Plastics Films

Plastics Films

Second edition

J. H. Briston, B.Sc., C.Chem., M.R.S.C., F.P.R.I., F.Inst.Pkg.

With two chapters by

Dr L. L. Katan, B.Sc., Ph.D., F.I.Chem.E., D.I.C., A.R.C.S., C.Eng, F.P.R.I.

George Godwin
in association with
The Plastics and Rubber Institute

George Godwin
an imprint of:
Longman Group Limited
Longman House, Burnt Mill, Harlow,
Essex CM20 2JE, England
Associated companies throughout the world

Published in the United States of America
by Longman Inc., New York

First published in 1974 by Illife Books, an imprint of the Butterworth Group
Second edition 1983

British Library Cataloguing in Publication Data
Briston, J. H.
Plastics films—2nd ed.
1. Plastic films
I. Title
668.4'95 TP1183.F5

CN ISBN 0-7114-5734-4

Library of Congress Cataloging in Publication Data
Briston, J. H. (John Herbert)
Plastics films
Bibliography: p.
Includes index
1. Plastic films. I. Katan, Leonard L. II. Plastics and Rubber Institute. III. Title.
TP1183.F5B74 1983 668.4'95 82-11808

ISBN 0-7114-5734-4

Printed in Hong Kong by
Wah Cheong Printing Press Ltd

Contents

Preface to the Second Edition

Since the first edition of this book there have been many advances in the field of plastics films. There is also a greater awareness of the socio-economic consequences of the use of plastics films, particularly in packaging.

The unified approach of the original book seemed to meet with general approval so this has been retained but a great deal of new material has been added to keep it up-to-date and to deal with some of the issues now being raised by governments and consumer groups. Three new chapters have been added, for example, covering Choice Criteria, Plastics and Energy, and Degradability, Recycle and Re-use. In addition, the two chapters on Health Safety and Odour and Taint, have been largely re-written. In terms of Odour and Taint there have been remarkable strides in the assessment of organoleptic effects. There have also been some developments in the science of Health Safety but the main changes have been in legislation, particularly on the subject of monomers. This emphasis on monomers arose from the vinyl chloride monomer storm that blew up during the two editions of this book.

New material has also been added on the more recent plastics (including linear low density polyethylene, polyvinylidene fluoride, parylene and acrylonitrile/methyl acrylate copolymer), stretch wrapping, and the ultra-violet and infra-red drying of printing inks.

The sections on extrusion and calendering have been expanded and summaries have been added to the chapters on Properties in order to emphasise the importance of the various properties on the packaging operation and the end-use performance of the film.

As in the first edition the needs of students in both the plastics and the packaging industry have been kept clearly in mind but the book should also be valuable to the converter and user of plastics films.

1 Introduction

Although the title of this book is 'Plastics Films' the subject of regenerated cellulose film has also been included because of its historical and present day importance. Regenerated cellulose is not a 'plastic' because at no stage in its manufacture is it capable of being moulded to shape. However, as a film it has a great deal in common with plastics films and it acts as an excellent reference point when properties and applications of plastics films are discussed.

The borderline between film and sheet may also need some clarification. There appears to be no hard and fast rule, but a convenient and widely accepted definition is that film comprises material up to 250 μm (0.010 in) in thickness whereas thicknesses above that figure refer to sheet. Some people tend to look on rigid material as sheet and flexible material as film but this is not so satisfactory as a practical definition. Materials such as unplasticised PVC and cellulose nitrate are fairly rigid even at thicknesses down to 70 or 80 μm (around 0·003 in), whereas low-density polyethylene is not really rigid even at thicknesses around 300 μm (0·012 in).

The subject of plastics film is already vast and is still growing rapidly so that a great deal of selection and condensation has been necessary to keep the subject within the bounds of this volume. Its aim has been to provide as full a background as possible to the student of plastics so that he may be in a position to consult the extensive literature on the subject. It is also intended to be a guide to the user or prospective user of plastics films, and to the worker in other fields of polymer technology.

1.1 HISTORICAL BACKGROUND

The first commercially successful plastics film was cellulose nitrate. It had some excellent properties, such as clarity, strength, resistance to moisture and dimensional stability but it also had one serious drawback, namely, flammability. As a film it has now been displaced by newer materials although as a sheet it still has at least one large application, namely, table tennis balls.

The most important development in films was the production of regenerated cellulose. The first process for dissolving cellulose and subsequently regenerating it was discovered by Schweitzer and formed the basis of his cuprammonium rayon fibre process. Cuprammonium films have also been made but are not so important commercially as those made by the Xanthate process. This was discovered in 1892 by three English chemists, Cross, Bevan and Beadle. A method for coagulation and regeneration of xanthated cellulose was patented by Stearn in 1898 and small commercial quantities of film were produced by John Chorley in Manchester a year later. However, the person chiefly responsible for the commercial development of equipment for the continuous production of cellulose film was J. E. Brandenberger, a Swiss chemist, and patents were first issued to him in 1911.

First applications for regenerated cellulose film were as decorative wrappings for luxury or semi-luxury goods. The development of heat sealable and moisture-proof grades (by means of coatings) meant that the film could be used as a protective wrapper and from then on its growth rate increased by leaps and bounds.

Cellulose acetate film was developed first, in 1909, as a 'safety' photographic film because of the dangers attached to the use of cellulose nitrate. Since then it has developed as a wrapping film where 'breatheability' is required (as in fresh produce packaging), in print lamination for the covers of journals, maps, etc., and in the manufacture of window cartons.

Of the films so far mentioned, regenerated cellulose was the almost unchallenged leader up to about 1950 since cellulose acetate was too expensive for the large tonnage packaging uses. The first film to challenge the supremacy of cellulose film was polyethylene but other, more recent, films such as polypropylene, PVC and polyethylene terephthalate have also either eroded cellulose film's markets or developed new ones of their own.

A calendered grade of polyethylene mixed with polyisobutylene was produced as early as World War II and was used for the packaging of Mepacrine tablets for the Services. In 1946, an extrusion process was developed, followed by a chill roll casting method in 1948. The latter had a better gloss and transparency than the extruded grades then available but was unsuitable for bag making purposes because of a serious tendency to block. A year later, a blown tubular film was made available which could be converted into bags although its clarity was poor by today's standards. Initial applications were mainly confined to industrial packaging, such as covers for equipment packed inside wooden crates, the packaging of electrical resistors and as drum liners. The development of prepackaging for fresh produce gave a boost to the use of polyethylene film because of its combination of transparency and high strength. When, eventually, high clarity grades of polyethylene film were developed, new markets opened up in the fields of display packaging such as textiles, woollens, soft toys and similar items. During this time, polymer costs were steadily falling and eventually polyethylene film became the cheapest transparent film available. Today, one of the largest tonnage outlets for polyethylene film is the heavy duty sack for fertilisers, peat, polymer granules and various chemicals. The strength of polyethylene is exemplified by the fact that a sack for 50 kg (1 cwt) of fertiliser has a thickness of only 200 μm (0.008 in).

Although it has many advantages, low-density polyethylene film is limp compared with regenerated cellulose film and special machinery had to be developed before it could be used for high-speed packaging. A more challenging competitor to regenerated cellulose was polypropylene with its good clarity and stiffness as well as good barrier properties. This film was introduced around 1959.

Other plastics films have made their debut since then but none of them has had the impact of polypropylene, mainly because of their high price relative to polypropylene. However, many have established their own special niche in the market place by virtue of some special property or other. The importance of plastics films today can be seen from the *Table 1.1* which gives consumption figures for some of the main films in the U.K. in 1979.

Up to the beginning of the 1970s the story of plastics films was one of steadily reducing prices, leading to a gradual replacement of more traditional packaging materials such as paper. Since 1973 the continuing heavy increases in oil prices have sharply reversed the downward spiral of polymer prices and have greatly reduced the rate of increase in plastics consumption. The growth of plastics films has not been halted completely, however, because there are valid technical reasons, in addition to economic ones, for using them in a wide variety of applications. When considered from a performance

Table 1.1

Film	*Consumption (tonnes)*
Low-density polyethylene	300 000*
High-density polyethylene	22 000
Polypropylene	34 000†
PVC	40 000
Regenerated cellulose	47 000

* includes heavy duty sacks (65 000 tonnes).
†*includes film tape used for woven sacks (9000 tonnes).

point of view plastics are still competitively priced in relation to other materials.

1.2 SOCIO-ECONOMIC CONSIDERATIONS

Reference has already been made to the economics of plastics films but there is also a socio-economic consideration which is becoming more important as time goes on. This is the use of energy in manufacturing processes. More strictly this should be looked at in terms of a total energy equation involving the energy used to produce the article to be packaged, the possible percentage loss if the product was sold in the unpackaged state and the energy used to produce the particular protective package. Such equations tend to show that in most cases the energy used in producing the packaging material is fully justified. In addition, the energy used in producing a plastics film bag or wrap is an order of magnitude less than that of a glass container (first use), a tinplate can or an aluminium container (to package an equivalent weight of foodstuff). As an example of the justification for packaging, 1 kg (2·2 lb) of potato crisps is estimated to consume 10 MJ (0.095 therms) of energy in its manufacture. Coated, oriented polypropylene bags, or bags made of coated cellulose film (to contain 1 kg—2.2 lb of crisps), would probably consume only about 0·3 to 0·35 MJ (0·003 to 0·0035 therms). The spoilage of unpackaged crisps would be 100% within a very short space of time and thus the packaging is fully justified. The energy consumption for a glass bottle (also able to hold 1 kg—202 lb of foodstuff) would be nearer 9 MJ (0·085 therms) on initial manufacture but this drops to about 0·5 MJ (0·005 therms) for subsequent reuse (reflecting the energy used in washing).

Another factor which has a bearing on economics is availability. Most plastics are derived from oil and, as this is a non-renewable resource, prices must obviously rise as the resources dwindle.

However, not all plastics are derived from oil. Cellulose (from trees and other plant life) forms the starting point for a range of plastics including regenerated cellulose and cellulose acetate films. Attention is also being directed increasingly towards a wide range of renewable resources including seaweed, sewer waste, field crops, etc. One important monomer—ethanol—can be derived, via the fermentation of sugars, from plant juices, cellulose and starch, while methanol (an important source of formaldehyde) can be obtained from agricultural residues or from municipal solid waste. Sugar itself has also been investigated as a feedstock for plastics. Economics is still a major barrier but this will change as the non-renewable sources become even more scarce.

It should not be forgotten that oil is not the only non-renewable resource available for plastics manufacture. Coal was widely used once and may well be again, so giving more time to develop renewable resources.

1.3 LEGISLATION

A large percentage of plastics films is used in packaging and this means that plastics films are widely affected by an increasing amount of legislation. One aspect of legislation is dealt with in Chapter 11 but there is also increasing pressure from consumer groups and governments concerning the impact of packaging on the environment. In the E.E.C., the Environment and Consumer Protection Service has set up a Packaging Working Group. This has already shown a very anti-packaging bias and although its initial attention has been concentrated on the use of returnable and one-trip containers for beverages it will probably move to other areas of packaging (including films) in due course. In the U.K. a Waste Management Advisory Council (WMAC) was set up in 1974 and went on to form a Packaging and Containers Working Group. WMAC was disbanded late in 1980 but not before the Working Group had produced an excellent report on returnable and non-returnable beverage containers. In order to present a coherent reply from industry to the growing criticism, leading U.K. packaging manufacturers, producers of packaged goods and retail organisations formed the Industry Committee for Packaging and the Environment (INCPEN) in 1974. Later, in 1976, the Waste Management Advisory Council indicated to INCPEN that, to avoid legislation, industry should consider producing a draft code of practice for packaging. INCPEN eventually produced a draft U.K. Code which was then discussed with the CBI, various consumer and environmentalist organisations, Pira (Research Organisation for the Paper, Printing and Packaging Industries) and observers from various government departments. The final Packaging Code was published in May 1978. The Code has eight points, the last of which reads: 'The package should be designed with due regard to its possible effect on the environment, its ultimate disposal and to possible recycling and reuse where appropriate.' The operation of the Code is supervised by the Packaging Council which includes members representing manufacturers of packaging, packers of consumer goods, distributors, consumer organisations, environmentalists and trade unions, meeting under an independent chairman.

Some of the main factors concerned with the packaging film/environment interface are discussed in Chapter 25.

Part 1
Film-Forming Materials

2

Polyolefins

The polyolefins form an important class of thermoplastics from any standpoint but they are particularly important as film formers since they include low and high density polyethylene and polypropylene. Low density polyethylene film constitutes the major portion of the total film market but high density polyethylene and polypropylene are by no means minor components.

The polyolefins which will be dealt with in this chapter are low density polyethylene, high density polyethylene, polypropylene and poly(4-methyl pentene-1). Ethylene–vinyl acetate copolymers will also be dealt with here because the film forming polymers contain ethylene as the major constituent. Ionomers, although based on ethylene, are discussed in a separate chapter since their structure and properties are sufficiently different to merit separate treatment.

In the case of polyethylene, the low density and high density polymers will be considered separately although a broad spectrum of polyethylenes is now available, due to co-polymerisation of ethylene with small amounts of other olefins such as butene-1, or by blending of the two types of polyethylene.

2.1 LOW DENSITY POLYETHYLENE

The first high-molecular-weight polymer of ethylene was produced by Imperial Chemical Industries Limited in 1933, during a pure research programme devoted to the effects of extremely high pressures on chain reactions. Special equipment was developed to withstand these pressures which were of the order of thousands of atmospheres. During an investigation of the reaction between ethylene and benzaldehyde, a white powder was produced and was identified as a polymer of ethylene.

The experiment was repeated using ethylene alone but this led to explosive decomposition of the ethylene and the apparatus was badly damaged. Two years passed before any further work was carried out but in the meantime improved apparatus had been developed.

What happened next owed a great deal to chance but was turned to good account. A fall of pressure occurred during a repeat of the experiment using ethylene alone and fresh ethylene was added to restore the pressure to its original value. At the end of the experiment about 8 g of a white powdery solid was obtained and identified as polyethylene. The success of this particular experiment was traced to the fact that the ethylene added to restore the pressure had, quite by chance, contained just the right amount of oxygen to initiate the polymerisation reaction. The element of chance had, in fact, entered at a still earlier stage because the pressure drop was apparently greater than could be accounted for by polymerisation and there must, therefore, have been a leak in the apparatus.

Low density polyethylene was first used in the electrical industry, particularly as an insulating material for underwater cables and, later, for radar. The development of polyethylene film came during World War II by calendering of a polyethylene–polyisobutylene blend and later, in 1946, by extrusion.

The polymerisation of ethylene can occur over a wide range of temperatures and pressures but most commercial high pressure processes utilise pressures between 1000 and 3000 atmospheres and temperatures between 100 and 300°C. Temperatures higher than 300°C tend to cause degradation of the polyethylene. The first initiator, as we have seen, was oxygen but other initiators and modifiers have also been used. The concentration of oxygen is critical. The process is highly exothermic and one of the earliest difficulties was the removal of excess heat from the reactors.

The ethylene is highly purified and then led over a reduced copper catalyst in order to remove any traces of oxygen. The precise amount of oxygen necessary to act as an initiator is then added and the gases are compressed in multi-stage compressors. They are then pumped into the reaction vessel by specially designed compressors. Two main types of reactor are in use, namely, the autoclave and the tubular. In either case, provision must be made for the close control of catalyst concentration, pressure and temperature. Any unconverted ethylene is separated from the molten polymer and is recycled. The polymer is then extruded as a continuous ribbon, solidified by cooling (usually in a water bath) and cut into granules. Film grades are usually given a further homogenisation by processing in an internal mixer, a refiner or a screw extruder.

The simplest structure for the polyethylene molecule is a completely unbranched chain of $—CH_2—$ units as shown.

$$—CH_2—CH_2—CH_2—CH_2—CH_2—CH_2—\text{etc.}$$

The vigorous nature of the high pressure process, however, militates against the straightforward process of chain growth and a great deal of chain branching occurs which has an important bearing on the properties of low density polyethylene. Both short and long branch chains are produced, examples of which are shown below.

$$\begin{array}{l} \quad CH_2—CH_2—CH_2—CH_2—CH_2—CH_2—CH_3 \\ \quad | \\ —CH_2—CH—CH_2—CH_2—CH_2—CH_2—CH_2—CH—CH_2—CH_2—CH_2—CH_2— \\ \qquad\qquad\qquad\qquad\qquad\qquad\qquad\qquad\quad | \\ \qquad\qquad\qquad\qquad\qquad\qquad\qquad\qquad\quad CH_2—CH_2—CH_3 \end{array}$$

It will be noticed that each branch chain contains a terminal methyl ($—CH_3$) group. A convenient way of characterising branching is by the number of methyl groups per 100 carbon atoms and this can be done by infra-red spectroscopy.

The occurrence of these branch chains prevents a close packing of the main polymer chains and this accounts for the fact that the process just described produces low density polyethylene.

2.1.1 PROPERTIES

The structure of the low density polyethylene molecule also affects properties other than density. One important property is that of crystallinity. The great length of polymer chains means that a certain amount of entanglement normally occurs and this prevents complete crystallisation on cooling; there are thus disordered areas between the crystallites. The structure of a partially crystalline polymer is shown in *Figure 2.1*.

Areas where the chains are parallel and closely packed are largely crystalline while the disordered areas are amorphous. These crystalline areas are known as crystallites. When the polymer melt is cooled slowly, the crystallites may form spherulites consisting of spherically symmetrical aggregates of crystallites plus amorphous polymer.

Figure 2.1 shows molecules passing through both ordered and

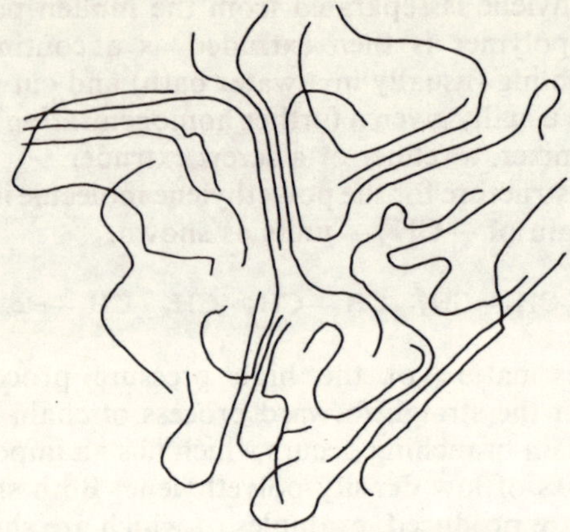

Figure 2.1

disordered states and this was once widely believed to be the case. More recent work seems to point to the fact that the molecules fold on themselves in a parallel arrangement to form lamellae. Crystallisation spreads as other molecules align themselves into position and also start to fold. The spherulites mentioned earlier occur due to irregularities in molecular structure which cause the crystalline growth to occur in more than one direction. The presence of chain branching will tend to reduce the possibility of an ordered arrangement and so reduce the crystallinity. The crystallinity of low density polyethylene usually varies between 55 and 70% (compared with 75–90% of high density polyethylene). The other property of importance which is affected by the chain branching is softening point. The fact that the chains cannot approach so closely to each other means that the attractive forces between them are reduced and less energy, in the form of heat, is necessary to cause them to move relative to each other and thus flow. The softening point of low density polyethylene is slightly below the boiling point of water and so the material cannot be used where boiling water or steam sterilisation are involved.

Low density polyethylene is a tough, slightly translucent material and is waxy to the touch. The density can vary between about 0·916 g/cm³ and 0·935 g/cm³. It can be blow extruded into tubular film or extruded through a slit die and chill-roll cast. The chill-roll casting process gives a clearer film but even blown film can be produced with good clarity with improvements in cooling of the bubble. Low density polyethylene film has a good balance of properties such as tensile strength, burst strength, impact resistance and tear strength. In addition, it retains its strength down to quite low temperatures (around −60° to −70°C). It is a good barrier to water and water vapour but is not so good a barrier to gases. It is not suitable, therefore, for the packaging of products susceptible to oxidation.

It has excellent chemical resistance, particularly to acids, alkalis and inorganic solutions. It is, however, sensitive to hydrocarbons and halogenated hydrocarbons and to oils and greases. The effect is one of absorption followed by swelling. The swelling is less for polyethylenes of high molecular weight. Certain polar organic chemicals can cause environmental stress cracking of low-density polyethylene. This is a phenomenon which can be caused by chemicals which normally would not attack or dissolve the polyethylene. In the presence of stresses, however, these same chemicals cause surface cracks or even complete failure of the material. Typical stress cracking agents are detergents, some essential oils, vegetable oils, benzaldehyde and nitrobenzene. The incidence of stress cracking can be greatly reduced by the use of high-molecular-weight grades of polyethylene.

By the use of suitable additives it is possible to produce low density polyethylene film with high slip (low coefficient of friction) and low blocking tendencies. Build-up of static electricity can be a problem but again the use of additives can give film with much improved properties in this respect. It is odourless and tasteless and is widely used in food packaging applications. One disadvantage of low density polyethylene film is its relatively low softening point and the material cannot, therefore, be steam sterilised.

Low density polyethylene film is easily heat sealed and gives good, tough welds. It cannot be sealed by high frequency methods since it has a very low loss tangent. Sealing by means of adhesives is difficult because of the low absorption characteristics of low density polyethylene and the use of water or solvent based adhesives is restricted. Hot melt adhesives (especially those based on polyethylene and polyisobutylene blends) can be used but are expensive and show little advantage over direct heat sealing.

Printing of low density polyethylene film can be carried out successfully by a variety of methods provided that the surface is treated beforehand. This pre-treatment is necessary because of the inert, non-polar nature of the film surface which makes it very difficult to achieve a good key, and is described more fully in Chapter 14.

Flexographic printing methods are the most popular for low density polyethylene film but gravure and silk screen methods are also used. Although low density polyethylene has excellent dimensional stability with respect to changes in ambient humidity conditions, it is easily stretched so that 'snatching' should be avoided while running through the printing machine otherwise bad register will be obtained.

The addition of starch to low density polyethylene produces a material of some interest in the field of biodegradable plastics. Such polyethylenes normally contain between 10% and 40% of starch by volume. In moist soils they are said to lose half their strength in about 12 months and to break down totally in 5 years but the latter is strictly true only for the starch content. The remaining framework of polyethylene will probably embrittle and break up. Dispersal or absorption into the soil could well follow.

Starch-filled polyethylenes can be processed on conventional equipment. The main use at present is for carrier bags although another application for degradable film is for mulching of some agricultural crops such as tomatoes.

2.1.2 USES

The high tonnage usage of low density polyethylene film is a good indication of the very large number of applications which have already been developed for this material. The overwhelming proportion of these uses are in the fields of packaging, building and horticulture and agriculture and are therefore dealt with in the appropriate chapters.

2.2 LINEAR LOW DENSITY POLYETHYLENE (LLDPE)

In recent years there has been a spate of process development aimed at producing low density polyethylene either by *low* pressure gas phase polymerisation or by liquid phase processes similar to those used for producing high density polyethylene. Although the result of these newer processes is indeed a low density polyethylene there are still differences between the conventional low density polyethylene and the new polymers. These latter are known as linear low density polyethylenes (or LLDPE) and are similar in structure to high density polyethylene but with slightly longer and more numerous side chains.

A typical LLDPE gas phase process is that of Union Carbide Corporation. The process is based on a fluidised bed of polyethylene. Ethylene is fed in at the base of the fluidised bed and the resultant polymer is taken out continuously, keeping a constant fluidised bed level within the reactor. The temperature is relatively low (about 100°C) while pressures are markedly lower (689–2068 kN/m^2—100–300 lb/in^2) than is the high pressure gas phase process developed originally by ICI. The efficiency is lower than in high pressure gas phase processes, being about 2% polymerisation per pass, compared with 15–30%. However, plant size can be greatly reduced by eliminating much of the equipment necessary in high pressure operation and the capital costs are also much lower. A somewhat similar process is the stirred bed system. A Ziegler type catalyst is used and a higher percentage polymerisation is achieved.

There are basically two types of liquid phase process, namely, slurry and solution. One slurry process is that of Phillips Petroleum which uses a chromium catalyst and operates at a temperature of 100°C and pressures between 689 and 4826 kN/m^2 (100 and 700 lb/in^2). The polyethylene is produced in particle form. Solution processes have been developed by Du Pont and by Dow Chemical, using a Ziegler type catalyst. Operating conditions are around 180–250°C and 2757–4137 kN/m^2 (400–600 lb/in^2). Additives can be incorporated directly in these processes.

The properties of LLDPE will vary according to the particular process employed in the same way that the properties of LDPE or HDPE vary. However, a major feature of LLDPE is that its molecular weight distribution is more narrow than that of LDPE. In general, the advantages of LLDPE over LDPE are improved chemical resistance, improved performance at both low and high temperatures, higher surface gloss, higher strength at a given density and a greater resistance to environmental stress cracking. In film form, LLDPE shows improved puncture resistance and tear strength. Typical figures for puncture resistance at a thickness of around 75 μm (0·003 in) are 0·4 ft lb/in^2 (834 J/m^2) for LDPE and 0·9 ft lb/in^2 (1877 J/m^2) for LLDPE. Similarly, the tear strengths about 350 gf (0·77 lb.f) for LDPE and 10500 gf (2·31 lb.f) for LLDPE (mean of machine and cross directions). At a density of 0·92 g/cm^3 the melting points of LDPE and LLDPE are 95°C and 118°C respectively.

LLDPE has been used for the manufacture of heavy duty sacks where its higher percentage elongation and ultimate tensile strength make it particularly suitable. Its higher melting point also opens up the possibility of hot fill applications, one suggested use being for cement sacks.

LLDPE is also making inroads into the stretch wrap market because of cost/performance improvements. One disadvantage is that LLDPE has considerably less inherent cling than other stretch films like PVC and EVA. One way of overcoming this is by the incorporation of cling additives. Another is to eliminate the need for cling by such devices as closure by mechanical crimping. LLDPE is also being used in stretch wrapping in the form of a laminate with other plastics films. One example is a triple laminate with conventional LDPE as the outer layers and an inner core of LLDPE. The advantages claimed here are a greater tear and puncture resistance and a higher tensile strength and elongation, compared with LDPE. The overall thickness can, therefore, be reduced and the stretchability is greater. The film weight per pallet load can, therefore, be reduced with a consequent cost saving. Other films have been developed that are based on blends of LLDPE with other polymers such as EVA. It should be noted that in shrink wrapping (as opposed to stretch wrapping) the lower hot melt strength of LDPE makes it the more suitable material.

2.3 IRRADIATED POLYETHYLENE

Work on the irradiation of polyethylene has shown that hydrogen gas (with small amounts of methane, ethane and propane) is liberated and the polymer becomes increasingly insoluble due to cross-linking of the molecule by C—C bonds. Increasing irradiation leads to yellowing and, eventually, the polyethylene becomes a dark ruby red colour. At first the polyethylene becomes more flexible but later it turns harder and eventually becomes quite brittle.

Irradiated polyethylene film has been produced by passing ordinary low density polyethylene film continuously through a high energy electron field. This converts it to an infusible film, which, with appropriate stabilisers, will withstand prolonged ageing at 105°C or short term exposure at 230°C.

The process slightly reduces gas and water transmission rates and the film has good clarity. Tear strength is good, both tear initiation and tear propagation being high.

2.4 HIGH-DENSITY POLYETHYLENE

Following ICI's discovery of low density polyethylene, the next most important step came in the early 1950s when Professor Ziegler, in the course of work on organometallic compounds, discovered catalysts which enabled the polymerisation of ethylene to be carried out at near atmospheric pressures and temperatures. At about the same time, two other low pressure methods were developed in the

U.S.A. by Phillips Petroleum Company and Standard Oil of Indiana.

These discoveries were significant not only because of the different manufacturing techniques involved but because the products themselves were found to be different from the conventional polythylene in several important respects.

The catalysts used by Professor Ziegler stemmed from his work on organo-aluminium compounds and the first to give satisfactory results was a combination of aluminium triethyl and a titanium derivative such as the tetrachloride. The active catalyst is actually a reaction product of the two materials since neither will polymerise ethylene to high molecular weight products if used alone.

The Phillips process utilised such catalysts as partly reduced chromic oxide supported on steam activated silica–alumina, while the Standard Oil of Indiana process made use of nickel oxide on charcoal.

The principles involved can be seen by reference to the Ziegler process. The catalyst system is suspended in a liquid hydrocarbon through which the ethylene gas is passed. The pressure is near atmospheric and the temperature is around 50–75°C. Polyethylene settles out as a granular powder and the resultant slurry is stirred until the viscosity becomes so high as to interfere with efficient dispersion. The mixture is then passed through working up and solvent recovery stages. In general, the sequence is one of catalyst deactivation, catalyst decomposition and removal, solvent recovery, drying, extrusion and granulation. The efficient removal of catalyst residues is extremely important as the electrical properties of the polymer may otherwise be severely affected. One method used is to add dry hydrochloric acid gas which forms alcohol soluble complexes with the titanium (this being the more difficult metal to remove). After preliminary washing with alcohol, the polymer is well washed with water. The resultant slurry is filtered or centrifuged and then dried.

The Phillips process operates at higher pressures (2750–3450 kN/m^2—400–500 lb/in^2) and higher temperatures (100–175°C). The liquid medium is usually cyclohexane and at these temperatures the polyethylene dissolves and is removed as an approximately 10% solution. The catalyst is removed quite simply by centrifuging. The polyethylene is obtained by cooling, pelleting and drying.

More recently (during the late 1960s) Union Carbide Corporation introduced a gas phase process for the manufacture of high-density polyethylene, based on a catalyst system developed by them. Ethylene, a small quantity of hydrogen, catalyst and comonomer (if used) are fed continuously to a gas phase reactor where polymerisation takes place at about 1960 kN/m² (286 lb/in²) pressure and at a temperature within the range 85–100°C. The reactor is discharged to a product receiver where any unreacted ethylene is recovered and recycled to the reactor. The polymerised product is transferred from the receiver to a product purge tank and then through a gas lock valve to a conveying system which takes it to the storage silos. From the silos the product is withdrawn for mixing, extrusion and granulation. Because gas phase polymerisation requires no solvent, there is no need for separation of solvent from the polymer, nor for solvent recovery. The catalyst efficiency is also claimed to be sufficiently high for residual catalyst removal to be unnecessary. There is, therefore, no need to wash and dry the polymerisation product and so there is no recovery of washing solvent and no effluent.

2.4.1 PROPERTIES

The linear structure mentioned earlier is closely approached by the polyethylenes produced by low pressure processes. Some branch chains are formed but these are short and are not so many in number. Copolymers of ethylene with, for example, butene-1 are also produced by low pressure processes in order to introduce controlled amounts of branching into the essentially linear molecule. Densities of the copolymers are around 0·945–0·95 g/cm³ whereas those of the linear homopolymers are around 0·96 g/cm³.

High density polyethylene film is stiffer, harder and has a less waxy feel than low density polyethylene. It can be made by blow extrusion or slit die extrusion (with chill roll casting or water quench) but with blow extrusion a rather milky, translucent film is obtained.

The softening point of high density polyethylene is higher (about 121°C) and so it can be steam sterilised. The low temperature resistance is about the same as that of the low density polymer. Tensile strength is higher than that of low density polyethylene as is the bursting strength but impact strength and tear strength are both lower. Because of their linear nature the high density polyethylene molecules tend to align themselves in the direction of flow and the tear strength of the film is much lower in the machine direction. This difference between machine direction and transverse direction tear strength can be accentuated by orientation to give a built-in tear tape effect.

The permeability of high density polyethylene is lower than that of low density polyethylene by a factor of about 5–6 times, its moisture vapour barrier properties being particularly good. In fact, of the films in common use its moisture vapour barrier properties are only surpassed by those of the vinyl chloride/vinylidene chloride copolymers.

The chemical resistance of high density polyethylene is also superior to that of low density polyethylene and, in particular, it has a better resistance to oils and greases. The solubility in organic solvents is reduced with increasing density, as is the permeability to solvents.

High density polyethylene is subject to environmental stress cracking in the same way as low density polyethylene but the effect can be reduced, again, by using high molecular weight grades where there is likely to be a problem.

2.4.2 USES

As with low density polyethylene the applications for high density polyethylene are largely in the fields covered by the applicational chapters, namely, Packaging, Film Fibres and Synthetic Papers.

2.5 POLYPROPYLENE

Early attempts to polymerise propylene using the high pressure process gave only oily liquids or rubbery solids of no commercial value. Later work by Professor Natta, in Italy, was directed towards the use of Ziegler-type catalysts. He found that if in the Ziegler catalyst system, aluminium triethyl plus titanium tetrachloride, the titanium tetrachloride was replaced by titanium trichloride, a stereospecific catalyst was formed which yielded crystalline high molecular weight polymers of propylene. A stereospecific catalyst is one which controls the position of each monomer unit as it is added to the growing polymer chain, thus allowing the formation of a polymer of regular structure from an asymmetric monomer unit such as propylene. This work constituted a major breakthrough in polymer technology since this was the first time that polymers with a regular spatial structure had been produced synthetically.

Basically Natta's polymerisation of propylene is similar to the Ziegler process for the preparation of high density polyethylene. The gas, under a pressure of about 100 atmospheres, is led into a reaction vessel in which is a well-stirred dispersion of catalyst in a liquid hydrocarbon. The temperature is kept low enough to ensure precipitation of the polypropylene as it is formed. Stirring is continued until the polymer content is between 35 and 40%, when the slurry is pumped to a flash drum where the unreacted propylene is removed and recycled. The slurry is then centrifuged to remove the liquid hydrocarbon.

Catalyst removal from the polymer is carried out by extracting with a weak solution of hydrochloric acid gas in methyl alcohol. After removal of the extractant, the polymer is washed with water to remove the acid, steam distilled to remove solid traces, dried, extruded and pelletised.

More recent work has made it possible to produce a type of polypropylene by gas-phase polymerisation methods.

The basic structural unit of polypropylene is:

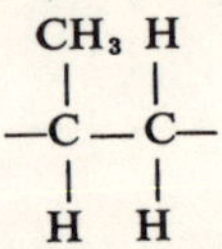

and is thus similar to that of polyethylene with a methyl group substituted for one of the hydrogen atoms. The polymerisation of

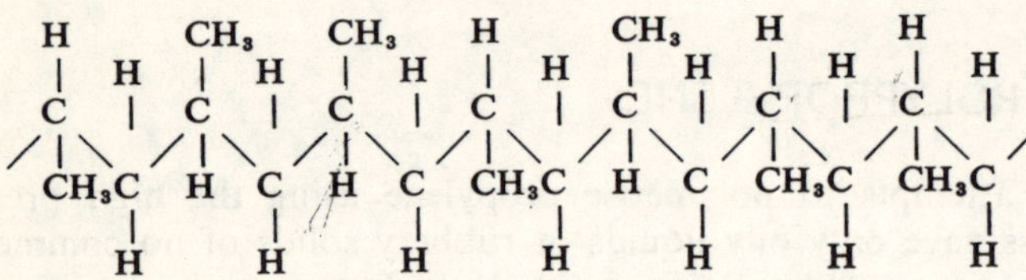

Figure 2.2. Atactic polypropylene

propylene without the aid of stereospecific catalysts produces haphazard linkages to give a rubbery or oily polymer with what is known as an *atactic* structure with a random distribution of the methyl groups on either side of the chain, as shown in *Figure 2.2*.

The regular crystalline polymer produced by stereospecific catalysts is known as the *isotactic* form, the name stemming from the original idea that the methyl groups were always situated along the same side of the polymer chain. In fact, the carbon atoms

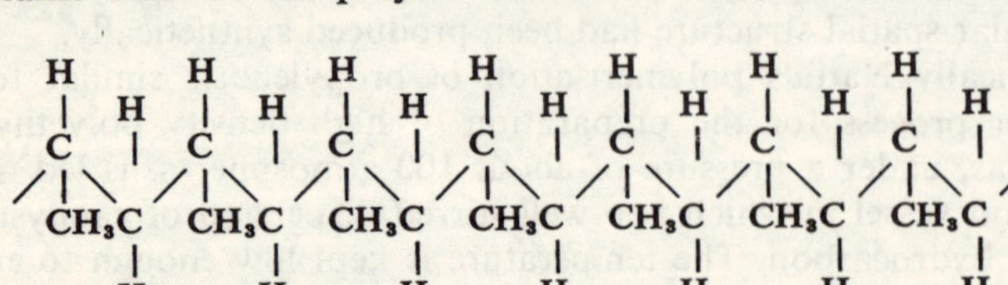

Figure 2.3. Isotactic polypropylene

arrange themselves into a helical chain with the methyl groups on the outside. *Figure 2.3* shows a 2-dimensional representation of the isotactic form. Two other forms are the *syndiotactic* and *stereoblock* polymers (*Figures 2.4* and *2.5* respectively).

The regular helices of the isotactic form can pack closely together whereas the atactic molecules will have a more random arrangement. Stereoblock polymers with infrequent inversion will behave very much as an isotactic polymer.

Some atactic polymer is produced even using a stereospecific catalyst system and the degree of isotacticity can be measured by extraction of the atactic molecule with n-hexane in which it is soluble. Both isotactic and stereoblock polymers are insoluble.

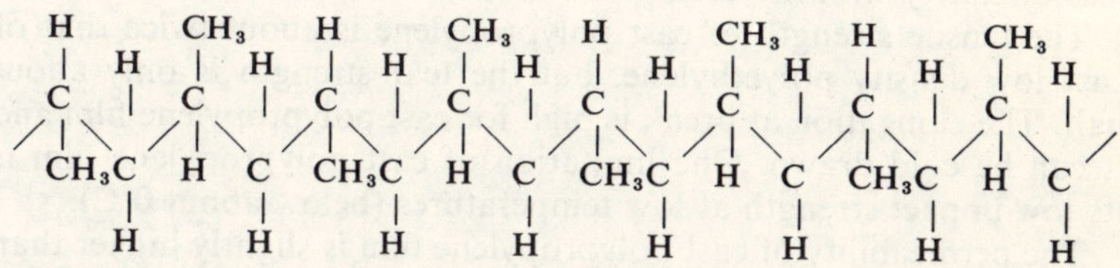

Figure 2.4. Syndiotactic polypropylene

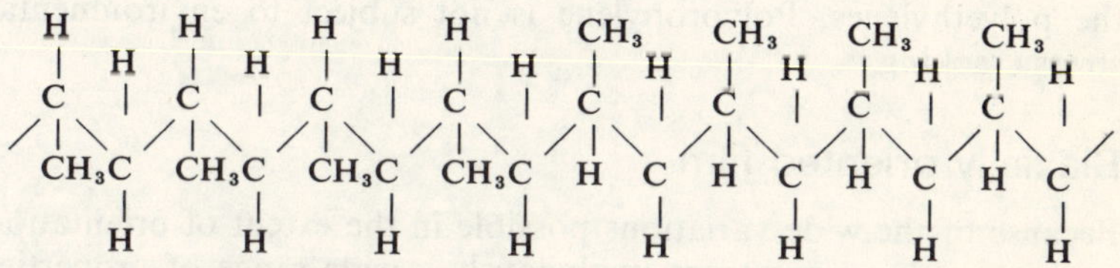

Figure 2.5. Stereoblock polypropylene

2.5.1. PROPERTIES

Polypropylene has a lower density than low density polyethylene (0·90 g/cm^3), is harder and has a higher softening point. Polypropylene film can be manufactured either by blow extrusion or by slit-die extrusion with subsequent cooling by chill-roll or water bath quench. The position is complicated by the fact that both methods have been used to produce a range of films which vary in the amount of orientation built into the film. Commercially, the films of most interest are slit-die extruded cast film with no orientation, biaxially oriented film with more or less equal orientation in the machine and transverse directions, and biaxially oriented films with unbalanced orientation. It is also possible to produce uniaxially oriented polypropylene, but this has a special significance in the manufacture of fibres, and is dealt with in Chapter 24. The various methods for producing biaxially oriented films are described in Chapter 8.

Cast film

Cast polypropylene film has good transparency and gloss but increasing thickness makes it more and more difficult to effect rapid cooling and thus prevent the growth of larger spherulites which cause haze (see Chapter 10). There is, therefore, a consequent loss of clarity with increasing thickness.

The tensile strength of cast polypropylene is about twice that of cast low density polyethylene, but the tear strength is only about half. The elongation at break is high for cast polypropylene film and it can be cold drawn. One limitation of cast polypropylene film is its low impact strength at low temperatures (below about 0°C).

The permeability of cast polypropylene film is slightly higher than that of high density polyethylene, but much lower than that of low density polyethylene. The chemical resistance of polypropylene is good and, in particular, it is more resistant to oil and greases than the polyethylenes. Polypropylene is not subject to environmental stress cracking.

Biaxially oriented film

Because of the wide variations possible in the extent of orientation in the two directions there is obviously a wide range of properties available. The balanced film has a tensile strength, in each direction, roughly equal to four times that of cast polypropylene film. Tear initiation is difficult, but tear resistance after initiation, is low. The elongation at break is also low because it has already been fully drawn. Biaxial orientation improves the gloss slightly and greatly reduces haze so that the oriented film has a greatly improved clarity over that of the cast film.

Biaxial orientation also improves the barrier properties of polypropylene film and its low temperature impact strength. Although both moisture vapour and gas barrier properties are improved by biaxial orientation, the gas barrier properties are still not as good as those of regenerated cellulose. However, coated grades of oriented polypropylene are available which are very much better in this respect. Coated grades were originally developed to improve heat sealability because shrinkage tends to occur when the highly stretched film is heated. The provision of a lower melting point polymer as a coating is one way of solving this problem.

Highly transparent film

Highly transparent blown polypropylene films have also been developed as direct competitors to regenerated cellulose film in areas of retail packaging where the prime need is consumer appeal, i.e. good surface gloss, high clarity, crackle, etc. The strength of such films is not so high as that of normal orientated polypropylene but this is seen as an advantage in some retail packaging where complaints are often made that plastics packs are too strong and difficult to open.

The process consists in extruding a tube downwards from a rotating spreader die and water cooling it after a short (300–500 mm—12–20 in) passage through air. The tube passes through a calibrator just below the water surface. The blow-up ratio is very low (less than 1:2). The preferred polymers are those with a higher than usual atactic and heptane soluble content.

2.5.2 USES

Many of the uses of polypropylene film (both cast and biaxially oriented) are in the field of packaging while the uniaxially oriented film is used for fibres. There is one field in which oriented polypropylene film is being increasingly used and that is print lamination. Included here is the covering of maps, charts, covers of technical journals and covers of paper-back books. Cellulose acetate is also used in this field, but polypropylene is cheaper because of its lower density and is also dimensionally stable in conditions of changing humidity. One disadvantage of polypropylene film is that special adhesives are usually necessary for good adhesion.

2.6 POLY(METHYL PENTENE) (TPX)

Methyl pentene polymers are a class of polymer based on 4-methyl pentene-1 as the principal monomer. This is copolymerised with other monomers using Ziegler type catalysts when an isotactic polymer is formed. Unlike polyethylene and polypropylene it is a very transparent polymer having a clarity nearly equal to that of acrylic sheet. It has the lowest density of any commercially available plastic (0·83 g/cm^3) and a softening point higher than that of polypropylene. The melting point is 240°C but it is still form stable up to about 200°C. It has good chemical resistance but its permeability to moisture vapour and gases is high.

No uses have yet been developed commercially for the unsupported film, but it has been used as a coating on paperboard to give heat resistant disposable trays suitable for cooking and serving foodstuffs.

2.7 ETHYLENE/VINYL ACETATE COPOLYMERS (EVA)

Theoretically the proportion of vinyl acetate in ethylene/vinyl acetate copolymers can vary from 1 to 99% but commercially available products to date have normally had less than 50% of vinyl acetate. Of these, the ones with 21–50% vinyl acetate are used as additives for waxes and adhesives and the copolymers with useful film forming properties lie in the range 1–20% vinyl acetate. It seems logical therefore, to treat the film forming EVA copolymers under the heading of polyolefins.

EVA films can be made by either blow extrusion or chill-roll casting methods. Cast film has the better optical properties but blow extrusion gives a tougher film and is the simpler method for the production of shrinkable film. Because it has a high elasticity, excessive tension should be avoided during haul-off.

2.7.1 PROPERTIES

The properties of EVA films vary according to the percentage of vinyl acetate in the molecule but in general the material compares with low density polyethylene as follows:

(1) Its heat seal temperature is lower.

(2) Its impact strength is higher.

(3) Its elasticity is greater.

(4) It has a greater resistance to environmental stress cracking.

(5) Its permeability to moisture vapour and to gases is higher.

(6) Its flexural life is greater.

(7) Its low temperature properties are better.

(8) EVA is tacky and has a greater tendency to blocking than has low density polyethylene.

(9) EVA's slip properties are worse, i.e. its coefficient of friction is greater.

(10) EVA can be sealed by high frequency heating but more power is needed than for PVC.

(11) Its printability is better.

(12) Its filler retention is higher.

Films with up to about 7–8% vinyl acetate may be considered as modified low density polyethylene, but the films with around 15–20% vinyl acetate are more like flexible PVC, with the added advantage that their flexibility is inherent and does not depend on added plasticisers which can later be leached out.

2.8 POLY(BUTENE-1)

Butene-1 is the next step up from propylene in the olefin family and is polymerised with the aid of Ziegler type catalysts. Its behaviour on crystallising from the melt is unusual inasmuch as the tetragonal crystalline Form II which is first obtained changes over a period of three to six days to the more stable hexagonal Form I. The transformation is accompanied by an increase in density, rigidity, strength and hardness. There is also an increase in crystallinity from about 30% to about 50%.

Polybutene film is normally prepared by blow extrusion but it can also be chill-roll cast. The unusual behaviour on cooling does not adversely affect film production.

2.8.1 PROPERTIES

Polybutene gives a tough film with good tear resistance, tensile strength, puncture resistance and impact strength. Its performance in these respects is superior to that of low density polyethylene which it resembles in density (0·915–0·92 g/cm³). The melting point of polybutene is similar to that of high density polyethylene (125°C). Polybutene retains its strength properties at elevated temperatures to a greater extent than does low density polyethylene and this makes it particularly suitable for hot-fill packaging applications. At the other end of the temperature range, polybutene embrittles at about −35°C.

Polybutene has similar chemical properties to those of low density polyethylene, in general, but it has a much better resistance to environmental stress cracking. It is a good barrier to moisture vapour, but is not a particularly good barrier to gases.

Polybutene film can be heat sealed to give strong seals whether the seals are made in-line after extrusion, when the film is in Form II or later when the film has changed to Form I. Polybutene can be printed with the type of flexographic equipment used for polyethylene film, after pre-treatment.

2.8.2 USES

Polybutene film has been used in the U.S.A. for milk pouches but its toughness makes it a strong contender for industrial packaging or for tarpaulins.

2.9 MELT FLOW INDEX (MFI)

The concept of melt flow index is so widely used that no discussion of the polyolefins is complete without it. The MFI is an indirect measure of the molecular weight of the particular polymer and is important to the converter because of the differences in processability between polymers of the same chemical type but with different molecular weights. We have also seen that the liability to environmental stress cracking is dependent on molecular weight.

The melt flow index is a measure of the melt viscosity of the polymer which in turn is related to the molecular weight. It is measured using the apparatus shown in *Figure 2.6* which is basically an extrusion plastometer with a standard orifice. The weight, in grammes, of the material extruded in 10 min under a constant dead weight and at a constant temperature is called the Melt Flow Index. A low melt flow index corresponds to a high melt viscosity and since melt viscosity is directly related to the molecular weight of the polymer a low MFI corresponds to a high molecular weight and vice versa. A strict relationship obviously only applies to polymers of the same chemical constitution, but it also is restricted to polymers of the same density since density also affects melt viscosity. In the case of the polyethylenes and EVA the standard temperature is 190°C but for polypropylene the temperature is 230°C in order to allow for its higher melting point.

The relationship between melt viscosity and MFI is a logarithmic one. Thus, for a MFI of 0·2 the melt viscosity of a low density

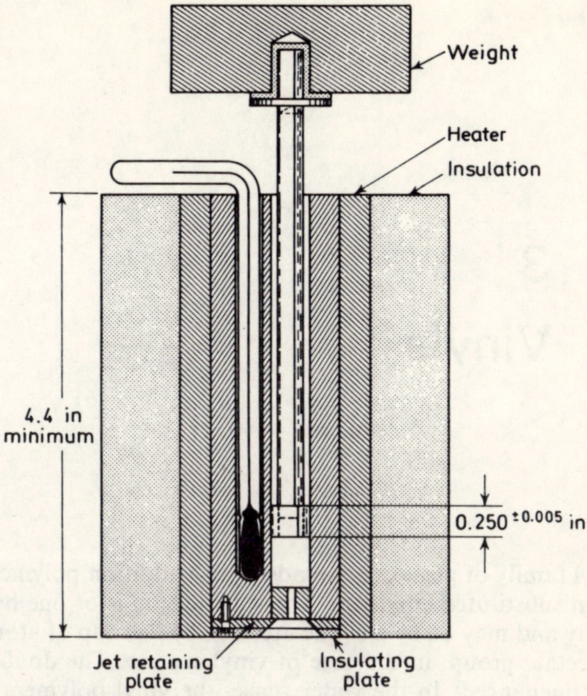

Figure 2.6. Apparatus for determining Melt Flow Index

polyethylene (density 0·92 g/cm³) is 300 000 P, while for a MFI of 2, it would be 30 000 P. Similarly a MFI of 20 corresponds to a melt viscosity of 3000 P. Extrusion processes, such as film making, require high melt viscosities and so low MFI grades are used.

FURTHER READING

BOENIG, H. V., *Polyolefins, Structure and Properties*, Elsevier, Amsterdam (1966)
RENFREW, A. and MORGAN, P., *Polythene*, Iliffe, London (1960)
RITCHIE, P. D., *Vinyl and Allied Polymers, Vol. 1.*, Iliffe, London (1968)

3
Vinyls

The vinyl family of plastics are made by the addition polymerisation of certain substituted ethylenes. The substitution is of one hydrogen atom only and may be by a single atom or by a group of atoms such as the acetate group, in the case of vinyl acetate. The double bond remains unchanged. In the wider sense, the vinyl polymers include such materials as polystyrene (based on styrene —CH_2—CH. C_6 H_5) but in general the term is taken to mean polyvinyl chloride (PVC), vinyl chloride/vinyl acetate copolymers, polyvinyl alcohol and vinylidene chloride/vinyl chloride copolymers. Polyvinyl acetate is also one of the common vinyls but is used only in dispersion form as an adhesive or as the basis for some emulsion paints.

3.1 POLYVINYL CHLORIDE (PVC)

PVC is made by the addition polymerisation of vinyl chloride. The overall reaction is shown below:

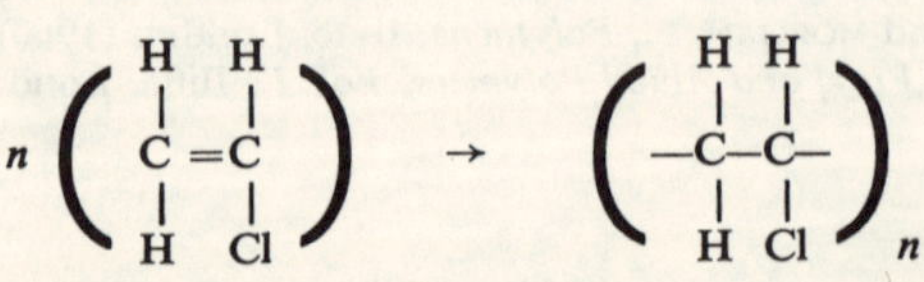

The most widely used polymerisation process is a suspension one. The vinyl chloride is stirred with water to which has been added a suspension agent such as methyl cellulose, gelatine or polyvinyl alcohol. The suspending medium (water) helps to dissipate the heat formed during the polymerisation reaction. The reaction is initiated by a catalyst which is soluble in the vinyl chloride but insoluble in the water. Benzoyl or lauroyl peroxide are both suitable catalysts. Strong agitation is necessary to maintain the suspension in droplet form and the speed of stirring is carefully controlled to control droplet size. Polymerisation is carried out for between 6 and 24 hours and the polymer is then in the form of a slurry in water. This is passed to a stripping tank to remove any unreacted vinyl chloride, filtered and then dried by passing the material through a continuous rotary drier. Increasing use is also being made of bulk polymerisation processes. These give a product particularly suitable for high clarity, low colour film.

It can be seen by looking at the structure of the vinyl chloride molecule that addition of molecules to the growing chain can take place either head to head (*Figure 3.1*), head to tail (*Figure 3.2*) or

$$(CH_2—CHCl)—(CHCl—CH_2)—(CH_2—CHCl)—$$

Figure 3.1. Head to head arrangement

$$—(CH_2—CHCl)—(CH_2—CHCl)—(CH_2—CHCl)—$$

Figure 3.2. Head to tail arrangement

else there can be a completely random arrangement. Materials polymerising in either of the two regular forms could be expected to be crystalline but those containing the random arrangement will be amorphous. PVC is one example of a head to tail arrangement but is mainly an amorphous polymer because it polymerises in the atactic form (see Chapter 2).

PVC can be made into film by the blow extrusion technique or by slit-die extrusion and these processes are widely used for producing the thinner gauges of film, particularly in unplasticised or lightly plasticised grades. One of the difficulties with the extrusion of PVC is its thermal instability at processing temperatures, combined with a high melt viscosity. With polystyrene or the polyolefins, viscosity can be reduced if necessary by increasing the processing temperature but the thermal instability of PVC prevents this solution being used. Die heads have to be carefully designed, therefore, to prevent stagnation of flow. One solution is to eliminate the use of a cross-head die by using a horizontal extruder with an in-line tubular die and horizontal take-off.

The most widespread method, however, of converting PVC into sheet or film is by calendering. A wide range of PVC films, with extremely varied properties, can be obtained from the basic polymer. The two main variables are formulation and orientation. Changes in formulation (chiefly plasticiser content) can give films ranging from rigid crisp films to limp, tacky, stretchable films. The degree of orientation can also be varied from completely uniaxial to balanced biaxial.

3.1.1 PROPERTIES

Unplasticised

As mentioned earlier, unplasticised PVC tends to degrade and discolour at temperatures fairly close to those used in its processing so that suitable stabilisers have to be included in the formulation. If the stabiliser is efficient then extremely clear and glossy films can be produced. The film is stiff and has a high tensile strength. The density is high (1·35–1·41 g/cm³) compared with those of polypropylene (0·90 g/cm³) and the polyethylenes (0·92–0·96 g/cm³) and so the yields are lower.

The water vapour permeability of PVC is higher than that of the polyolefins in general but is still adequate for many purposes. The gas permeability is lower, however, so that good protection against rancidity can be given to oils and fats.

Unplasticised PVC film has excellent resistance to oils, fats and greases and is also resistant to acids and alkalis. It is, however, softened by certain solvents, particularly chlorinated hydrocarbons and ketones. The films have little tendency towards blocking but static charge pick-up may cause trouble unless anti-static formulations are used.

Plasticised

To some extent the properties of plasticised PVC film depend on the type of plasticiser used, as well as on the amount. In general, increasing amounts of plasticiser increase the limpness and softness of the film and also improve its low temperature properties. Plasticised PVC is more likely to have an odour and is more liable to attack by solvents. Plasticised PVC films can be obtained with excellent gloss and transparency provided that the correct stabiliser and plasticiser are used.

Both plasticised and unplasticised films can be sealed by high frequency welding techniques but the speed of welding can be increased by the addition of certain plasticisers, such as organic phosphates. Printing is also possible on both types of film but some plasticisers and lubricants tend to migrate to the surface in flexible grades and may cause lifting of the inks. It is not necessary to pre-treat the surface of the film, and in this respect PVC film differs from polypropylene and the polyethylenes.

3.1.2 APPLICATIONS

Most of the applications for unplasticised PVC film lie in the field of packaging and are, therefore, discussed in Chapter 20.

A great deal of plasticised PVC is used in thicknesses over 250 μm and so should really be classified as sheet but there is no clear cut demarcation line. Plasticised PVC is used for curtains (particularly shower curtains), tablecloths and lightweight rain-wear. It is also used in the home in the form of decorative sheeting. This is adhesive backed, with a removable paper protective backing, and is used for covering shelves, whitewood furniture, etc.

Industrially, plasticised PVC film is used for the protective covering of equipment as well as for the lining of metal drums, to prevent corrosion when carrying certain liquids.

Applications for thin, plasticised PVC film are mainly in the field of packaging and are also dealt with in Chapter 20.

3.2 POLYVINYLIDENE CHLORIDE (PVDC)

The films normally used are, in fact, copolymers of vinylidene chloride and vinyl chloride but they are usually referred to as PVDC. Both casting and blow extrusion methods are used for the production of PVDC films although blow extrusion is preferred for the production of oriented film. If the PVDC film is produced by extrusion from a slit-die then it must be quenched, either by extruding into a cold water bath or by casting onto a chilled roller, in order to inhibit crystallisation. Minimum crystallinity is required in order to obtain a stretchable film. Because PVDC film shows an appreciable crystallisation rate at room temperature it must be oriented immediately after extrusion. The preferred method, however, for producing biaxially oriented PVDC film is by blow extrusion which produces transverse and longitudinal orientation simultaneously and easily.

3.2.1 PROPERTIES

Oriented PVDC film is a clear film with good strength characteristics, particularly burst strength. It is heat sealable at fairly low temperatures (120–158°C) but it is not particularly stable when heated for any length of time above 60°C. It has a high resistance to tear propagation but is difficult to handle on packaging equipment because of its limpness and 'cling'.

PVDC's outstanding property is its low permeability to water vapour and gases. It is likely to be used increasingly widely as a component in coextruded film laminates because it can often supply adequate barrier properties at gauges too thin to be handled in unsupported film form. It is already used to coat a wide variety of substrates, such as paper, regenerated cellulose and polypropylene but this entails a further manufacturing operation which coextrusion would eliminate.

The applications of PVDC film are dealt with in the chapter on 'Packaging'.

3.3 VINYL CHLORIDE/VINYL ACETATE COPOLYMERS

In these copolymers the acetate grouping is bulkier than the chlorine atom and serves to prevent close contact of the polymer chains. It acts, therefore, as an internal plasticiser. If extra flexibility is required, normal plasticisers can also be added. The material is used more in the form of sheet than film, a particularly important application being the manufacture of long playing records.

3.4 POLYVINYL ACETATE

Vinyl acetate can also be polymerised alone to give a material having some similarities to PVC but with a much greater solubility in organic solvents. Polyvinyl acetate is not used as a film but is important as an adhesive in laminating.

3.5 POLYVINYL ALCOHOL

Vinyl alcohol does not exist and polyvinyl alcohol is made by the hydrolysis of polyvinyl acetate. It is an unusual polymer in one other respect, being soluble in water. It is often utilised, therefore, in the manufacture of film sachets used to give controlled dosage in water. The sachet, plus contents, are simply added to the required amount of water, the sachet dissolves and the contents are released. This method is particularly valuable where the contents are toxic or where there are other reasons for not touching them.

FURTHER READING

MATTHEWS, G. A. R., *Vinyl and Applied Polymers, Vol. 2*, Iliffe, London (1972)

NASS, L. I. (Ed), *Encyclopaedia of PVC, Vol. 1* (1976), *Vol. 2* (1977), *Vol. 3* (1977), Marcel Dekker, New York

PENN, W. S., *PVC Technology*, MacLaren, London (1966)

PRITCHARD, J. G., *Polyvinyl Alcohol: Basic Properties and Uses*, MacDonald, London (1970)

4
Ionomers

The word ionomer was coined to describe a new family of polymers containing both covalent and ionic bonds. Covalent bonds are the usual ones between the separate atoms on each polymer chain while the ionic bonds exist between the chains. Ionic bonds are more powerful than the normal attractive forces between polymer chains and they modify a number of polymer properties.

The ionomers were developed by the du Pont Company and are sold under the tradename 'Surlyn' A (a range of 'Surlyn' D materials is also sold but these are dispersions and are used for coatings and primers rather than as films).

'Surlyn' A is based on ethylene and so resembles polyethylene in many of its properties. However, carboxyl groups are also located at intervals along the polymer chain by means of copolymerisation

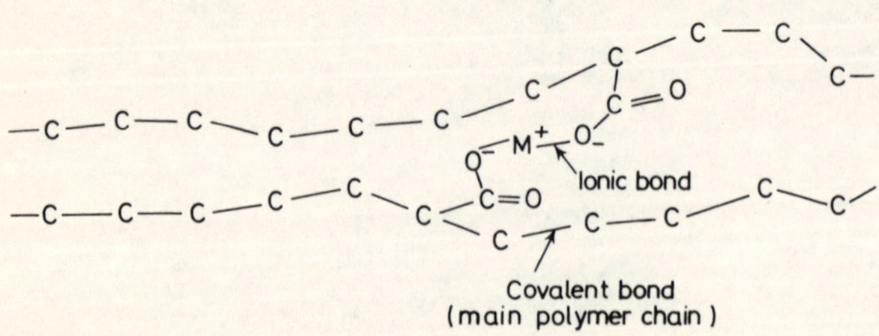

Figure 4.1. Ionic link in ionomer

and these provide the anionic portion of the ionic cross-links. The cationic portion of the links is supplied by metal ions. Typical metal ions used are sodium, potassium, magnesium and zinc. A typical ionomer has been found to contain 2.8 % of sodium which is equivalent to 17 sodium atoms per 1000 carbon atoms. A schematic representation of the structure of an ionomer is given in *Figure 4.1*.

The ionic links between the chains strengthen, stiffen and toughen the polymer without, however, destroying its melt processability. This is completely different from the conventional process of cross-linking because even though the ionic links are powerful, they are not as powerful as conventional covalent cross-links. The latter, of course, set up a permanent three-dimensional network which prevents the occurrence of melt flow. With ionic links, the application of heat causes them to weaken and processing is possible by all the conventional thermoplastic techniques. The residual ionic linkages are strong enough, however, to increase the melt strength markedly. The drawing characteristics of ionomers are extremely good, therefore, and they perform particularly well in extrusion coating and skin packaging applications.

The ionic forces also have a pronounced effect on the crystalline morphology. Low density polyethylene itself is a semi-crystalline polymer and can only be given good clarity by rapid quenching of the film, with a consequent repression of spherulite growth. The ionic forces in ionomers virtually eliminate all traces of visible spherulites thus giving excellent clarity. This is not accompanied by a reduction in impact strength as is the production of high clarity low density polyethylene film.

One other difference brought about by ionic linkages is the fact that, unlike other transparent polymers, ionomers are completely unchanged in appearance by exposure to any organic solvent at room temperature. No crazing or other type of surface attack is observable and ionomers cannot, in fact, be completely dissolved in any commercial solvent even at elevated temperatures.

Properties

Ionomers are flexible, tough materials with extremely good clarity. Their toughness is even more noticeable at low temperatures where they are able to outperform ordinary low density polyethylene. Under equivalent test conditions, namely, a steel ball rolling down an inclined plane against a refrigerated sample, a commercial ionomer withstood the impact down to −99°C whereas low density polyethylene failed at −68°C. The question of the increase in melt strength has already been mentioned and is important for a number of reasons.

(1) It determines how much draw down can be achieved in a given cycle time, for a given film thickness.

(2) How puncture resistant the film will be during processing.

(3) How tightly the product will be secured to the base in skin packaging.

From plots of melt strength against draw rate, it has been shown that ionomers are approximately ten times tougher than low density polyethylene in the molten state at any given draw rate. This means that sharper objects can be packaged in ionomer films, at higher speeds, without puncture. In addition, less pinholing or tearing is likely to occur during deep draw applications.

The additional polar groups in the molecule absorb more infra-red radiation than does a polyethylene molecule and thus ionomer films heat more quickly under infra-red heaters. This factor, taken in conjunction with the greater melt strengths, means faster heating plus resistance to excessive sagging in the frame of a skin packaging machine and so results in faster cycles. Ionomers are extremely tough and flexible and have tensile strengths above those of either low density or high density polyethylene. They also exhibit good abrasion resistance, comparable with that of materials such as nylon, polycarbonate and polyacetal. Tabor abrasion test results showed 'Surlyn' A to have lost 3·9 mg/1000 cycles compared with a polycarbonate sample which lost 6–7·1 mg in the same number of cycles.

Ionomers have good filler acceptance and can be compounded with high proportions of inexpensive fillers without a serious loss of physical properties.

Ionomer films have a greater resistance to oils and greases than has low density polyethylene at room temperature but the differences in behaviour are much reduced at elevated temperature. Chemically they are resistant to weak and strong alkalis but they are slowly attacked by acids. Other chemical properties are similar to those of low density polyethylene, the ionomers being resistant to ketones, esters and alcohols but swelling slightly in hydrocarbon solvents. Stress cracking resistance is also good and is claimed to be higher than that of low density polyethylenes of equivalent melt flow index. For outdoor exposure, the use of carbon black as an antioxidant is usually necessary, as in the case of low density polyethylene.

The gas permeability of ionomer films is similar to that of low density polyethylene but the water vapour permeability is somewhat higher, as is the water absorption. The polar nature of ionomer films means that printing is easier than on low density polyethylene but some pre-treatment is still necessary.

Electrical properties are similar to those of low density polyethylene but the polar nature again has an effect and the volume resistivity is lower (10^{17} ohms cm as against 10^{17}–10^{19} ohm cm for low density polyethylene) although it is still higher than that of high impact polystyrene or polycarbonate. The dielectric strength is higher than that of low density polyethylene, however, while the dielectric constant is the same.

As already mentioned, the film is extremely clear and figures for haze as low as 1% for 30 μm thick film have been reported. Ionomer films can be made by either the blow extrusion or slit-die, chill-roll casting process but no indication was given as to which method was used for the film giving 1% haze.

Strong heat seals can be obtained with ionomer films and the combination of high melt strength and relatively high viscosity of the ionomer resin reduces squeeze-out or package distortion during heat sealing operations. On the other hand, these properties dictate a firm control of jaw pressure during heat sealing in order to obtain optimum strength. At low pressures a seal may be obtained, but it is peelable. This can be a desirable property in some instances but not when complete seal integrity is essential.

5
Styrene Polymers and Copolymers

5.1 POLYSTYRENE

Polystyrene is made by the addition polymerisation of styrene, 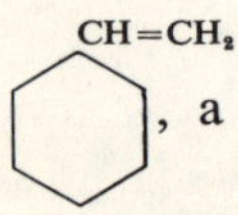, a mobile liquid, boiling at 145°C. The simplest method of polymerisation is to heat styrene but the reaction itself gives off heat and this is difficult to remove. If the process is a batch one the solid reaction product is difficult to handle. One way round this problem is to trickle a partially polymerised solution of polystyrene in styrene down a heated tower. The molten polymer so formed is collected at the base of the tower then extruded and granulated.

Another polymerisation method is to suspend the styrene in water by stirring rapidly, together with a stabiliser, and a styrene soluble catalyst, such as benzoyl peroxide. If the stirring is carried out correctly, the styrene is dispersed as small droplets which polymerise to give polystyrene beads. The presence of large amounts of water helps to reduce the build-up of heat which occurs in the bulk process. Other processes which have been used for the polymerisation of styrene include emulsion and solution ones.

Polystyrene film can be manufactured either by blow extrusion or by slit-die extrusion. It is intrinsically a brittle film and is unsuitable for packaging applications, without modification. It has, however, been used for electrical applications in the form of a dielectric in capacitors. Biaxial orientation of polystyrene by the methods outlined previously overcomes the problem of brittleness and gives a clear, sparkling film.

5.1.1 PROPERTIES

In addition to possessing excellent clarity and sparkle, biaxially oriented polystyrene is rigid and has a high tensile strength. Its impact strength is much improved by orientation and it has good resistance to low temperature. As one would expect from a pre-stretched film, its elongation is low. It has a high refractive index (1.59) and a softening point of around 90–95°C.

It has a medium permeability to gases (being more permeable than polypropylene but less permeable than low density polyethylene) but a high permeability to water vapour. The moisture vapour transmission rate decreases rapidly at temperatures below 0°C, however, so that it is quite suitable for use as a food packaging material at low temperatures.

It has outstanding electrical properties including a very low power factor, high dielectric strength and a high volume resistivity. Chemically it is resistant to strong acids and alkalis and is insoluble in aliphatic hydrocarbons and the lower alcohols. It is soluble, however, in aromatic hydrocarbons, the higher alcohols, esters and chlorinated hydrocarbons.

Oriented polystyrene film can be thermoformed into quite intricate shapes although special techniques have to be used because of its orientation which leads to a tendency to shrinking on heating. Applications, apart from the use in capacitors already mentioned, lie mainly in the field of packaging and so are dealt with in Chapter 20.

5.2 HIGH IMPACT POLYSTYRENE

As stated earlier in its unmodified state, polystyrene is a brittle material and its impact strength is not always adequate for certain applications. Orientation of polystyrene film is one way of improving the impact strength but even this is not always good enough and the material is also rather expensive.

Another approach is the incorporation of synthetic rubbers in the polystyrene to give a toughened polymer. There are two main methods of achieving this:

(1) By mechanically mixing the rubber and the polystyrene on a two-roll mill or in an internal mixer.

(2) Inter-polymerisation of a solution of rubber in styrene.

If a suspension method is used, the rubber and a catalyst are dissolved in the styrene, followed by dispersion by stirring in water with a suitable suspension agent, such as polyvinyl alcohol. The polymerisation is completed by heating the suspension. Solvent methods are also used, toluene being a suitable solvent. The viscous liquid resulting from polymerisation is transferred to a batch still, where it is heated under reflux to complete the polymerisation. The solvent is then distilled off and the high impact polystyrene is extruded into strands and chopped into granules. Bulk processes are also possible, using a tower reactor.

The incorporation of synthetic rubber to increase the impact strength and flexibility of polystyrene also, unfortunately, affects its clarity and even in thicknesses down to 100 μm it is no more than translucent. The film is produced by slit-die extrusion and is sometimes given a post-extrusion process to improve the gloss. In heat glazing, infra-red heaters are situated immediately in front of the extruder die lips, while an alternative method is roll polishing where the gap between the rollers is set at less than the film thickness, thus impressing the gloss from the rollers.

5.2.1 PROPERTIES

Toughened, or high impact, polystyrene is more flexible and has a higher impact strength than has the unmodified material, but this is accompanied by a reduction in tensile strength and thermal resistance. The surface hardness is also reduced and the material has a high elongation at break. Chemical properties are much the same as those for unmodified polystyrene.

High impact polystyrene is an excellent material for thermoforming and draw ratios of up to 3:1 have been obtained with a good control of wall thickness. Applications for toughened polystyrene film lie either in the field of packaging or in that of synthetic papers. These are dealt with in Chapters 20 and 23 respectively.

5.3 EXPANDED POLYSTYRENE

The manufacture of expanded polystyrene film/sheet is described, along with that of other foamed materials in Chapter 8.

5.3.1 PROPERTIES

Expanded polystyrene film has a non-communicating cell structure and so it does not pick up water. It is a good cushioning material (when used in sufficient thickness) and has good thermal insulation properties. It has an attractive, non-abrasive surface and it can be thermoformed to give trays, interleaves or containers. Because of its thermal insulation properties, double-sided heaters are essential when thermoforming. The film can be embossed by passing it through cold embossing rollers and this opens up applications in fancy wrappers and high quality labels.

When originally reeled up, the film has a density of about 128–160 g/l (8–10 lb/cu. ft.) but this can be reduced to 40–48 g/l (2½–3 lb/cu. ft.) by a post-expansion process involving reheating with steam or under infra-red heaters. The attainment of low densities is important economically, of course, because thicker film can be made from the same weight of material and there is thus an overall advantage in rigidity per unit weight (see Chapter 8).

5.4 STYRENE/ACRYLONITRILE COPOLYMER (SAN)

Styrene/acrylonitrile copolymers are usually produced by suspension or emulsion polymerisation techniques similar to those used in polystyrene.

5.4.1 PROPERTIES

These copolymers have better chemical resistance, surface hardness and more resistance than has the polystyrene homopolymer. The natural material has a yellowish tinge and it is usual to add blueing to counteract this. Resistance to weathering is good and this allows the copolymer to be used in outdoor applications, such as glazing (usually as a component in do-it-yourself double glazing).

5.5 ACRYLONITRILE/BUTADIENE/STYRENE (ABS)

There are several methods of making these co—or terpolymers but the general principles can be seen from the following examples.

(1) Styrene and acrylonitrile are added to a polybutadiene emulsion and the mixture stirred and warmed to 50°C. A water

soluble initiator such as potassium persulphate is then added and the mixture is polymerised.

(2) A butadiene/acrylonitrile rubber latex is added to a styrene/acrylonitrile latex. The mixture is then coagulated and spray-dried.

5.5.1 PROPERTIES

As can be imagined the properties vary widely, according to the composition and the particular manufacturing method. In general, however, ABS has good impact strength, good chemical resistance and is very tough. It is attacked by methyl ethyl ketone and by esters. Applications for film and sheet lie mainly in the higher gauges but the thinner material is thermoformed into items such as tubs and trays. ABS sheet can also be formed into pots by cold forming techniques. Manufacturing cycles are much shorter because of the elimination of the heating and cooling cycle while another advantage is the fact that there is less distortion during forming so that printing can be carried out on the flat sheet.

FURTHER READING

BOUNDY, R. M. and BOYER, R. F., *Styrene, its Polymers, Copolymers and Derivatives*, Reinhold, New York (1952)

TEACH, W. C. and KIESSLING, G. C., *Polystyrene*, Reinhold, New York (1960)

6 Cellulose and Cellulose Derivatives

6.1 REGENERATED CELLULOSE

The manufacture of regenerated cellulose film is described in Chapter 8 on manufacturing methods and this chapter only covers its properties. Before dealing with these, however, it will be helpful to consider the chemical structure of cellulose. Cellulose is a carbohydrate and has the chemical structure shown below:

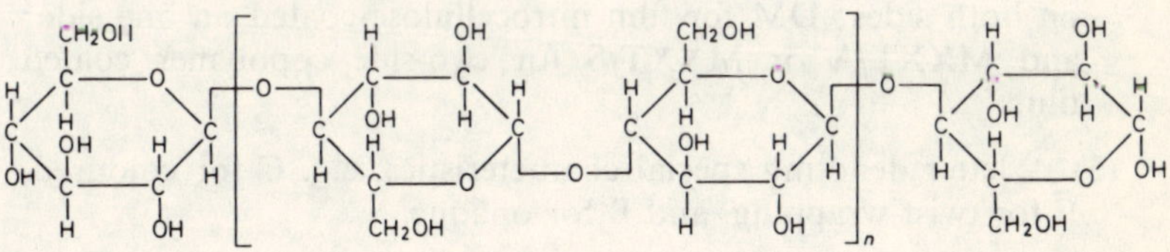

Since cellulose has a structure consisting of long chain molecules with no cross-linking it would appear that it ought to be a thermoplastic. However, the presence of a large number of hydroxyl groups in each molecule (three-OH groups per six carbon atom group) leads to a high degree of hydrogen bonding and the chains are thus very strongly attracted to each other. The application of heat, therefore, leads to degradation (charring) before the point is reached at which the hydrogen bonding could be overcome sufficiently to allow relative movement of the molecules to take place. The insolubility of cellulose is also due to the hydrogen bonding which must be overcome before solubility can be achieved.

6.1.1 PROPERTIES

There are very many types of regenerated cellulose film, including moistureproof and non-moistureproof grades, heat sealable and non-heat sealable grades, single and double side coated films and so on. Some of the more important grades are given here together with the coding which has now been agreed among the major suppliers.

The codes used are based on four elements which should be written in the following order:

(1) A code number to indicate the approximate weight in grammes per 10 m^2 of film.

(2) One or more letters denoting the basic types of film: P for non-moistureproof; M for standard film nitrocellulose-coated on both sides; DM for film nitrocellulose-coated on one side; and MXXT/A or MXXT/S for two-side copolymer coated film.

(3) A letter denoting special characteristics, e.g. C for coloured, F for twist-wrapping, and B for opaque.

(4) A number used as a suffix where an additional description is required to cover special features of manufacture and certain end-uses.

The following are the code names and suffixes that have been agreed by the U.K. makers of cellulose film.

Code Letters

M — nitrocellulose-coated on both sides. Coating is heat-sealable only when the letter 'S' is included.

QM — nitrocellulose-coated on both sides but less moisture-proof than M.
PS — nitrocellulose-coated on both sides but less moisture-proof than QM.
DM — nitrocellulose-coated on one side.
P — non-moistureproof.
MXXT/A — copolymer-coated by aqueous dispersion on both sides.
MXXT/S — copolymer-coated by solvent process on both sides.
MXDT/A — copolymer-coated by aqueous dispersion on one side.
MXDT/S — copolymer-coated by solvent process on one side.
B — opaque.
C — coloured (name of colour follows, e.g. 325PFC red).
F — for twist-wrapping.
S — heat-sealable.
U — for adhesive tape manufacture.
V — has high surface slip.

Suffixes

11 — for tobacco.
12 — for tobacco in areas of high humidity.
21, 24 — for release-agent use.
30, 36, 39 — films with increased flexibility (30 for frozen foods: 36 for biscuit wrapping).
71 — anti-oxidant.
81 — matt.

Some examples of the use of these codings are given below.

350MS is a moistureproof film, nitrocellulose-coated on both sides and is heat sealable. Weight of 10 m^2 of film equals 350 g.

345QMS is a somewhat less moistureproof film, nitrocellulose-coated both sides and is heat-sealable. Weight of 10 m^2 of film equals 345 g.

325P is a non-moistureproof film, uncoated. Weight of 10 m^2 of film equals 325 g.

The more general properties of regenerated cellulose film include a transparency equal to that of glass in the visible region (about 90% transmission) and greater than glass in the u.v. region. It has a high tensile strength and bursting strength and is a good barrier to oils, greases and odours. Dry cellulose film is practically impermeable to gases, but it becomes more permeable when wet. The basic film has a very high moisture vapour permeability but this, as we have seen, can be markedly reduced by the addition of suitable coatings.

Regenerated cellulose film is comparatively free from static electricity problems and can be handled easily on automatic wrapping machines. It is readily printed with good results especially on the basic film.

It should be noted that creasing of moistureproof cellulose film adversely affects its water vapour permeability because of damage to the coating.

The applications of regenerated cellulose film lie mainly in the field of packaging and so are dealt with in Chapter 20.

6.2 SUBSTITUTED CELLULOSES

We have seen that the reason why cellulose is not a thermoplastic is that hydrogen bonding, due to OH groups, is so strong that the application of heat leads to degradation before the bonding can be overcome sufficiently to allow relative movement of the molecules. If enough of these OH groups are substituted by other chemical groups the hydrogen bonding can be reduced sufficiently to produce a thermoplastic. The substitution must be uniform, however, or local hydrogen bonding may occur. Before any substitution treatment is carried out, therefore, the cellulose is given a pre-treatment to open up the structure so that the substituting reagent can penetrate it uniformly and rapidly. Two main types of cellulose derivatives are commercially available, the esters and the ethers. In esterification the hydroxyl groups are replaced by ester groups such as nitrate or acetate, while in etherification, the hydrogen atom of the hydroxyl group is replaced by groupings such as ethyl, methyl or benzyl, the oxygen atom acting as a bridge. As far as film uses are concerned the esters are more important than the ethers. Methyl cellulose is a water soluble powder and is used as a thickening agent and emulsifier while benzyl cellulose is expensive, has a low softening point and is unstable to heat and light. Ethyl cellulose is used as a solvent-based strippable coating and also forms a film which is tough and flexible at low temperatures.

Of the esters, cellulose nitrate is important as the forerunner of our modern plastics industry. The other cellulose esters which will be described here are cellulose acetate, cellulose acetate/butyrate and cellulose propionate.

6.3 CELLULOSE NITRATE (CELLULOID)

Cellulose nitrate is made by reacting cellulose with mixed nitric and sulphuric acids. An average of 2 out of 3 hydroxy groups are replaced by nitrate ($O.NO_2$) groups. It has to be plasticised before it can be processed, the usual plasticiser being camphor. Cellulose nitrate film is made by a solvent casting process as described in Chapter 8.

6.3.1 PROPERTIES

Cellulose nitrate is a clear, tough, rigid film with a slightly yellow tint. This basic colour does not, however, interfere with the colourability and celluloid can be obtained in an unlimited range of colour effects. It is dimensionally stable and is resistant to dilute acids and alkalis. It is highly flammable, burning with a hot, white flame. It tends to discolour on ageing and becomes brittle due to loss of camphor. It is soluble in methanol, ether/alcohol mixtures, esters and hydrocarbons.

6.3.2 APPLICATIONS

Thin celluloid sheet is still the main material used for table tennis balls and a great deal is still used for decorative purposes such as covering various musical instruments. Its main use as a film was in cinematography and photography but it has now been replaced by cellulose acetate or polyethylene terephthalate because of its flammability.

6.4 CELLULOSE ACETATE

The normal product of acetylation of cellulose is one in which substantially all the hydroxyl groups are replaced by acetate ones, i.e. the product is cellulose triacetate. This is soluble only in expensive or toxic solvents such as chloroform or tetrachloroethane and this held up its use for some time. It was then shown that a mild acid hydrolysis of the normal acetylation product gave a cellulose acetate (known as secondary cellulose acetate) which was soluble in acetone. The hydrolysis is carried out until an average of around one out of three of the acetyl groups is hydrolysed back to hydroxyl.

The acetylation is carried out by pre-treating the cellulose with glacial acetic acid to open up the structure, then reacting with a mixture of acetic anhydride, sulphuric acid (as a dehydrating agent) and methylene chloride as solvent. The fully acetylated material is then stirred with dilute sulphuric acid until the desired degree of hydrolysis is achieved. Sodium acetate solution is then added to prevent further deacetylation. The methylene chloride is distilled off and recovered. The cellulose acetate is finally washed and dried. If required as cellulose triacetate, of course, the reaction is stopped after acetylation. Both types of cellulose acetate now have their particular uses. It should be noted that where the triacetate is required as a fibre, the fibrous form of the original cellulose is preserved by carrying out the acetylation in a non-solvent medium.

6.4.1 PROPERTIES

Secondary cellulose acetate

The properties of cellulose acetate vary with the degree of acetylation, and the degree and type of plasticisers used but in general, it is a hard, tough material with good electrical properties when dry, combining good insulating properties with good corrosion and arc resistance.

Secondary cellulose acetate film is sensitive to moisture absorption and so is not dimensionally stable in conditions of changing relative humidity. It possesses crystal clarity and good gloss and has a high tensile strength when dry. It is easily printed and is often reverse printed and used as the outer layer in a laminate in order to give scuff-proof decoration. Plasticisers are incorporated so that care must be taken when choosing a grade for food packaging.

Although the film is softened by heat it is not readily heat sealable but is usually sealed with adhesives. It can also be coated with a heat sealable composition. Its permeability, both to water vapour and gases is high. Weak acids and alkalis have only a slight effect but strong acids and alkalis cause decomposition. Secondary cellulose acetate is soluble in acetone and 80/20 acetone/water mixtures but is insoluble in methylene chloride whereas the triacetate has exactly opposite solubilities. In thicker gauges it is easily fabricated by cutting and creasing, and by thermoforming.

6.4.2 APPLICATIONS

Cellulose acetate film has many uses, particularly in the fields of packaging and display, which are dealt with in Chapter 20. It has also found use as a glazing material for greenhouses, chicken coops and cold frames, since it is generally transparent to u.v. light (u.v. light-transmission through 250 μm film—about 85%: visible light—transmission through 250 μm film—about 93%). Reinforcement is sometimes given to the film by placing cotton scrim or wire mesh between two layers of film and laminating in a hot press, or by passing the reinforcement through a cellulose acetate casting solution. Cellulose acetate is also used as a dielectric in capacitors and for electrical tape.

6.4.3 CELLULOSE TRIACETATE

Properties

The triacetate film is less susceptible to moisture and is more dimensionally stable. It is also less permeable to gases and moisture vapour than the secondary acetate. As mentioned under the properties of secondary cellulose acetate, the triacetate is soluble in methylene chloride and insoluble in acetone and 80/20 acetone/water mixture.

Applications

Triacetate film is used mainly for cinematography or X-ray film although it is now experiencing competition from the physically tougher polyethylene terephthalate.

6.5 CELLULOSE ACETATE/BUTYRATE (CAB)

The manufacture of CAB is similar to that of cellulose acetate, using mixed acetic and butyric acids as the esterifying agent.

6.5.1 PROPERTIES

CAB has a higher softening point than has cellulose acetate and a better resistance to acids. Although it is discoloured by oxidising acids such as 10% nitric acid, it resists 10% hydrochloric acid and 30% sulphuric acid. The moisture absorption of CAB is lower than that of cellulose acetate and this, plus the fact that less plasticisers need to be used to attain any particular performance, makes it more dimensionally stable and gives it better resistance to weathering. It also has a good resistance to petrol and oil.

One drawback is that on storage it generates the smell of butyric acid, which closely resembles that of rancid butter.

6.5.2 APPLICATIONS

An interesting application is the use of CAB as a strippable film to protect tools, gauges and similar items during storage. A plasticised formulation is heated and the object dip-coated. When the object is required for use the coating is easily removed by stripping.

Acetate/butyrate film can be thermoformed and is used for a range of transparent boxes and blisters. It is not suitable for food packaging because of the odour problem mentioned above.

6.6 CELLULOSE PROPIONATE

The manufacture of cellulose propionate is similar to those of the other cellulose esters but using propionic acid. It is dimensionally stable, hard but not brittle and has good flexural strength. It is resistant to weak acids and alkalis and to petrol. It is used for blister packs to hold bulky objects because of its improved toughness.

FURTHER READING

YARSLEY, V. E., FLAVELL, W., ADAMSON, P. S. and PERKINS, N. G., *Cellulosic Plastics*, Iliffe, London (1964)

7
Miscellaneous

7.1 NYLONS (POLYAMIDES)

The first nylons were prepared by the condensation of di-acids with di-amines and were characterised by a number derived from the number of carbon atoms in the parent compounds. Thus, nylon 6.6 is the condensation product of adipic acid, $COOH(CH_2)_4COOH$ and hexamethylene diamine, $NH_2(CH_2)_6NH_2$

$$n.\ COOH(CH_2)_4\ COOH + n.\ NH_2\ (CH_2)_6\ NH_2 \rightarrow$$
$$\left[CO(CH_2)_4\ CO.NH(CH_2)_6\ NH \right]_n + 2n.\ H_2O$$

Later, methods were developed for the manufacture of nylons by the condensation of certain ω-amino acids. Nylons prepared by this route are characterised by a single number derived from the number of carbon atoms in the parent amino acid. Nylon 11, for example, is prepared from ω-amino -undecanoic acid, $NH_2(CH_2)_{10}COOH$.

Of the wide range of nylons now available, the ones most commonly used in film form are nylons 11, 12, 6 and 6.6. In addition to unsupported nylon films (produced by casting or blow extrusion) there is a range of laminates, including coaxially extruded ones such as polyethylene/nylon. A more recent development is a thin, biaxially oriented film, produced in Japan and Germany.

7.1.1 PROPERTIES

In general, nylons are tough materials with high tensile strength and good resistance to abrasion. They also have high softening points and can withstand steam sterilisation (up to about 140°C) and dry heat to even higher temperatures. They retain their flexibility well at low temperatures so that they can be used over a wide temperature range. Nylons, however, have rather high water absorption and their mechanical properties can consequently be affected by water. The effect is not permanent and the full properties are restored on drying. Nylons 11 and 12 are better in this respect and have a lower water absorption than nylons 6 and 6·6. Nylon films have good impact and burst strength and are fairly readily heat sealed. They are polar in nature and can also be sealed by high frequency methods.

Nylon films have fairly high moisture vapour permeabilities but they are very good gas barriers and can be used for vacuum packaging. They also form good barriers to odour. Because of their polar nature, nylon films can be printed without pre-treatment.

Chemically, nylon films are resistant to weak acids but are attacked by concentrated mineral acids. They are resistant to alkalis, even at high concentrations and are particularly resistant to organic solvents, oils and greases.

The transparency of nylon films is excellent, especially when biaxially oriented but their gloss is only fair although this, too, is improved by orientation.

Electrical properties of nylons are affected by the ambient humidity in the same way as their mechanical properties. At low humidities, nylons have fairly good electrical insulating properties but these deteriorate as the humidity increases. As with the mechanical properties, the effect is not permanent and the electrical properties are restored to their full value when the material is dried.

Biaxial orientation of nylon films improves certain properties such as mechanical strength, stiffness and transparency. The permeability to water vapour and gases is also reduced.

7.1.2 APPLICATIONS

Many of the uses for nylon films lie in the field of packaging and are, therefore, dealt with in Chapter 20 on 'Packaging'.

Non-packaging uses include blown nylon film for tapes and cable insulation while thin gauges of the oriented film have been vacuum metallised and used for stamping foils and metallised yarns.

7.2 POLYCARBONATE

Polycarbonate is a linear polyester of carbonic acid and is made by an ester interchange reaction between bisphenol A and diphenyl carbonate:

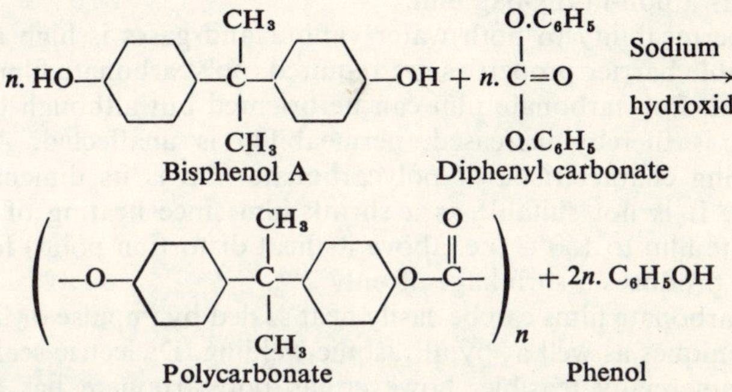

Polycarbonate film is usually made by slit-die extrusion onto a polished metal roller. The temperature of the 'chill' roller is quite high (about 130°C) because the softening point of polycarbonate is over 200°C. Solvent casting is also used to produce polycarbonate films but the high cost limits this method to the production of photographic film base in which optical properties are important.

7.2.1 PROPERTIES

Polycarbonate is outstanding in its combination of high temperature resistance, high impact strength and clarity. It also retains its properties well with increasing temperature. At 125°C, for example polycarbonate film still exhibits a tensile yield strength of around 31 MN/m² (4,438 lb/in²) which is about that of high density polyethylene at room temperature. Low temperature properties are also excellent, the brittle point being below −135°C. Tear initiation and tear propagation resistances are fairly high while fold endurance is only moderate. Bursting strength is high.

Chemically, polycarbonate is resistant to dilute acids but is strongly attacked by alkalis, and bases such as the amines. It is resistant to aliphatic hydrocarbons, alcohols, detergents, oils and greases but is soluble in chlorinated hydrocarbons, and methylene chloride is used in the solvent cementing of polycarbonate. It is also partially soluble in aromatic hydrocarbons, ketones and esters and these materials act as stress-crazing agents at elevated temperatures or under conditions of stress. Immersion in water at room temperature for several weeks causes no significant changes in tensile properties or elongation but boiling water does affect elongation. After only four hours, elongation drops from 100% to 50%. However, ultimate and yield tensile strengths are not affected even after one week in boiling water, and the material is perfectly adequate for use as a boil-in-the-bag film.

The permeability to both water vapour and gases is high and if appreciable barrier properties are required, polycarbonate film must be coated. Polycarbonate film can be oriented but although tensile strength is thereby increased, permeability is unaffected. A distinguishing characteristic of polycarbonate film is its dimensional stability. It is not suitable as a shrink film since heating of polycarbonate film to 150°C (i.e. above its heat distortion point) for ten minutes produces a shrinkage of only 2%.

Polycarbonate films can be easily heat sealed by impulse or heated jaw techniques as well as by ultrasonic welding. Dielectric sealing is not commercially feasible, however, as polycarbonate has a low power factor. Because of its high softening point, normal heat sealing is not very rapid and sealing times of 7 s for 18 μm film and 8 s for 15 μm films have been quoted.

Thermoforming of polycarbonate film is readily carried out and deep draws with good mould detail are obtainable. Rapid gel times are obtainable with polycarbonate because it sets at a high temperature so that cooling times are much reduced.

Good printing results are obtainable with silk screen, flexographic and gravure processes but as with other films the ink used must be correct in terms of compatibility and performance. In particular, polycarbonate films are preferable at high temperature because of their heat resistance and the inks must be similarly heat resistant. Vacuum metallising gives good results because of the polycarbonate's transparency and the finished material has a high gloss.

7.2.2 APPLICATIONS

The majority of polycarbonate film applications are packaging ones and so are dealt with in Chapter 20 on 'Packaging'.

7.3 POLYETHYLENE TEREPHTHALATE (POLYESTER)

This is a linear polyester and, like Nylon 6.6, was first prepared as a fibre. It was first formed by the reaction between ethylene glycol

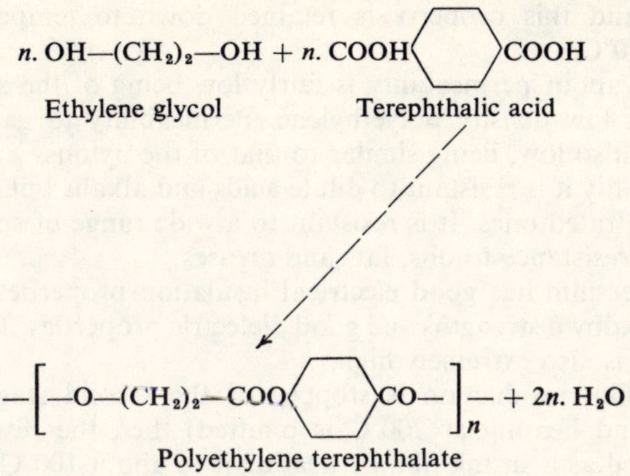

and terephthalic acid but is commonly made commercially by ester interchange of dimethyl terephthalate.

The process for making the film itself is rather more complicated than for some other films and deserves special mention. The film is quenched immediately after extrusion to prevent it from crystallising and is then biaxially oriented by stretching it at a temperature above 80°C. This has the usual effect of improving the mechanical properties and reducing brittleness. If the film is subsequently re-heated to above 80°C it will tend to contract considerably. A third manufacturing stage is added, therefore, to improve heat stability. The oriented film is firmly held to prevent contraction and is then heated to about 200°C. This causes a considerable amount of crystallisation without appreciably affecting the orientation and, when subsequently cooled, the film is stable dimensionally up to 200°C.

7.3.1 PROPERTIES

Polyester films are very tough and strong and have excellent transparency. However, slip characteristics are poor unless slip additives are incorporated, and these make the film slightly hazy. No other additives are present and the film is inert towards foodstuffs. Heat sealing is difficult because of shrinkage and embrittlement (due to crystallisation—see Chapter 15 on 'Sealing of Film') which occurs at the heat seal. Adhesives are sometimes used for sealing but the best results are obtained by laminating with low density polyethylene film and utilising the excellent heat seal properties of the latter.

Polyester film has a high resistance to tearing and has a high fold endurance. It is also resistant to abrasion. The impact strength is good and this property is retained down to temperatures of about −70°C.

Water vapour permeability is fairly low being of the same order as that of low density polyethylene. Permeability to gases and to odours is also low, being similar to that of the nylons.

Chemically it is resistant to dilute acids and alkalis but is attacked by concentrated ones. It is resistant to a wide range of solvents and has good resistance to oils, fats and greases.

Polyester film has good electrical insulation properties including high breakdown strengths and good dielectric properties. The volume resistivity is also extremely high.

If the film production is stopped at the second stage (i.e. the holding and heating at 200°C is omitted) then the resultant film can be used as a shrink film by heating it to about 100°C.

7.3.2 APPLICATIONS

Polyester film which is taken after the first stage of manufacture (i.e. before orientation) can be thermoformed although the amount of draw in all directions, and the thermal conditions during forming are critical because of a tendency to split and become opaque due to premature crystallisation. In the U.S.A., thermoformable laminates have been developed, consisting of 12.5 μm polyester plus 50 μm low density polyethylene. The polyethylene gives the material heat sealability and better barrier properties.

Second stage, or shrinkable, films have been used for the wrapping of cooked meat products where vacuum retention is essential because of the deleterious effects of oxygen. Polyester bags have also been used for the long-term storage of frozen poultry as well as a number of other packaging applications, mentioned in Chapter 20.

In the U.K., the high cost of polyester film has inhibited its use in packaging although its very high strength allows it to be used in thicknesses down to 6 μm. However, at this thickness it is not easy to handle.

There are a great number of non-packaging uses for polyester films, including magnetic recording tapes, pressure sensitive tapes and cable wrapping. The use of polyester film for recording tapes allows longer playing time per given reel size because it can be used in thinner gauges due to its high strength. There are many other electrical uses such as the manufacture of small capacitors. Polyester film's higher breakdown strength means that smaller capacitors can be made for a given voltage when it is used instead of paper as the dielectric. Vacuum metallisation can lead to even further miniaturisation. Electric motors, too, can be insulated with a single layer of polyester film. Its dimensional stability, allied to its insulation value, makes it suitable as a base for flexible circuits when laminated to copper foil.

The surface of polyester film can be roughened mechanically, or otherwise modified by laquering, so as to accept pencil and ink marking. It is then used in the manufacture of drafting materials for original drawings or for tracing. The properties which make it particularly suitable for this sort of work are its dimensional stability, durability and high tear resistance. It has also been used in book binding and as a base for carbon 'papers'. A similar application to the latter is its use for electrical typewriter ribbons where a clean release of the ink and sharp impression of the type is important.

Polyester films have a number of decorative uses, especially when metallised by vacuum deposition methods. Metallised polyester film is laminated to PVC, for instance, and used in the manufacture of handbags, shoes and decorative trim (in cars, etc.). Lacquered metallised film is slit into very narrow strips (about 1 mm wide) and then used as textile threads. Christmas decorations and hot stamping foils are other decorative uses. An interesting use in this field is for transferring the gold embossing on to U.K. commemorative postage stamps. Finally, polyester film is used in the manufacture of ultra-lightweight mirrors. The film is metallised and stretched over a framework attached to a backing board in such a way that the film is clear of the backing board. In addition to being light in weight, the extremely low heat content of the film means that condensation rarely occurs because the film rapidly attains thermal equilibrium with the atmosphere. These lightweight mirrors have been used in aircraft washrooms.

7.4 ACRYLIC MULTIPOLYMER

This is a terpolymer, one monomer being an acrylic. The other two monomers are not disclosed. The material is known as XT Polymer in the U.S.A. but is imported into the U.K. under the name Cyanacryl. Extrusion is carried out by slit-die methods, the minimum thickness available being 125 μm.

7.4.1 PROPERTIES

The transparency of acrylic multipolymer is good and is improved still further when in contact with liquids. It has a low gas permeability (comparable with that of PVC) and this, together with its transparency had led it to compete with PVC in the U.S.A. The price in the U.S.A. is only slightly greater than that of PVC but acrylic multipolymer has a 20% higher yield. In addition, it has greater thermal stability than has PVC and so a greater use of re-ground scrap is possible. The price situation in the U.K. is not so favourable, however, as acrylic multipolymer is almost double the cost of PVC.

One useful characteristic of acrylic multipolymer is its strength and rigidity in thin wall sections. Reductions in wall thickness of about a third have been claimed in the U.S.A. as compared with a similar container in high density polyethylene. It is a tough material and has a very good impact strength. It maintains its toughness at low temperatures (down to about −40°C) and it also has a high heat distortion temperature which enables containers to be hot filled.

Chemically it is resistant to aliphatic hydrocarbons, acids and alkalis and detergents. It is highly resistant to oils, fats and greases. However, resistance to aromatic hydrocarbons and chlorinated solvents is poor and high concentrations of alcohol should also be avoided.

Gas barrier properties of acrylic multipolymer are good and aroma transmission is also low. Water vapour transmission, however, is higher than that of polyethylene or PVC.

Acrylic multipolymer can be decorated by hot stamping or silk screen printing and no pre-treatment is necessary. The material will also accept pressure sensitive or heat sensitive labels. The uses for acrylic multipolymer films are still mainly those for packaging and are mentioned in Chapter 20.

7.5 ACRYLONITRILE/METHYL ACRYLATE COPOLYMER (BAREX)

This material is also an acrylic, consisting mainly of acrylonitrile copolymerised with methyl acrylate and a small percentage of a butadiene/acrylonitrile rubber. It has an interesting combination of properties including good clarity and high barrier properties. It also has good impact strength and creep resistance. It was developed in the U.S.A. and was initially intended as a bottle blowing material for carbonated drinks. However, it is now produced as a film and when laminated with other materials such as low density polyethylene it is suitable for thermoforming containers for cheese and meat. In these applications it will compete with nylon/PVDC laminates by virtue of its low gas permeability, its resistance to UV light and its toughness.

7.6 PROPYLENE/VINYL CHLORIDE COPOLYMER

These copolymers are, essentially, modified PVC's, the propylene content being up to about 10% by weight. Copolymerisation with propylene improves the thermal and processing stability and makes possible a more rapid extrusion rate. Although these copolymers were originally developed as bottle blowing and injection moulding compounds, they are claimed to be equally suitable for film and sheet extrusion.

7.6.1 PROPERTIES

A range of flexibilities can be obtained, from rigid to very flexible. The rigid materials are claimed to retain the desirable properties of PVC homopolymer resins such as high modulus and strength properties, chemical inertness and heat distortion temperatures up to 80°C. In addition, though, they are said to offer superior gloss, colour and clarity. Their thermal stability is better than that of PVC and they can be processed at lower temperatures. The risk of degradation during processing is thus very much reduced.

One factor arising from the greater thermal stability is that less efficient stabilisers can be used to produce satisfactory results. This is important from the point of view of producing formulations approved by the American Food & Drug Administration (F & DA) and other bodies, for use in contact with foods. It seems to be almost a fact of life that stabilisers approved for use in contact with food are less efficient than stabilisers which are not approved. The applications for these films lie mainly in the field of packaging and are thus dealt with in Chapter 20.

7.7 FLUOROPOLYMERS

7.7.1 POLYCHLOROTRIFLUOROETHYLENE (PCTFE)

PCTFE is formed by the addition polymerisation of chlorotriflu-roethylene, under pressure, in the presence of excess water to carry off the large amount of heat evolved. Commercial films are usually copolymers.

Properties

PCTFE film is non-flammable and is resistant to a wide range of chemicals. Its water vapour permeability is the lowest of any polymer film and its water absorption is almost nil. Gas barrier properties are also good being nearly as good as those of the nylons.

It retains its flexibility down to temperatures of around −195°C and has a softening point between 185° and 205°C according to its grade and crystallinity. It can be vacuum metallised and laminated to paper, aluminium foil or a range of other plastics films.

Applications

The main use for PCTFE film, either alone or as a laminate, is the strip or blister packaging of pharmaceutical capsules and tablets where the product is very hygroscopic and requires a high degree of protection from water vapour. It has also been used, in conjunction with low density polyethylene, as the soft dome blisters of 'push-through' packaging for tablets.

In the U.S.A., PCTFE film has been used for the packaging of aerospace hardware and valves which have subsequently to work in contact with liquid oxygen. There is always some chance of particles being abraded from the bags and these present a potential explosion hazard in contact with liquid oxygen. PCTFE film has been found to be stable under impact in these conditions. Unsupported PCTFE films may be sealed by impulse sealers, dielectric heating or by ultrasonics, while laminates can be sealed on heated jaw sealers.

7.7.2 POLYVINYL FLUORIDE

This film could have been classified with the vinyls but has more points of resemblance with other fluorocarbon polymers. It is formed by the addition polymerisation of vinyl fluoride which is a gas at normal temperature and pressure. Polymerisation is normally carried out at a temperature of 80°C and a pressure of 300 atmospheres using a peroxide initiator.

Properties

Polyvinyl fluoride film has excellent resistance to solvents, acids and alkalis, and can even be boiled in strong acids and alkalis without losing its film form or strength. It is unaffected by boiling in carbon tetrachloride, acetone, benzene and methyl ethyl ketone for two hours and is impermeable to oils and greases.

It is strong, flexible and is extremely resistant to failure by flexing.

It retains this toughness and flexibility over a wide temperature range and is highly resistant to thermal embrittlement.

Water vapour permeability of polyvinyl fluoride film is low, as is its permeability to gases and to most organic vapours. The exceptions in the case of organic vapours are esters and ketones.

Electrical insulation properties are good, including high dielectric constant and volume resistivity. This, coupled with its excellent thermal ageing properties and chemical resistance makes PVF films of use in insulation and cable manufacture.

One of PVF's outstanding properties is its weatherability. It is extremely resistant to degradation by sunlight and retains its clarity and flexibility for many years in strong sunlight. Coupled with this are its good surface properties. It is stain resistant and has good scratch and abrasion resistance. PVF film can be printed or decorated by embossing. Heat sealing can be carried out by impulse and dielectric methods.

Applications

PVF's weatherability, surface properties, toughness and chemical resistance have led to its use as a paint substitute. In many ways it is superior to paints for outdoor protection. In industrial areas, dirt is more easily washed off it and its surface does not chalk, craze or erode appreciably even after long exposure. Because the PVF film is clear, it has also been used for glazing. It is readily laminated to aluminium, galvanised steel, plywood, etc., and can be painted, both when freshly applied or after weathering for some time. When laminated to polyethylene film, it has been used as a substitute for roofing felt, giving better weatherability, freedom from maintenance and reduced application costs. Laminates with PVC can be vacuum formed and PVF film has also been used in skin packaging, giving close conformity to the shape of the product.

7.7.3 POLYVINYLIDENE FLUORIDE

Polyvinylidene fluoride is more flexible than PVF and is much more extensible (300% instead of 100%). It has similar gas barrier properties to PVF but is a better water vapour barrier. It is also softer than PVF and so is not suitable for use as a paint substitute. This limits the range of possible uses and it is unlikely to compete in packaging markets because of its high cost.

Its useful temperature range is from −68°C to 150°C and it has good electrical properties, including a high permittivity. The latter property makes it very suitable for use in capacitors.

7.8 POLYURETHANE

Polyurethane film is a comparative newcomer although polyurethanes are well-known in other forms, including foams, elastomers and surface coatings. As a film it is soft but very tough. It can be made by blowing, extrusion casting, solution casting or by calendering. It has a very high tear strength, high abrasion resistance and is extremely resistant to oils and greases. This combination of properties has led to its use for specialised packaging in the industrial and military fields, including the short term storage of potable water. Other applications include the packaging of parts requiring storage in grease or oil where the film's grease resistance, toughness and clarity are important.

7.9 POLYIMIDES

The polyimide molecule has a very rigid, stable structure consisting of heterocyclic rings (rings containing more than one type of atom) bracketed with aromatic rings, as shown:

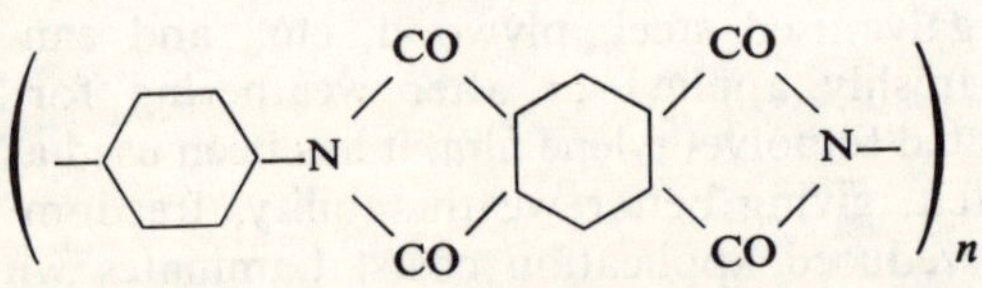

The structure shown above is a typical one but many variations are possible.

Polyimide films are made by first synthesising long chain percursor molecules in which the heterocyclic rings are not closed. These precursors are soluble in certain solvents and a solution is cast on to a smooth surface. The solution is heated to drive off the solvent and complete the heterocyclic ring structure. The film is then stripped off. In some cases the film may be deposited *in situ*.

Melting points range as high as 600°C and polyimide films have been used in applications where temperatures range from −270°C to 400°C. Abrasion resistance is good and the films are also noted for resistance to electrical and nuclear radiation. Chemically, they are extremely resistant to organic solvents and dilute acids. They are attacked by alkalis and concentrated acids, however.

Polyimide film has been used for flexible printed circuits because of solder resistance and its dimensional stability at high temperatures. Although polyimide films are very expensive, they find many electrical applications because of the weight to space savings made possible by their high dielectric strength, high radiation stability and high temperature resistance. They are used in wire and cable insulation and the U.S. railroad and aerospace industries use ultra-thin polyimide film for its space saving and for its capacity to withstand temperatures high enough to melt or even decompose most conventional plastics materials.

Polyimides are used in electric motor insulation, especially for those used in electric locomotives which generate large surges of power (and hence heat) when starting up.

7.10 POLY(p-XYLENE) (PARYLENE)

This polymer also has an unusual method of film formation, being deposited from the vapour phase in thicknesses as low as $\frac{1}{4}$ μm (0·01 mil). When the vapour condenses on to a cold surface it gives an adherent coating or it can be stripped off as a film. One example of parylene's use as a coating is in the micro-encapsulation of granules and powders. A very thin coating of parylene is sufficient to protect the very reactive lithium metal granules from water.

Parylene film is a very good barrier against essential oils. Low density polyethylene bottles, coated internally with parylene, lost less than 2 g of lemon oil (out of half litre) in 3 years. Uncoated low density polyethylene would normally present no barrier at all to lemon oil. Parylene's maximum temperature of use is around 265°C for short term use and around 220°C for long term use.

The polymer is expensive and its use can only be justified when its special properties are essential.

Part 2

Manufacture and Properties of Films

8
Manufacturing Methods

8.1 EXTRUSION OF FILM

There are fundamentally two different methods of extruding film, namely, blow extrusion and slit die extrusion. The former method produces tubular film, which may be gussetted or lay-flat while the latter results in flat film. If desired, lay-flat film can be slit to give flat film. The equipment for film extrusion consists of an extruder, fitted with a suitable die, equipment to cool the molten film, haul-off machinery and a wind-up unit. Blow extrusion and slit die extrusion vary in the design of die used and in the type of cooling. The haul-off and wind-up equipment is also different.

The design and operation of the extruder up to the die is the same for both methods, however, and will be briefly described here before considering the different types of film manufacture. The basic extrusion process is designed to convert, continuously, a thermoplastics material into a particular form, in this case film. The basic sequence of events is as follows:

(1) Plasticisation of the raw material in granule or powder form.

(2) Metering of the plasticised product through a die which converts it to the required form (i.e. tubular or flat).

(3) Solidification into the required form.

(4) Winding into reels.

Processes (1) and (2) are carried out in the extruder, whilst (3) and (4) are ancillary processes.

A typical extruder is shown in *Figure 8.1* and consists essentially of an Archimedean screw which revolves within a close fitting, heated barrel. The plastics granules are fed through a hopper mounted at one end of the barrel and carried forward along the

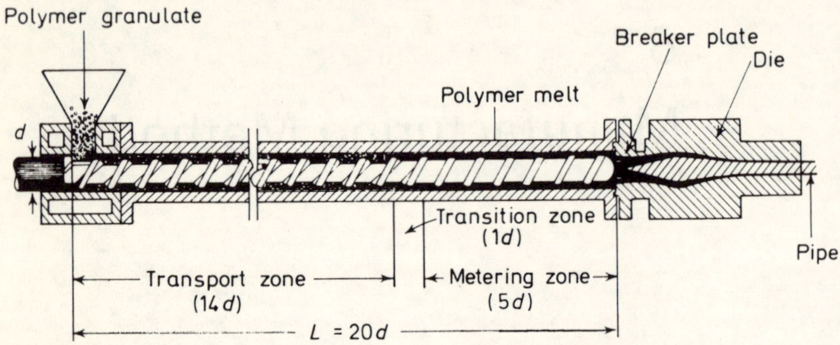

Figure 8.1. Scheme for a typical single-screw extruder ($l/d = 20$) *shown extruding pipe*

barrel by the action of the screw. As the granules move along the screw, they are melted by contact with the heated walls of the barrel and by the heat generated by friction. This frictional, or exothermal, heating is appreciable and in modern high speed machines it can supply the whole of the heating required for steady running, external heating being required only to prevent the machine from stalling at the start of the run when the material is cold. The screw then forces the molten plastic through the die which determines its final form. The most important component of the extruder is the screw and different designs of screw are used for extruding different polymers. Extruder screws are characterised by their length to diameter ratio (commonly written as L/D ratio) and their compression ratios. The compression ratio is the ratio of the volume of one flight of the screw at the hopper end to the volume of one flight at the die end. L/D ratios most commonly used for single screw extruders are between about 15:1 to 30:1 while compression ratios can vary from 2:1 to 4:1. An extruder screw is usually divided into three sectors, namely, feed, compression and metering. The feed section transports the material from under the hopper mouth to the hotter portion of the barrel. The compression section is that section where the diminishing depth of thread causes a volume compression of the melting granules. The main effect of this is an increase in the shearing action of the molten polymer due to the relative motion of the screw surface with respect to the barrel wall. This improves the mixing and also leads to an increase in frictional heat and a more uniform heat distribution throughout the melt. The function of the final section of the screw is to homogenise the melt further, meter it uniformly through the die and smooth out pulsations.

Just prior to the die is fitted a breaker plate supporting a screen pack consisting of a number of fine or coarse mesh gauges. The screen pack filters out any contamination which might be present in the raw material. This is particularly important in the case of thin film extrusion where even the smallest of contaminating particles could cause holes or even breaks in the film. The screen pack also increases the back pressure in the extruder and this improves the

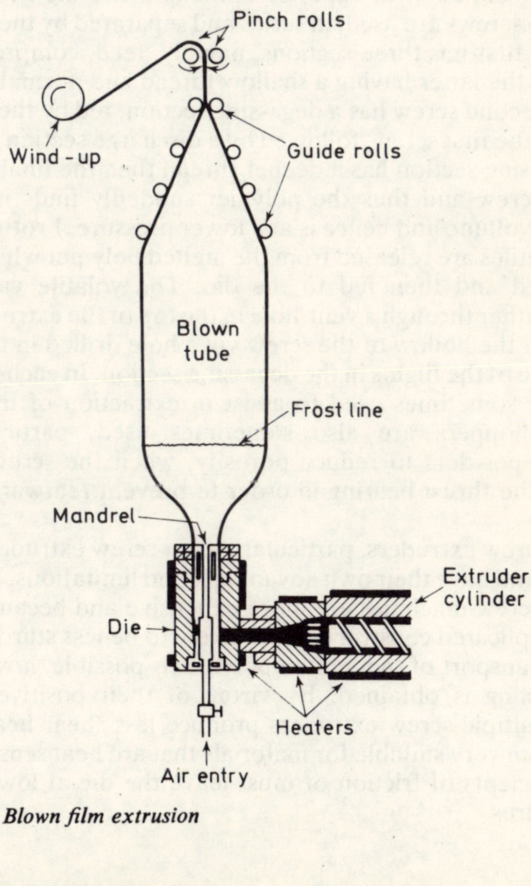

Figure 8.2. Blown film extrusion

mixing and homogenisation of the melt. The screw is usually cored for steam heating or water cooling. When the maximum amount of compounding is required, the screw is cooled. This improves the quality of the extrudate but slightly reduces the output.

Correct design of the head is important and should promote streamline flow of the material with no 'dead spots' where material could stagnate and so decompose due to overheating. This is particularly important in the case of PVC where the degradation point of the material is close to the temperature necessary for adequate flow.

A fairly recent development in extruder design has been the use of venting or degassing zones in order to remove any volatile constituents from the extrudate before it emerges from the die. This is achieved by releasing the plastic melt from the state of compression to which it has been subjected, whereupon the water and other volatiles will vaporise and cause the melt to froth. In effect, two screws are used, in series and separated by the degassing zone. The first has three sections, namely, feed, compression and discharge, the latter having a shallow thread and normally running full. The second screw has a degassing section, fed by the discharge section of the first screw, followed by a discharge section of its own. The degassing section has a deeper thread than the final section of the first screw and thus the polymer suddenly finds itself in an increased volume and hence is at a lower pressure. Frothing occurs as the volatiles are released from the melted polymer which is again compressed and then fed to the die. The volatile vapours are released either through a vent-hole in the top of the extruder barrel, or through the hollow of the screw via a hole drilled in the trailing edge of one of the flights in the degassing section. In each instance, a vacuum is sometimes used to assist in extraction of the vapour. Vacuum hoppers are also sometimes used, particularly for dry-blend powders to reduce porosity, when the screw must be sealed at the thrust bearing in order to prevent rearward polymer leakage.

Multi-screw extruders, particularly twin-screw extruders are also available and have their own advantages and limitations. In general, multiple-screw machines are more expensive and because of their more complicated construction are likely to be less sturdy. A more positive transport of the molten polymer is possible, however, and better mixing is obtained. By virtue of their positive pumping action, multiple-screw extruders produce less shear heat and this makes them very suitable for materials that are heat sensitive, have low coefficients of friction or must leave the die at low extrusion temperatures.

8.1.1 BLOWN FILM EXTRUSION

A typical set-up for blown film extrusion is shown in *Figure 8.2*. In this instance the molten polymer from the extruder enters the die from the side but entry can also be effected from the bottom of the die. Once in the die, the molten polymer is made to flow round a mandrel and emerges through a ring shaped die opening, in the form of a tube. The tube is expanded into a bubble of the required diameter by an air pressure maintained through the centre of the mandrel. The expansion of the bubble is accompanied by a corresponding reduction in thickness. Extrusion of the tube is usually upwards but it can be extruded downwards, or even sideways. The bubble pressure is maintained by pinch rolls at one end and by the die at the other. It is important that the pressure of the air is kept constant in order to ensure uniform thickness and width of film. Other factors that affect film thickness are extruder output, haul-off speed and temperatures of the die and along the barrel. These must also be strictly controlled.

As with any extrusion process, film blowing becomes more economical as speeds are increased. The limiting factor here is the rate at which the tubular extrudate can be cooled. Cooling is usually achieved by blowing air against the outside surface of the bubble. Under constant air flow conditions an increase in extrusion speed results in a higher 'frost' line (the line where solidification of the extrudate commences) and this leads to bubble instability. Increasing the air flow gives more rapid cooling and lowers the 'frost' line but this is limited in its application because too high a velocity of the air stream will distort the bubble. Various designs of air cooling rings have been worked out in order to produce improved cooling without these attendant difficulties and one such design (designed and patented by Shell) is shown in *Figure 8.3*. It consists of a conically shaped ring provided with three air slits, the airstreams from which are so directed and regulated that the space between the bubble and the ring decreases gradually towards the top of the ring. This gives improved cooling by increasing the speed of the airstream. The design also results in a zone of under-pressure at the top of the ring and this greatly improves the bubble stability.

Blown film extrusion is an extremely complex subject and there are many problems associated with the production of good quality film. Among the many defects which can occur are variations in film thickness, surface defects such as 'orange peel', 'apple sauce' or 'fish eyes', low tensile strength, low impact strength, hazy film, blocking and wrinkling. Wrinkling is a particularly annoying problem because it can be costly, leading to scrapping of a roll of

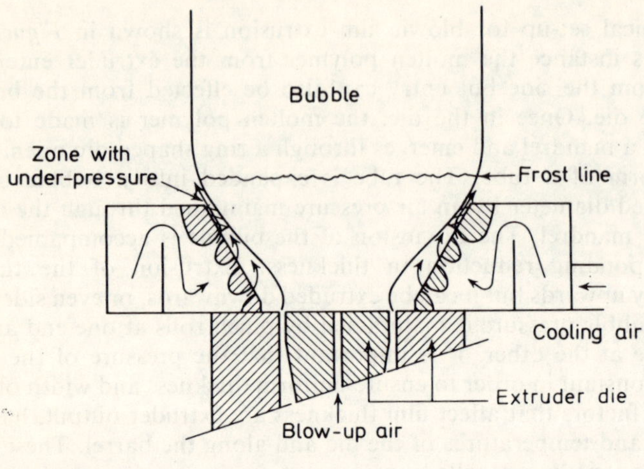

Figure 8.3. Shell cooling ring

film, and because it can arise from such a wide variety of causes that it is likely to occur even in the best regulated extrusion shop. If the film is too cold when it reaches the pinch rolls, for instance, it will be stiff and this may cause crimping at the nip and wrinkling. One way of raising the film temperature at the nip rolls is to raise the melt temperature but this can lead to other troubles such as blocking. In fact, this is illustrative of the whole subject of film blowing inasmuch as compromises are often necessary to achieve the best balance of properties. Wrinkling can also be caused by the die gap being out of adjustment. This causes variations in film thickness and can lead to uneven pull at the pinch rolls. Another cause of wrinkling may be surging from the extruder or air currents in the extruder shop. Both of these factors can cause wobbling of the film bubble and thus wrinkling at the wind-up stage. The film bubble may be stabilised by supporting it with horizontal stationary guides or the whole extruder may be protected from stray air currents by a film curtain. Other causes include non-alignment of the guide roll and the pinch rolls, or non-uniformity of pressure across the face of the pinch rolls.

Among the surface defects mentioned earlier, 'fish eyes' are due to imperfect mixing in the extruder or to contamination. Both of these factors are controlled by the screen pack which not only screens out contaminating particles but improves homogeneity by increasing the back pressure in the extruder. 'Orange peel' or 'apple sauce' are also surface defects caused by inhomogeneity of the molten polymer.

Since low density polyethylene forms by far the greatest percentage of all film made, it will be useful to consider the influence of the various polymer parameters such as melt flow index and molecular weight on the film properties. Impact strength, for instance, increases with molecular weight (i.e. with decreasing melt index) and with decreasing density. Heavy duty sacks, for instance, are normally made from polyethylene grades having densities between 0.916 and 0.922 g/cm^3 and melt indices between 0.2 and 0.5. For thinner technical film as used in building applications or waterproof lining of ponds, higher melt indices have to be used because of the difficulty of drawing down very viscous melts to thin film. Melt indices of between 1 and 2.5 are more usual, therefore, and impact strengths are less than for heavy duty sacks. Clarity is, however, improved. Where a good balance of properties is required as in the medium clarity/medium impact grades, slightly higher densities are used (0.920 to 0.925 g/cm^3) and the melt index is varied between 0.75 and 2.5. For high clarity, a high density and a high melt index are required since increases in both these properties cause an increase in see-through clarity, a decrease in haze and an increase in gloss. High clarity film will, of course, have a relatively poor impact strength because of the high melt index and such film should not be used for packaging heavy items.

8.1.2 SLIT-DIE EXTRUSION (FLAT FILM EXTRUSION)

In flat film extrusion the molten polymer is extruded through a slit-die and thence into a quenching water bath or on to a chilled roller. In either case the essence of the process is rapid cooling of the extruded film and cooling is, therefore, applied within a very short distance of the die lips (usually between 25 and 65 mm—1·0 in and 2·6 in). This short distance is also dictated by the necessity to reduce 'necking' of the film web, with consequent loss of width. In the chill roll casting method, the melt is extruded onto a chromium plated roller, cored for water cooling *(Figure 8.4)*. The rapid cooling leads to the formation of small crystallites and this gives a clearer film.

Where the quench bath method is used, the water temperature should be kept constant for best results. At constant extrusion temperature, lower quench temperatures improve slip and anti-blocking properties while higher quench temperatures give film that is easier to wind without wrinkles and with better physical properties.

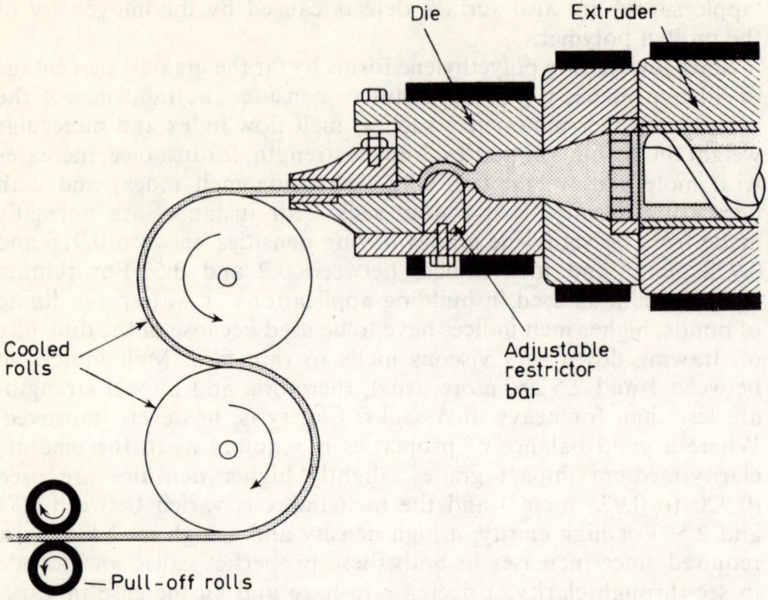

Figure 8.4. Film casting (slit-die extrusion)

Slit-dies for flat film are wide in comparison with the diameter of the extruder head and this means that the flow path to the extreme edges of the die is longer than to the centre. Flow compensation

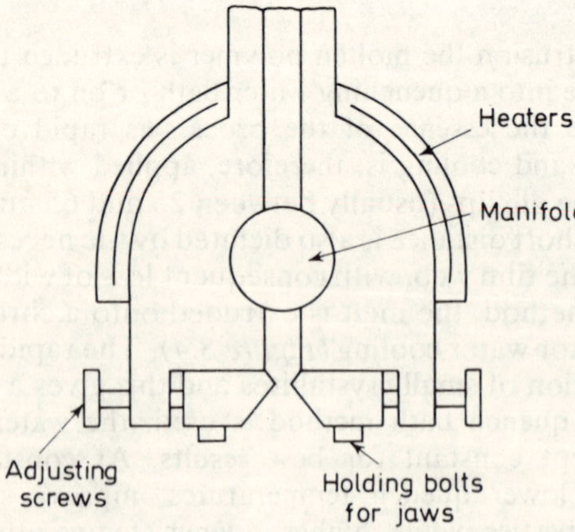

Figure 8.5. Cross-section of manifold-type die for film

is usually obtained by a manifold die, a cross-section of which is shown in *Figure 8.5*. It consists of a lateral channel (or manifold) of such a diameter that the flow resistance is small compared with that offered by the die lips. The manifold can only be efficient in its task of flow compensation if the viscosity of the melt is fairly low so that higher temperatures are necessary for flat film extrusion. This limits the use of the manifold die to materials of good thermal stability while another consequence of the higher extrusion temperature is the necessity for a heavier screen pack in order to maintain satisfactory back pressures. The inside surface of a flat die has to be precision machined and well polished since the slightest surface imperfection will result in striations or variations in gauge.

8.1.3 COMPARISON OF BLOW AND CAST FILM PROCESSES

Some of the advantages of the tubular film process are as follows:

(1) The mechanical properties of the film are generally better than those of cast film.
(2) The width of lay flat film is easily adjustable and there are no losses due to edge trimming. This latter is necessary for flat film because of the thickening of the film edge due to necking-in.
(3) Lay flat film is more easily converted into bags since it is only necessary to seal one end of a cut length to make the bag.
(4) The cost for making wide blown film is much lower than for wide cast film because the cost of chill rollers increases rapidly with width due to the difficulty of precision grinding longer rollers.
(5) A tubular film die is more compact and is cheaper than a slit-die producing film of comparable width.
(6) The tubular process is easier and more flexible to operate.

These advantages must be balanced against the advantages of the slit-die process which are as follows:

(1) Very high outputs can be obtained by slit-die extrusion units.
(2) Slit-die film normally has superior optical properties but it should be noted that special rapid cooling processes have been developed for tubular film, particularly in the case of poly-

propylene film. One example, the Shell Tubular Quench Process, will be described in more detail later.

(3) Thickness variation is usually less with slit-die extruded film.

8.1.4 WATER COOLED POLYPROPYLENE FILM

Among the advantages of the air-cooled tubular process mentioned above were cheapness, ease of conversion into bags and flexibility of operation. These factors have been largely responsible for the large scale penetration of low density polyethylene film into packaging markets. Polypropylene cannot be processed on the same equipment since the rate of cooling is inadequate to prevent the formation of large 'spherulites' (crystalline aggregates) in the film. This leads to the production of a brittle film, having a matt, opaque appearance. Clear polypropylene film can be produced by chill roll casting techniques but the equipment is expensive and is not normally

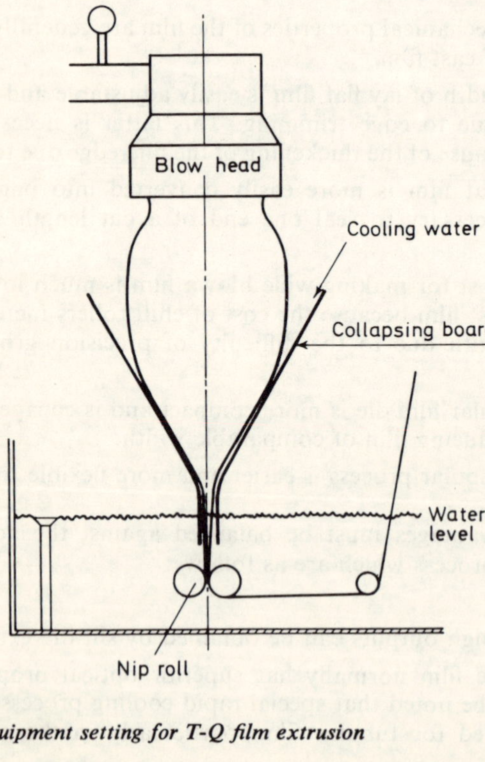

Figure 8.6. Equipment setting for T-Q film extrusion

economic at outputs below about 600 tonnes per annum. This has greatly hindered its penetration of the clear film packaging market.

Techniques of water cooling tubular polypropylene film, however, have opened up ways of producing clear film, with greater toughness and at no greater cost than cast polypropylene film. Among the different techniques commercially available is Shell's Tubular Quench (TQ Process) which involves downward extrusion of a tubular extrudate from an annular die followed by rapid cooling on water-covered converging boards. At the same time the tubular extrudate is inflated with air in the normal way to give film of the required lay flat width and thickness. The water film that runs down the converging boards shock cools the film and causes rapid crystallite formation and hence the formation of small spherulites with a consequent increase in clarity. The layout of the TQ Process is shown schematically in *Figure 8.6*. As with polyethylene, the blow up ratio influences the balance of molecular orientation between machine and transverse directions and this affects film impact strength, tensile strength and tear strength in the usual manner.

The properties of TQ polypropylene film are similar to those of the cast film. However, the ability to vary the blow up ratio allows a measure of control over the molecular orientation and this in turn can result in an improvement in mechanical properties compared with cast film where the orientation is essentially all in the machine direction of the film. The degree of orientation in the TQ Process is still low compared with that of a true biaxially oriented film and

Table 8.1

Property	*TQ film*	*Cast film*	*Blown film (air cooled)*
Haze (%)	3	4	>20
Gloss	75	70	25
Clarity (%)	4–6	5–9	—
Coefficient of static friction	0.15–0.40	0.15–0.40	0.30–0.50
Ultimate tensile stress (kg/cm²)			
MD	500	450	190
TD	470	380	205

TQ film does not compete with it in properties. The production of biaxially oriented film will be dealt with later but a comparison of polypropylene film produced by chill roll casting, the TQ process and biaxial orientation is given in *Table 8.1*.

The TQ Process is particularly valuable for outputs of film below the economic output of a cast film line (up to about 600 tonnes per annum). Costs are much lower and benefit from the great flexibility of the blown film process and the absence of edge trim waste.

Another process is the Dow-Taga process. In this the blown film goes through a hollow ring where it is coated with a film of water which flows from the ring and on to the film before it is collapsed. Different sizes of rings must be used to match changes in blow up ratios whereas with the TQ Prccess, no changes in equipment are needed when making different diameters of tubular film.

In a third process, developed by Kokoku Rayon and Pulp Co. Ltd. in Japan (now the Kohjin Co. Ltd.), the tube of polypropylene film, after it comes out of the die, is passed over a mandrel that extends down into a water bath. After cooling, the film comes out of the bath, still in tubular form. It is then dried and the tube collapsed between a set of rolls.

8.2 CALENDERING

In this process continuous sheet or film is made by passing heat-softened material between two or more rolls. Calenders were originally developed for processing rubber but they are now widely used for producing themoplastics (mainly flexible PVC) sheet and film.

In essence calendering consists in feeding a plastics mass into the nip between two rolls where it is squeezed into a film that then passes round the remaining rolls. It then emerges as a continuous film, the thickness of which is determined by the gap between the last pair of rolls. The surface of the film is determined by that of the last roll and can be glossy, matt or embossed. After leaving the calender the sheet is cooled by passing it over cooling rolls, and is then fed through a beta-ray thickness gauge (Figure 8.7) before it is reeled up.

The plastics mix fed to the calender may be a simple hot melt as in the case of, say, polyethylene but with PVC, a great deal of effort is put into the pre-calendering processes such as blending, mixing, gelling and straining. In addition to the polymer there may be inert mineral fillers (added to reduce the cost and modify the physical properties), pigments, lubricants (to assist in processing), stabilisers and plasticisers. The dry components (with the exception of the pigment) are loaded into a ribbon blender and agitated sufficiently to produce an even dispersion. If plasticisers are to be added they are sprayed into the powder mix during the initial period of

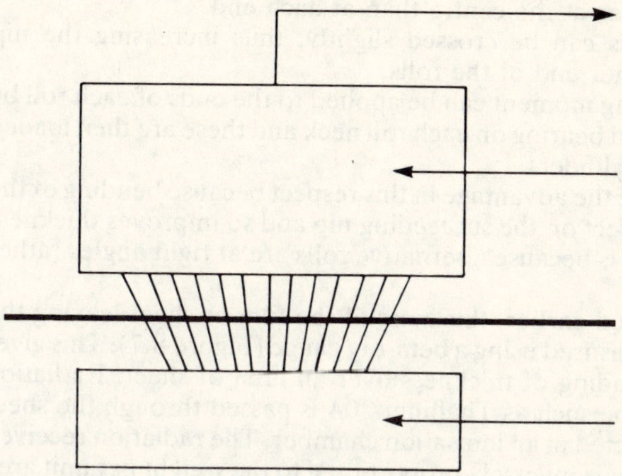

Figure 8.7 Beta-ray gauge

blending. When correct dispersion has been achieved the blend is removed via a valve at the base of the mixing chamber and weighed out into batch tins. If pigment is required it is added at this stage to each separate batch. The batch tins are then emptied into a primary mixer, such as a Banbury, and treated for about five to ten minutes at 120–160°C. The combination of heat and the 'kneading' action of the Banbury causes partial gelation of the mix. This partly gelled material is then passed to a two-roll mill and made to form a sheet round the front roll. This may be fed directly to the calender but for thin sheet or film a straining process is added to remove any coarse particles. A typical strainer consists of a single screw extruder with a filter screen directly in front of the screw. The screen consists of a fine stainless steel woven wire mesh backed up by a medium to coarse mesh for mechanical support, together with a perforated strainer plate.

Calenders may consist of from two to five hollow rolls (or bowls) arranged for steam heating or water cooling and are characterised by the number of bowls and their arrangement as, for example, I, Z or inverted L types. Four roll, inverted L and Z types are the most usual.

Very high forces have to be exerted on the rolls in order to squeeze the plastics to a thin film and these cause the rolls to bend giving sheet that is thicker in the middle than at the edges. Some ways in which the effect is counteracted are given below.

(1) The rolls can be 'crowned', i.e. they can be made with a greater diameter at the centre than at each end.

(2) The rolls can be crossed slightly, thus increasing the nip opening at either end of the rolls.

(3) A bending moment can be applied to the ends of each roll by having a second bearing on each roll neck and these are then loaded by hydraulic cylinders.

Z types have the advantage in this respect because bending of the rolls has no effect on the succeeding nip and so improves thickness accuracy. This is because alternative rolls are at right angles rather than in-line.

As mentioned earlier, thickness of the film or sheet leaving the calender is measured using a beta-ray gauge (Figure 8.7). This gives an average reading of thickness over an area of sheet. Radiation from an isotope such as Thallium 204 is passed through the sheet and then collected in an ionisation chamber. The radiation received by the chamber is inversely proportional to the weight per unit area of the material being measured.

Calenders are of massive construction because of the large forces necessary to squeeze the plastics mass into a thin film. They call for high temperatures, with little tolerance across the rolls and high pressures, again with low tolerances. A large floor area is also needed because of the associated plant such as mixers, blenders, temperature control systems, haul-off equipment and other ancillary plant. Calendering is, therefore, a capital intensive process and calenders tend to cater for wide width film (around 1.8m—6.0 ft wide) because the cost is obviously proportionally less. However, these large machines tend to be used only for plasticised PVC because unplasticised PVC has a much higher melt viscosity which makes it harder to handle on rolls of this width.

A special calendering process known as the Luvitherm process was developed, therefore, to produce unplasticised PVC films. The PVC is flash heated up to 220°C while in contact with a specially designed aluminium roll and the hot film produced is normally oriented immediately after calendering in the high temperature zone. The calender unit is narrower than the large units described earlier and the output is lower. Special grades of PVC are used, with special stabilisers, and the compounding stage can be an extruder compounder feeding direct to the calender.

Calendering usually produces film with a better uniformity of gauge compared with that obtained by extrusion. A number of factors contribute to this, one of which is the great care paid to the engineering of the calender bowls (as mentioned earlier). The final gauge is very much dependent on the gap between the final two bowls whereas in extrusion the gauge is dependent more on blow-up ratios (in the case of tubular film) or draw down speeds (in the case of slit-die film). In addition, in an extruder cross-head die there may be a range of path lengths which lead to preferential flow and induce variation in gauge. In the tubular die process there are complicating factors such as the structures supporting the mandrel in the die. This also has an effect on the melt flow, leading to gauge variations.

Another advantage of calendering is that better mixing is obtained. The amount of energy available in a calendering line is very much more than in an extrusion line, so calendered film is less dependent on the uniformity of the feedstock.

The main advantage of extrusion for PVC is the much lower capital costs involved, so that shorter runs can be made at economic rates.

8.3 SOLVENT CASTING

Solvent casting methods are expensive but cellulose nitrate film was made in this way because of the flammable nature of the film. A casting solution is prepared by dissolving cellulose nitrate and camphor (as plasticiser) in a 70:30 ether/alcohol mixture using paddle mixers. After 8–10 h mixing the product is filtered through cellulose wadding and de-aerated in vacuo at a temperature just under the boiling point of the solvent.

The film casting is carried out on an endless belt, a smooth surface being obtained by depositing a layer of hard gelatine on a

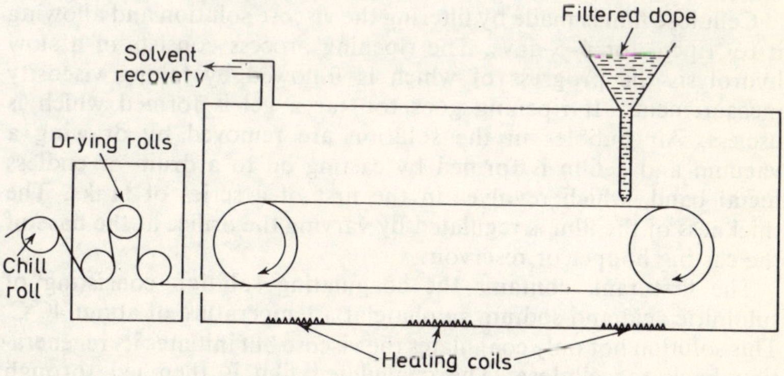

Figure 8.8. Endless belt method of casting cellulose nitrate film

copper support. After being stripped from the casting surface the film is seasoned in a heated drying cabinet where it passes over a zig zag pattern of rollers. After a final passage over chilled rollers to cool the film, it is wound into reels. The general lay-out of a film casting process is shown in *Figure 8.8*.

Some cellulose acetate film is produced by solvent casting, as well as by extrusion, and solution casting has also been used for producing vinyl chloride/vinyl acetate copolymer films.

8.4 CASTING OF REGENERATED CELLULOSE FILM

Methods of dissolving and then regenerating cellulose date back to 1857 but the one which is the basis for the modern cellulose film was discovered by Cross, Bevan and Beadle in 1892. It was not until 1911, however, that Brandenberger patented the continuous manufacture of cellulose film. Uses for the film were restricted at first because of its high moisture vapour permeability but the development in 1927 of moistureproof grades opened up many new markets and cellulose film is still an important packaging material.

The basic raw material, cellulose, is obtained from wood pulp or cotton linters. After soaking in caustic soda for an hour, the excess alkali is pressed out. It is then shredded and allowed to age for 2–3 days and absorbs oxygen from the air. This reduces the length of the cellulose molecule chain and so reduces the viscosity of the solution during the next stage. Here the pulp is transferred to rotating churns and then sprayed with carbon disulphide. The solution is discharged into a tank and dispersed by stirring with dilute caustic soda solution. The resultant solution is known as viscose.

Cellulose film is made by filtering the viscose solution and allowing it to 'ripen' for 4–5 days. The ripening process consists of a slow hydrolysis the progress of which is followed by taking viscosity measurements. If ripening goes too far a gel is formed which is useless. Air bubbles in the solution are removed by drawing a vacuum and a film is formed by casting on to a drum or endless metal band, which revolves in the first of a series of tanks. The thickness of the film is regulated by varying the orifice at the base of the casting hopper or reservoir.

The first tank contains the coagulating solution consisting of sulphuric acid and sodium sulphate at a temperature of about 40°C. This solution not only coagulates the viscose but initiates its regeneration back to cellulose. The coagulated film is then led through further tanks by means of guide rolls. The tanks contain various solutions designed to complete the regeneration, wash out acid carried over from the coagulating bath, remove any elemental sulphur, carbon disulphide, hydrogen sulphide, etc., and bleach the now transparent but still slightly coloured film. The film is then run through a bath containing glycerol or ethylene glycol which act as plasticisers and confer flexibility on the film. Finally it is passed through a drying oven and reeled up.

At this stage, the film is extremely permeable to moisture vapour and is not heat sealable and a subsequent coating treatment must be given if moistureproofness or heat sealability are required. A typical coating consists of dibutyl phthalate–plasticised cellulose nitrate, together with waxes and natural resins dissolved in organic solvents. Urea formaldehyde is also incorporated to anchor the coating and prevent it from floating off in damp conditions. The film is passed through the coating bath, dried to evaporate the solvent then passed through a high humidity chamber to restore the flexibility. Where even better barrier properties are required, a coating of polyvinylidene chloride is applied.

8.5 ORIENTATION OF FILM

Orientation of film by stretching it under heat is widely applied to films such as polypropylene, polystyrene, nylon and polyethylene

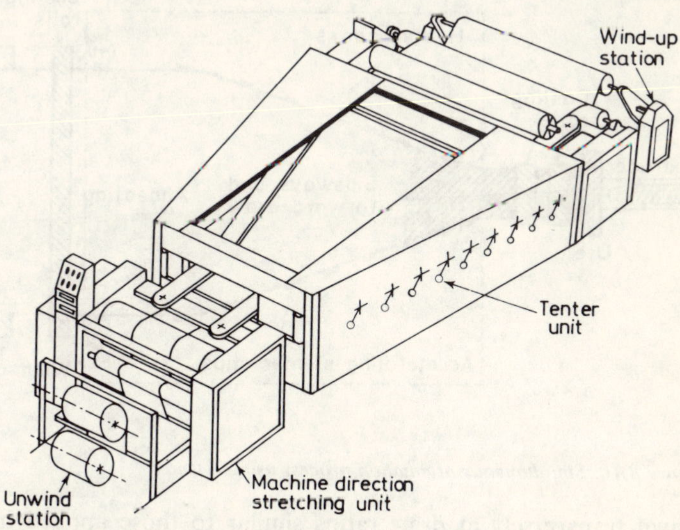

Figure 8.9. Two-stage orientation—flat film

terephthalate to improve clarity, impact strength and, particularly in the case of polypropylene, its barrier properties.

Basic polystyrene film, for instance, in its non-oriented form is very brittle and has only a limited use as a dielectric in capacitors. When biaxially oriented (i.e. oriented in two directions at right angles), the film is quite tough and can be thermoformed into crystal clear tubs, trays and larger items such as cake covers.

The largest application of orientation techniques, however, is in the manufacture of polypropylene film and the various processes will be illustrated mainly with respect to this film. The main processes can be divided into linear and tubular. The principle of the linear type can be illustrated by considering the two stage process shown in *Figure 8.9*. Thick cast film of around 500–600 μm thick is made by the chill roll casting process described earlier in the chapter. This is then fed to a system of differential draw rolls, i.e. rolls running at gradually increasing speeds. These rolls are heated sufficiently to bring the film to a suitable temperature (below the melting point). Under these conditions, the film is stretched in the machine direction at a draw ratio which is normally between 4:1 and 10:1. After leaving the draw rolls, the film is fed into a tenter frame which consists of two divergent endless belts or chains fitted with clips. These clips grip the film so that as it travels forward it is

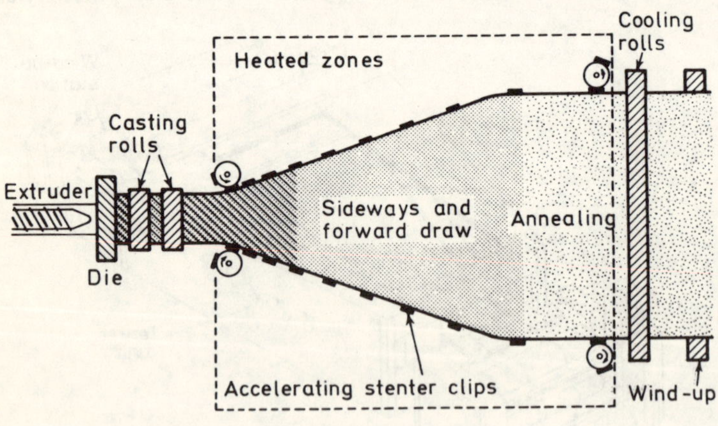

Figure 8.10. Simultaneous orientation process using a tenter

drawn transversely at draw ratios similar to those applied in the machine direction. The tentering area is also heated, with accurate control of temperature, usually by passing the film through an oven. On leaving the tenter frame, the film is cooled by passing over a cooling roller, then reeled up. Systems are also available in which this sequence of operation is reversed, i.e. the tenter frame comes first, followed by the differential draw rolls.

The two operations can also be carried out simultaneously, as in *Figure 8.10*. The film is gripped by its edges as it leaves the casting rolls and is then moved forward at an increasing speed while being stretched transversely by the diverging grips. Mechanically, this type of operation can be carried out by a tenter frame in which the clips are moved by a screw of increasing pitch. The magnitude of the transverse draw would then be controlled by the angle of divergence of the clips, in the usual way, while the magnitude of the forward draw would depend on the rate of increase of pitch.

As indicated, orientation can also be achieved by tubular processes. Molten polymer is extruded from an annular die and then quenched to form a tube. The wall thickness of the tube is controlled partly by the annular die gap and partly by the relative speeds of extrusion and haul-off. The tube passes through slow running nip rolls and is then re-heated to a uniform temperature. Transverse drawing is achieved by increasing the air pressure in the tube, the draw ratio being adjusted by adjustments to the volume of entrapped air. The air is trapped by pinch rolls at the end of the bubble remote from the extruder and these are run at a faster speed than the first pair, thus causing drawing of the film in the machine direction. The tubular process is thus another method of obtaining simultaneous transverse and forward orientation.

A range of films is obtainable from the various methods, from a biaxially oriented film with balanced properties, i.e. the same draw ratio in each direction, via biaxially oriented film with unbalanced properties (different draw ratios in the two directions), to completely uniaxially oriented film (either in the machine direction or transversely). Uniaxially drawn films tend to fibrillate when stretched at right angles to the direction of orientation and this reduces their usefulness as packaging film. The phenomenon is, however, utilised in the production of film fibres and this subject is dealt with in Chapter 24.

If oriented polypropylene film is heated to about 100°C immediately after drawing, it will shrink unless it is restrained in some way. This can be prevented by a process of annealing or heat setting. The film is heated, under controlled conditions, and while held under restraint. After cooling, the film will not shrink if heated to below the annealing temperature and the film is said to be heat set. The physical and optical properties of the film remain unchanged.

8.6 EXPANDED FILMS

Reducing the density of polymers by the creation of a cellular structure gives articles having a greater stiffness in bending for a given weight of polymer. The technique can be applied to extruded films as well as to moulded articles and has been so applied in the case of polystyrene and, more recently, the polyolefins.

The greater rigidity for a given weight of material is achieved because the rigidity of a beam is proportional to the cube of its thickness. For unit surface area, therefore, a film of double the thickness can be made from *X* g of material if the density is halved. Doubling the thickness means an increase of two cubed, = eight, times the stiffness for a given modulus but the modulus varies linearly with density so that halving the density halves the modulus, giving a net increase of four times the stiffness. This can have very important economic advantages.

8.6.1 EXPANDED POLYSTYRENE FILM

There are two main methods for the production of expanded polystyrene film. The first of these starts with polystyrene beads which have been impregnated under pressure with a liquefiable gas, usually pentane. The film is blow extruded, using a twin screw extruder and the addition of a nucleating agent such as a citric acid/sodium bicarbonate mixture. The nucleating agent has been found to be necessary in order to give a fine cell size. As the molten mass leaves the extruder die head and the pressure is released, the material starts to expand. The blown bubble technique is used in order to take up the corrugations which occur during the expansion process. The bubble is normally blown horizontally for ease of handling during start-up since the material is hard and rigid, unlike polyethylene film. When the bubble is collapsed, the tube is cut at each side to give two flat sheets which are reeled up separately. This is necessary because the folded edges of the tube are a source of weakness due to the rigidity of the polystyrene.

The other method utilises ordinary polystyrene beads as feedstock and gassing is carried out in the extruder. The film is then blown as before. The extruder has to be constructed specially for this type of operation and it is only economically viable for large tonnage offtakes—of the order of 400 tonnes/annum. The extrusion of pre-gassed beads is cheaper at lower tonnages and has the advantage of flexibility since the extruder can be used for other work.

Control of sheet orientation in either operation is achieved, as usual, by adjusting the haul-off speeds and the blow-up ratio. The orientation should be as balanced as possible as any appreciable difference in strength between machine and transverse directions could lead to splitting of the sheet during subsequent thermoforming operations.

The density of the sheet as it leaves the extruder is usually about 80 g/l (5 lb/ft^3) but subsequent operations, such as blowing-up, and any stretching that occurs during haul-off, compress the sheet and so increase its density. The collapsing of the bubble and subsequent passage of the film through nip rollers also increases the density.

Where low density sheet is required it is passed under an infra-red heater just prior to reeling up.

8.6.2 EXPANDED POLYOLEFINS

These are made by adding a masterbatch to the normal granules and then blow extruding as usual. The masterbatch contains a blowing agent and processing aids. The blowing agent is a compound which breaks down at the temperature of extrusion to give nitrogen which expands the molten mass on extrusion. When the melt leaves the high pressure zone in the die, the gas diffuses to discontinuities in the melt and bubbles are formed. In expanding the melt the gas has to stretch the polymer in order to create a cell. The melt temperature has to be carefully controlled at this point, therefore, because it affects the melt strength.

Films produced in this way have some of the attributes of paper, e.g. appearance, handle, stiffness and some retention of creasing. Densities are of the order of 600 g/l (37·5 lb/ft^3).

8.7 PLASTICS NET FROM FILM

Nets are usually made by the knitting of mono- or multi-filaments or by direct extrusion processes giving nets with the crossing strands fused together at the point of contact.

It is also possible to produce a net from film by embossing and then stretching. The film splits in a regular manner and an open work pattern is produced. The technique is applicable to a large number of polymers, while other variables are the embossed pattern and the loading conditions.

Outlets so far are in the surgical and medical fields, in dressings, and as adhesives between two layers of cloth which are fused together by heat. The regular spacing and porosity of the net remain while the laminated materials acquire excellent drape and resistance to washing or dry cleaning.

FURTHER READING

BEADLE, JOHN D. (Ed), *Plastics Forming*, Macmillan, London (1971)

ELDEN, R. A. and SWAN, A. D., *Calendering of Plastics*, Iliffe, London (1971)

FISHER, E. G., *Extrusion of Plastics*, Iliffe, London (1964)

MILES, D. C. and BRISTON, J. H., *Polymer Technology, 2nd edition*, Chemical Publishing Co. Inc., New York (1979)

9
Mechanical Properties

The properties of the various film forming materials have already been compared and contrasted. Selection of the best film for any particular use is a matter of matching these film properties against the end-use performance required. In packaging, for instance, it is the performance required by the product to be packaged in its particular marketing environment. The latter, it should be noted, includes the price at which the package can be sold.

In order to carry out this matching of properties against requirements, it is necessary to know what the various properties actually mean in practice and to have some method of quantifying them. The testing of film properties is also necessary for other reasons. Research laboratories must have some method of testing film properties in order to evaluate new products. The more efficient such an evaluation can be made, the more likelihood there is that the guesswork can be taken out of the decision as to whether the product is worth commercialising and, if so, for which market it is particularly suitable. Another objective of testing is to aid in process development. Test results can help in evaluating the results of changes in raw materials or processing variables, and can enable realistic manufacturing specifications to be set up.

There is a very wide range of criteria to be applied when considering the selection of a film for a particular purpose. In general, however, they fall into three broad groups. The first group covers those properties concerned basically with the strength of the film. These properties include tensile strength, impact strength, stiffness, bursting strength and tear strength. The second group of properties may broadly be classed as transmission properties and include permeability to gases, vapours and odours. Light transmission should also be included here and would cover properties such as 'see-through' clarity and haze. For convenience other optical properties like gloss are dealt with in this section.

Both of the groups already mentioned deal with those properties which are important to the final end-use performance. The last group includes properties which are more concerned with the performance of films on converting or packaging equipment. Examples of this sort of property are coefficient of friction, blocking, heat sealability and crease or flex resistance.

Some of the more important properties of plastics films are discussed in this and the next three chapters together with an outline of the various test methods used in their measurement.

9.1 TENSILE AND YIELD STRENGTH, ELONGATION AND YOUNG'S MODULUS

These four properties are considered under a single heading because the same equipment can be used for measuring each of them. To avoid possible confusion during the subsequent discussion, a few definitions are given below:

Stress This is the ratio of the force exerted on a body to its cross-sectional area,

$$\text{i.e.} \qquad = \frac{\text{applied force}}{\text{cross-sectional area}}$$

Strain This is a measure of change in the dimensions of the body when a force is applied to it and is calculated with reference to its original size. In tensile testing, strain is defined as:

$$= \frac{\text{total elongation}}{\text{gauge length}}$$

Gauge Length Gauge length is the original length of that part of the test piece over which the change in length is determined. For films, this is usually taken to be the original distance between the test specimen grips.

Tensile strength (or more accurately, ultimate tensile strength) is the maximum tensile stress which a material can sustain and is taken to be the maximum load exerted on the film specimen during the test, divided by the original cross-section of the specimen. Yield strength is the tensile stress at which the first sign of a non-elastic deformation occurs and is the load at this point (known as the yield point) divided by the original cross-section of the specimen.

Elongation is usually measured at the point where the film breaks and is expressed as the percentage of change of the original length of the specimen between the grips of the testing machine. Its importance is as a measure of the film's ability to stretch.

Young's modulus, or the modulus of elasticity, is the ratio of stress to strain over the range for which this ratio is constant, i.e. up to the yield point. It is a measure of the force that is required to deform the film by a given amount and so it is also a measure of the intrinsic stiffness of the film.

Yield strength is usually more important than ultimate tensile strength, especially during the passage of the film through wrapping or printing equipment where a sudden 'snatch' could cause a non-reversible distortion and, hence, out-of-register printing.

During the unwind operation both yield point and elongation are important properties. If the elongation is high there is the danger of uneven stretching of the film unless special handling techniques are used. Too low an elongation should also be avoided as any sudden unbalance in the unwind operation could lead to breaking of the film. A certain amount of tension is necessary during unwind so that films with a low yield strength are in danger of being stressed beyond their yield point.

9.1.1 TEST METHODS

There are many types of tensile testing machines available but the basic principle of them all is that a film strip is held at one end by a fixed clamp and at the other end by a moveable one. The clamps are then drawn apart and the tension in the strip is measured at various clamp separation distances. The load versus extension is measured by instruments and plotted on a chart. The film strip must be aligned vertically in the clamps so that the stresses are uniform across the width of the strip and parallel to the direction of loading. The upper clamp is usually suspended from a pivot and vertical alignment is more easily obtained if the strip is fastened in this clamp first. Self-aligning grips should be used in tensile tests although even these are not a complete answer to the problems of uneven loading. Both clamps should be tightened evenly so that no slipping occurs.

The preparation of the test strips is also important. They should be cut cleanly with no nicks or other imperfections in the edges, otherwise failure may occur by tearing. The thickness of the test strips should be checked prior to testing and any which vary more than about 10%, over their length, should be discarded. Since most plastic films have different properties in the machine and transverse directions, the tests should be carried out on samples cut from each of these directions. Five to ten samples will normally be sufficient to obtain reliable results. The differences in tensile properties often have to be taken into account during conversion of the film into bags or other packages.

The various types of testing machine are characterised by their differences in the application of the load. They usually fall into one of the following categories, namely, constant rate of loading, constant rate of elongation and constant rate of powered clamp separation.

Constant rate of load

One example of a constant rate of loading instrument works on the principle of the inclined plane. A tilting table has a carriage which moves down the table under the influence of gravity. The angle of the table is changed during the test to produce a constant rate of loading. This type of instrument is limited in its loading capacity and in the elongation it can accommodate and is now little used.

Constant rate of elongation

The principle of this type of test is illustrated in *Figure 9.1*. The strip is clamped between the upper (fixed) grip, which is attached to a load weighing cell, and a moveable grip driven at a constant rate, usually by an electric motor. The rate at which the powered grip moves can be preselected. Since the upper grip moves hardly at all, the rate of powered grip movement may be taken as the rate of extrusion of the film strip. Electrical signals proportional to the movement of the grip and to the applied load are fed to the axes of a recorder which then traces a load/elongation curve directly. Such machines are very versatile because of the combination of different drive speeds, recorder-chart magnification ratios and strain gauge sensitivities which can be used. They are also accurate since they are virtually free from frictional or inertial errors.

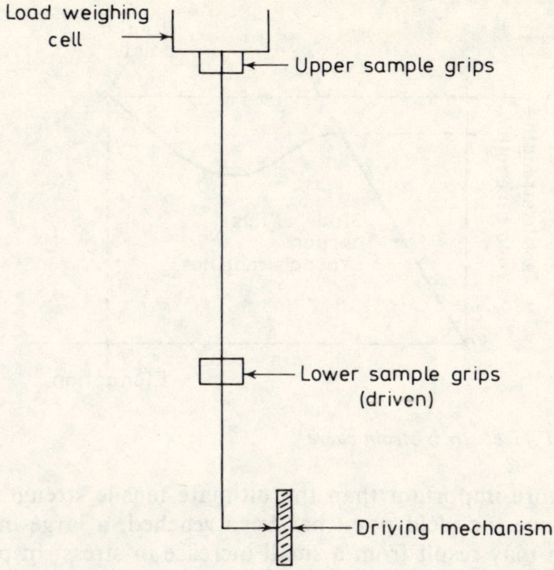

Figure 9.1. Tensile tester—constant rate of elongation

However, they are expensive and require expert maintenance and service.

Constant rate of powered grip separation

Before electronic tensile testers were available a popular type of machine was one which uses a pendulum to measure the applied load. The lower grip is powered, as before, and moves at a constant rate. In this type of instrument, however, the upper grip also moves since it is attached to a weighted pendulum. The rate of separation of the grips and the rate of straining are both variable, therefore. Although this type of instrument is somewhat less accurate it is cheaper and is still in widespread use, particularly for quality control purposes.

A typical plot of stress against strain is shown in *Figure 9.2*.

The importance of the various tensile properties depends on the specific end-use requirements but some general observations can be made. The tensile strength, for instance, must be great enough so that the film will not break when subjected to the sort of load likely to be encountered in practice. In many cases the yield strength

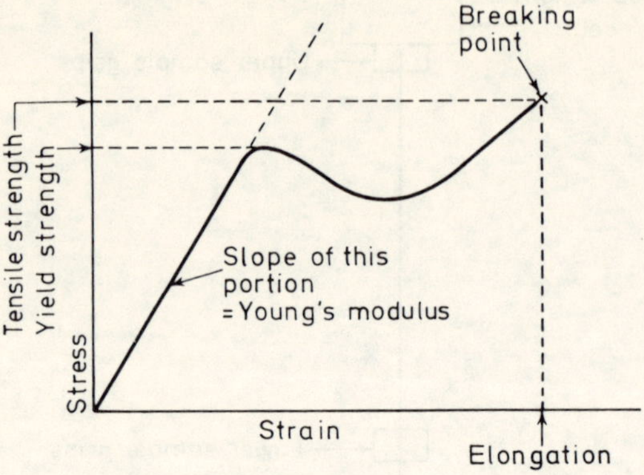

Figure 9.2. Typical stress/strain curve

may be more important than the ultimate tensile strength. This is because once the yield point has been reached, a large increase in elongation may result from a small increase in stress. In packaging uses, this can result in a tight wrap becoming loose. The yield strength is also important in handling on printing and laminating equipment. Up to the yield point, any elongation is reversible and the film will revert to its original strength when the stress is removed.

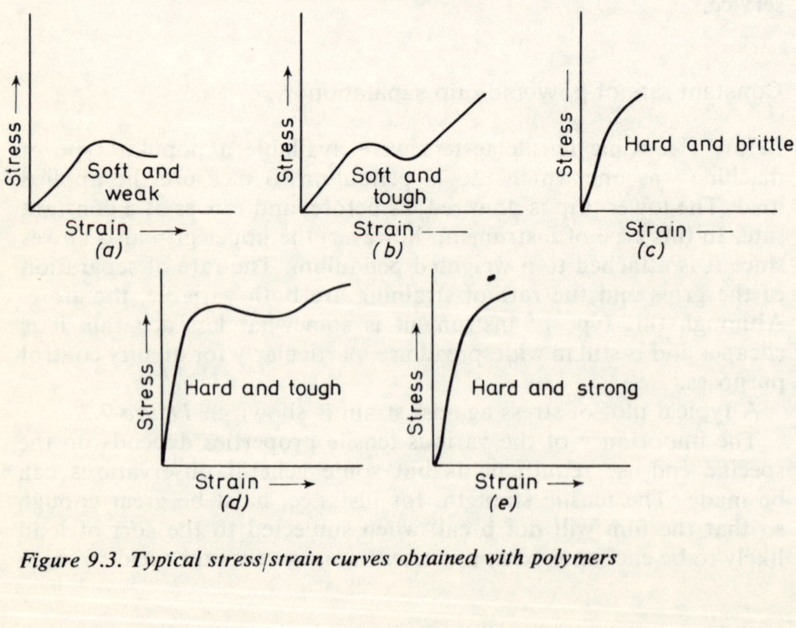

Figure 9.3. Typical stress/strain curves obtained with polymers

After the yield point, some of the elongation is non-elastic and there is always a residual elongation. A sudden snatch which leads to stresses greater than the yield strength will permanently stretch the film and lead to registration problems.

Films having a high elongation are sometimes used in order to relieve any stresses caused by an applied load and so reduce the risk of rupture. Under similar circumstances, films with low elongations would need to have fairly high tensile strengths to withstand any applied loading.

A great deal of information about the film can be obtained from the shape of its stress/strain curve. In addition to the numerical values for tensile strength, Young's modulus, elongation, etc., it is possible to obtain some idea of the toughness of the material by measuring the area under the curve. This area is a measure of the energy needed to break the test specimen and hence is directly related to toughness. Some typical curves are shown in *Figures 9.3* (*a–e*) and represent the following types of behaviour.

Soft and weak

Soft and weak films have a low value for Young's modulus, coupled with a low tensile strength. The elongation at break is only moderate. As can be seen, the area under the curve is low.

Soft and tough

These films have an appreciably higher area under the curve. The tensile strength is moderately high but Young's modulus is low as with soft and weak materials. The yield point is well marked and the elongation is high.

Hard and brittle

As one would expect, the tensile strength and Young's modulus are both high. Brittle materials have no distinct yield point, however, and the elongation is low.

Hard and tough

The high tensile strength, Young's modulus and elongation combine to give the large area under the stress/strain curve, thus justifying the adjective, tough.

Hard and strong

Such films are intermediate in character between the hard and brittle, and the hard and tough materials, Although there is still no clearly defined yield point, the elongation and tensile strengths are both greater than for hard and brittle materials.

9.2 BURST STRENGTH

The burst strength of a film is the resistance if offers to a steadily increasing pressure applied at right angles to its surface under certain defined conditions. The burst strength is taken to be the pressure at the moment of failure of the film and is essentially a measure of the capacity of the film to absorb energy.

Bursting tests are usually based on those developed for paper. One such test is the Mullen Burst Test. The test is carried out by applying pressure to one side of a clamped disc of standard dimensions and the pressure required to burst the disc is recorded. The pressure is normally applied using a liquid medium, such as glycerol or ethylene·glycerol, transmitted via a rubber diaphragm. Test results are affected by the speed at which the pressure is applied and by the diameter of the clamped disc. The lower the rate of pressure application and the smaller the diameter of clamped disc, the higher the bursting pressure.

9.3 IMPACT STRENGTH

Impact strength of a film is a measure of its ability to withstand shock loading. Such tests can, of course, be carried out on the finished package, one method being to fill, say, polyethylene bags with sand and subject them to controlled drop tests. The results of such tests cannot usually be applied to any other size or shape of pack, even if the same film is used. It is usual, therefore, to carry out impact tests on the basic film and some of these tests are described here.

One common method used to measure the impact strength of films is known as the Falling Dart method. The dart consists of a hemispherical head fitted with a shaft to which removeable weights can be added, together with a locking collar to keep the weights in place. The rest of the apparatus consists of a table, with a hole in the centre, over which are fitted grips designed to hold a sample of film without slipping. At the side of the hole a vertical pole is fitted. This supports an electromagnet which is used to drop the dart on to the film. The magnet can be moved up and down the pole to give any desired height of drop. A circular piece of film is placed in the grips, taking care that the film lies flat, without any creases or folds, and that it covers the whole of the surface of the grips. This is to avoid any slippage of the test sample which would result in too high a test result. The magnet is energised and the shaft of the dart (which is steel tipped) is attached to the magnet, via a centring hole. The magnet is then de-energised and the dart falls on to the film surface. If the film does not break, the operator should try to catch the dart on the rebound in order to avoid multiple impacts on the film. On the other hand, should the dart break the film and pass through the hole in the table, there should be some provision made for catching the dart and preventing damage to the head. This is because any nick or scratch on the dart head could cause anomalous results when dropped on to the next sample of film.

There are many ways in which the impact strength can be measured and expressed using the Falling Dart method. In one method the weight of the dart and the height of drop are adjusted to give rupture of all the film samples. The dart is decelerated by the film and the energy thus absorbed by the film sample is calculated. The original potential energy of the dart is known from its mass and its height above the sample while kinetic energy after impact can be calculated by measuring the time taken to pass two photoelectric cells placed immediately below the test samples. The photoelectric cells are connected directly with electronic timing devices. The energy absorbed is then found by subtraction of the residual kinetic energy from the potential energy of the dart and this is the impact strength of the film.

A simpler method is to drop the dart from a constant height and adjust its weight in equal increments from a minimum just too light to rupture the sample up to a maximum just heavy enough to cause rupture of all the samples. The weight at which 50% of these samples rupture, multiplied by the height through which the dart falls is taken to be the impact strength of the film. A convenient way of treating the results is the Staircase Method which needs a minimum of 20 samples.

Using some standard height, drop the dart and observe the result. If a failure is observed, the weight of the dart is reduced by a standard amount (say 5 g). If the drop does not rupture the film, then increase the weight of the dart for the next drop. Continue the drop impact testing, decreasing or increasing the weight of the dart by the standard increment depending on whether the previous drop caused a failure or not. Record for each drop, the weight of the dart and the results of the test, i.e. failure or non-failure. The results can be recorded as follows:

Let O represent a non-failure, F represent a failure. Draw up a column of dart weights on the left hand side of the page and record the results of consecutive drops as they are obtained.

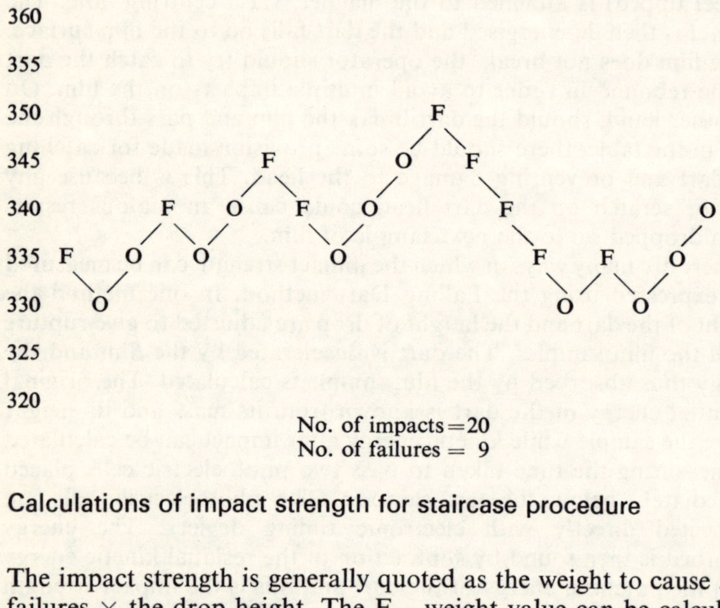

No. of impacts = 20
No. of failures = 9

Calculations of impact strength for staircase procedure

The impact strength is generally quoted as the weight to cause 50% failures × the drop height. The F_{50} weight value can be calculated from the following equation:

$$\text{Impact weight} = \frac{(n_0 w_0 + n_1 w_1 + n_2 w_2 + \cdots n_l w_l)}{n_0 + n_1 + n_2 + \cdots n_l} \pm e \text{ g}$$

Where n_0 = Number of least frequent events at weight w_0 g
n_1 = Number of least frequent events at weight w_1 g
n_l = Number of least frequent events at weight w_l g
e = Constant weight increment between weights w_0, w_1, etc.

If the least frequent event is a non-failure, use $+e$.
If the least frequent event is a failure, use $-e$.

In the example already shown, the calculations are based on failure as this was the least frequent event.

$$\text{Impact weight} = \frac{(3 \times 335) + (3 \times 340) + (2 \times 345) + (1 \times 350)}{3 + 3 + 2 + 1} - 5\text{ g}$$

$$= \frac{1005 + 1020 + 690 + 350}{9} - 5\text{ g}$$

$$= \frac{3065}{9} - 5\text{ g} = 336\text{ g} \ (= 335\text{ g to nearest 5 g})$$

Another approach to the matter of impact testing is the use of a pendulum instead of the falling dart. The film is clamped vertically and struck by a pendulum which has swung from a known height. The residual energy of the pendulum after it has ruptured the film, is measured by a pointer on a calibrated scale or by an electric timer activated by a pair of photoelectric cells. The loss in energy is a measure of the impact strength of the film.

9.3.1 IMPACT FATIGUE

The pendulum test is based on the application of a force more than sufficient to cause failure of the film and to a certain extent this is also true of the falling dart test. However, in use it is likely that the film will have to withstand a series of impacts which, singly, are not sufficient to cause rupture. These smaller impacts may eventually cause failure by fatigue and this would be shown up by some repetitive impact test. Such tests are time consuming but equipment has been designed to carry out impact fatigue testing. One such piece of equipment consists of an inclined plane which holds a number of stainless steel balls. The incline can be raised to allow the balls to be dropped at various heights on to the clamped film. Balls of different sizes can be used and this, with the variation in drop height, gives a great number of possible variables. When failure of the film eventually occurs, a gate prevents any more balls falling on to the film.

9.4 TEAR STRENGTH

The usual tests for measuring tear strength actually measure the energy required for tear propagation rather than for tear initiation, as it is difficult to start a tear in most plastics films. More specifically, what is measured is the energy absorbed by the test specimen in propagating a tear that has already been initiated by cutting a small nick in the sample, with a razor blade.

Tear strength is an important property of packaging films and a knowledge of both resistance to tear initiation and tear propagation, may be necessary. In heavy duty sacks, for instance, possible rough handling may demand that tears do not run from small snags or punctures incurred during transit. On the other hand, applications relying on a tear tape to give easy access to the contents, require ease of tear propagation in one direction.

One of the most common tests for measuring tear strength is the Elmendorf Test which, like so many other film tests, was originally developed for paper testing. The Elmendorf Test measures the energy required to propagate a tear through a specified length of film. The apparatus has two grips set side by side with only a small separation. One grip is stationary and is mounted on an upright on the instrument base. The second grip is moveable and is mounted on a pendulum or a heavy sector of a circle. This pendulum is mounted on an almost frictionless bearing and swings on a shaft fixed perpendicular to the upright pillar. The sample of film is clamped in the two grips and a slit of standard dimensions is introduced centrally into the film using a razor blade. When the pendulum is released it swings down and tears the sample along a continuation of the slit. The energy required to complete this tear

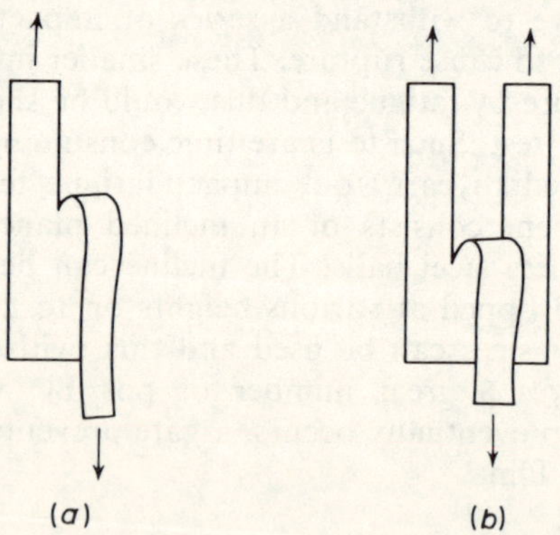

Figure 9.4. Tear test specimens: (a) 'trouser' tear, (b) 'tongue' tear

is measured on a scale attached to the pendulum by means of a pointer carried by the pendulum on its return swing. This indicates the residual energy in the pendulum and thus the amount of energy lost in tearing the film sample. The Elmendorf Tear Test is described more fully in ASTM D. 1922-61T.

Tear strengths can also be measured at slower rates of tear using a tensile testing machine. Two possible types of tear test samples are shown in *Figure 9.4*.

In this type of test, measurement is made of the force necessary to keep the tear moving at a fixed speed. ASTM D. 1938-62T describes such a test. Tear propagation can also be measured on a tensile testing machine using a film sample cut to the shape shown in *Figure 9.5*.

When the sample is pulled between the grips of a tensile testing machine, a tear starts at the root of the 90° notch. For fairly flexible films both the load necessary to initiate the tear and to continue it

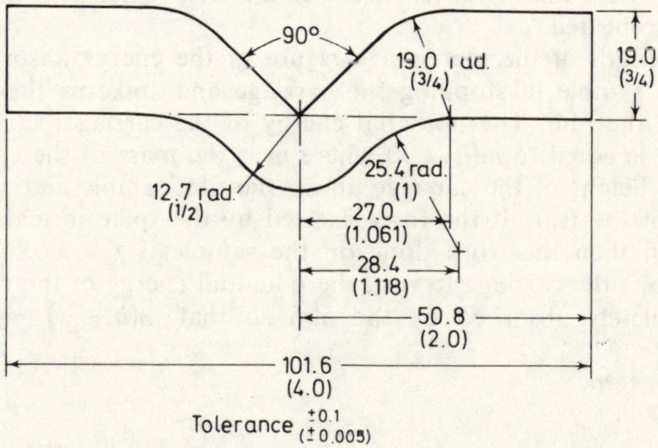

Figure 9.5. Cutting edges of punch for tear strength test specimen. (*Dimensions in millimetres with inch equivalents in parentheses*)

can be recorded. For brittle films, however, only the maximum tear initiation force is measured.

In all the tear tests described, samples should be cut from both machine and transverse directions as the tear strength can vary widely according to the direction of tear.

9.5 PUNCTURE PENETRATION TEST

Puncture propagation is also important, particularly in the handling or transit of shipping sacks. A test that measures the combined effects of puncture and tear is one developed by Patterson and Winn [*Mater.Res. Stand.2* (1962.396)]. A loaded carriage, fitted with a horizontally projecting spike, of standard tip and dimensions, falls from a fixed height on to a vertically mounted film sample.

The top of the film sample holder is curved away from the rails so that when the sample is draped over it and clamped in place, the spike snags the film at a point which is always the same distance away from the sample edge. The carriage is guided by a pair of rails during the drop so as to standardise the manner in which the spike engages the film sample. The spike is 3.2 mm ($\frac{1}{8}$ in) in diameter and extends 39.6 mm (1.56 in) from the carriage face. The point is conical with a 30° included angle. The probe penetrates the film within a very short distance and the major contribution to the absorption of energy is found to be the resulting tear propagation. Arbitrarily, a minimum tear length of 39.9 mm (1.57 in) has been selected and if this is not reached then a heavier carriage is used and the test repeated.

The length of the tear is a measure of the energy absorbed by the film sample in stopping the carriage and spike as they move through the film. The potential energy of the carriage before it is released is equal to $m(h + x)$ where m is the mass of the carriage, h is the height of the carriage above the film sample and x is the length of the tear. If the force exerted by the spike in tearing the film is f, then the work done on the sample is $f \times x$. Since the film brings the carriage to rest, the potential energy of the carriage is completely absorbed by the film so that $m(h + x) = fx$ and

$$f = \frac{mh}{x} + m.$$

9.6 STIFFNESS

Stiffness can be considered as the resistance of the film to distortion and, in particular, to bending. It is a compound property and depends on the thickness of the film, as well as the inherent stiffness of the material. Young's modulus is one measure of the inherent stiffness of a single ply film and the determination of this has already been described. There are also more direct methods of measuring stiffness, one of which is by means of the Handle-O-Meter Stiffness Tester. The test is a simple one and consists in measuring the force required to push a sample of film into a slot of a given width over which the film has been laid. The method is suitable mainly for thin film and the principle of operation is shown in *Figure 9.6*.

A sheet of film is placed on a metal plate containing a slot that extends the whole width of the plate. The width of the slot is usually 5 mm (0·2 in) but wider slots may have to be used with thick film. Machine direction stiffness is obtained by placing the film with the

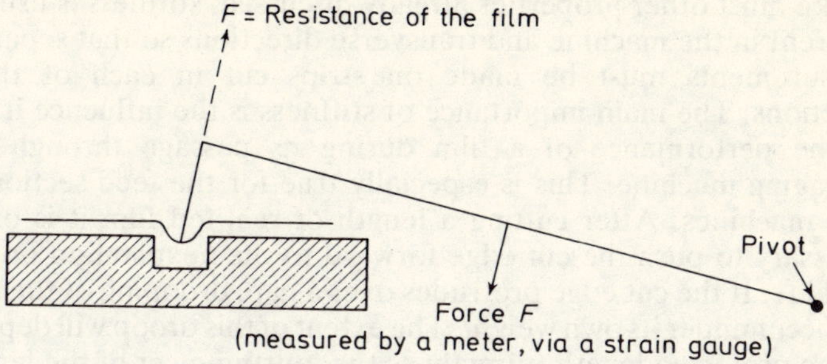

Figure 9.6. Principle of Handle-O-Meter stiffness tester

machine direction at right angles to the slot while transverse direction stiffness is obtained with the transverse direction of the film in a similar position. In operation, the equipment automatically lowers a horizontal bar through the slot against the resistance of the film. The force exerted on the bar is measured via a strain gauge and is indicated on a meter. The force first increases, reaches a maximum, and then decreases. The maximum value of the force, as indicated by the instrument, is taken as the Handle-O-Meter stiffness of the film.

The results may be affected by slip agents in the film and by electrostatic attraction (increasing the frictional resistance between the film and the base plate). Efforts have been made to reduce such effects by reducing the widths of the film sample so that it was not much wider than the width of the slot. However, the behaviour of the film in practice, on packaging machines, is probably more closely related to results obtained when they are affected by factors such as slip and electrostatic attraction. On the other hand, the dependence of stiffness on the cube of the thickness is obscured unless such extraneous factors are eliminated.

For thicker films, stiffness may be measured by treating a film strip as a beam. The test sample is placed on two supports, suitably spaced, and the resultant beam is loaded at the centre. The deflection produced for a given load is then measured. The test can be performed in other ways. Thus, the maximum load which the film strip/beam is able to support can be measured or the load necessary to produce a specified deflection can be determined.

One of the simplest tests for measuring stiffness is the cantilever test where one end of the film strip is clamped in a horizontal plane and the amount by which the film droops under its own weight is measured.

Like most other properties already discussed, stiffness is usually different in the machine and transverse directions so that separate measurements must be made on strips cut in each of these directions. The main importance of stiffness is the influence it has on the performance of a film during its passage through the packaging machine. This is especially true for the feed section of such machines. After cutting a length of reel-fed film it is often necessary to push the cut edge forward to the next station on the machine. If the cut edge protrudes over a gap, drooping of the film will occur under its own weight. The extent of this droop will depend on the projected length (directly as the fourth power of the length projected), on the thickness of film (inversely as the square of the thickness) and on the intrinsic rigidity of the film. If the gap is bridged by a metal plate then the problem of 'droop' is replaced by problems concerned with the absence of slip, particularly with soft, sticky films. Such problems can be solved by the use of grippers to pull the film through instead of pushing it.

Stiffness may also be important to the final end use. Thus, in the packaging of woollen goods a limp film is preferred so that the softness of the contents can be appreciated without opening the package.

9.7 FLEX RESISTANCE

The resistance to repeated flexure or creasing is important in use. Some films are highly resistant whereas others will fail by pinholing or total fracture after bending only a few times. In essence the resistance to flexing is measured by repeatedly folding the film, backwards and forwards, at a given rate. The number of cycles to failure is recorded as the flex resistance. One method of measuring is known as the Schopper Folding Endurance Test and was originally designed for measuring the folding endurance of paper. A film strip (15 mm wide × 100 mm long—0·6 in wide by 4·0 in long) is held at each end by grips which keep the sample under a constant tension. A slotted metal strip is fitted over the mid-point of the sample and is driven back and forth by an electric motor at a rate of 120 double folds per minute until failure of the film occurs. The number of double folds up to the point of failure is recorded on a counter. Details of the Schopper test are given in TAPPI (Technical Association for the Pulp and Paper Industry) T.423m-50 (TAPPI Standard and Suggested Methods).

With some of the more tough and flexible polymer films, even a large number of flexings may not lead to fracture. In such cases it may be worth running the test on various thicknesses since a thicker film may show failure at a relatively low number of flexings. Another point is that even if failure does not occur, certain properties of the film may be seriously impaired. Permeability may be increased, for example, or tensile properties may be reduced. The optical properties may also be seriously affected. One other way of testing crease or flex resistance, therefore, is to subject the film to a given number of cycles in the test equipment and then compare the relevant properties with those of the uncreased film.

9.8 COEFFICIENT OF FRICTION

The coefficient of friction is a measure of the ease with which the surface of one material will slide over another. Thus, films which are slippery and move easily over various surfaces have a low coefficient of friction. Referring to the diagram (*Figure 9.7*), F = frictional force and is the force, parallel to the surface that resists the sliding motion. The ratio of the frictional force, F, divided by the normal force, R, is the coefficient of friction, μ, when the forces are in equilibrium.

One of the simplest methods for measuring the coefficient of friction is by means of an inclined plane. The surface of the plane

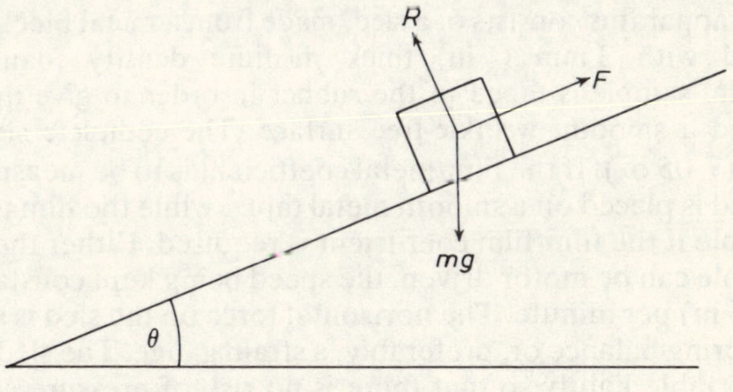

Figure 9.7. R=normal reaction=$mg\cos\theta$; F=frictional force=$mg\sin\theta$; m=mass of object on plane; g=acceleration due to gravity

is covered by a sample of film, as is the weight on the plane. The angle of the plane is slowly increased until the weight just starts to move when the various forces can be taken to be in equilibrium. The angle θ at this point is known as the angle of repose. The coefficient of friction μ then $= F/R = \tan\ \theta$. The coefficient of friction measured in this way is known as the static coefficient of

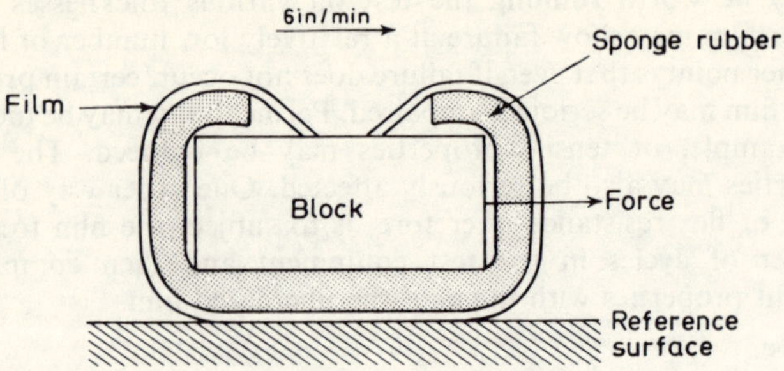

Figure 9.8. ASTM apparatus for determining coefficient of dynamic friction

friction. If the plane is set at an angle such that the weight moves easily, and the slope is then reduced until the weight just comes to rest, then the value measured for μ is known as the dynamic coefficient of friction. This is usually lower than the static coefficient although the two can also be equal. The value of the coefficient of dynamic friction is affected by the speed of movement of one surface over the other and the coefficient of friction may also be affected by humidity, temperature and the presence of static electricity.

A more satisfactory method of measuring coefficients of friction is by the procedure ASTM 1894-63, the principle of which is shown in *Figure 9.8*.

The apparatus consists of a sled, made from a metal block wrapped around with 3 mm ($\frac{1}{8}$ in) thick medium density foam rubber. The film sample is taped to the rubber in order to give the base of the sled a smooth, wrinkle-free surface. The complete sled weighs 200 g (7·05 oz). If the film/metal coefficient is to be measured, then the sled is placed on a smooth metal table, while the film is taped to the table if the film/film coefficient is required. Either the block or the table can be motor driven, the speed being kept constant at 150 mm (6 in) per minute. The horizontal force on the sled is measured by a spring balance or, preferably, a strain gauge. The sled is placed on the table lightly so that there is no risk of pressure causing an artificial bond between the surfaces. When the drive is started the original deflection of the load measuring device is noted and this is used to calculate the static coefficient of friction by dividing by the total weight of the sled. The average deflection is also noted and used to calculate the dynamic (or sliding) coefficient of friction.

Where the metal table is being used to measure film/metal coefficients, the metal surface must be thoroughly cleaned after each run. This is necessary because films which contain additives liable to diffuse to the surface may leave a smear on the metal.

The frictional properties of a film are important, both during its passage through printing or wrapping machines and after being made up into a bag, sack or overwrap. In general, one does not talk so much about high coefficients of friction as about low or high slip. High slip is normally required for a number of reasons. During the passage of a film, through packaging equipment of all sorts, it is subjected to a variety of forces including some which press the film tightly against flat metal surfaces. If the slip characteristics are low the passage of the film through the equipment can be arrested completely. The consequent strain in the film (particularly in a thin film) can lead to elongation or other distortion of the film which will affect register in the printing or bag making stages. High slip is also generally required once the package has been overwrapped since the finished packs have to slide over various surfaces as they are ejected from the machine.

The failure to slide during passage through a machine, mentioned previously, is known as seizure and the tendency towards this is not always shown by coefficient of friction data. This often arises because the coefficient of friction is determined at relatively low normal forces, compared with high localised forces which may occur in a packaging machine. Modifications of the standard friction test have been made, the main differences being the use of a sled modified to give a much greater loading normal to the surface, together with a reduction in the speed of movement. A satisfactory speed has been found to be 2.5 mm (0.1 in) per minute. In addition to measuring the magnitude of the tangential forces developed, provision is made for a pen trace to be made during the test. A trace which is fairly smooth and parallel to the surface indicates a surface with little tendency to bind. A sawtooth trace, however, indicates a surface which tends to stick and then slip and such a surface would be likely to give trouble on a packaging machine.

Frictional properties are also important in the manufacture of the film, especially when being wound up. Good roll formation, for instance, is dependent on the correct level of friction. Too much slip may cause telescoping of the roll during handling or transit whereas too little can cause buckling on the roll. There are, of course, occasions when a low slip value is a definite requirement. Heavy duty sacks, for instance, are often required to be stacked to heights of several metres and any tendency towards slip can be a possible danger.

9.9 BLOCKING

A property somewhat akin to friction is blocking, which is the tendency of two adjacent layers of film to stick together, particularly when left under pressure for some time, as when films are stacked in cut sheets. It can also make bags made from layflat film difficult to open. Blocking is particularly marked in the case of films with very smooth surfaces which can come into very close contact with consequent exclusion of air. Other factors which affect blocking are static charges, surface treatment (such as printing pre-treatment) and storage conditions. Anti-blocking additives are often added to the film to reduce the tendency to blocking. These act by diffusing to the surface and forming non-adherent layers.

The degree of blocking is determined by the force required to separate the two layers of blocked film, when the force is applied perpendicularly to the surface of the film. In BS.1763: 1956, rectangular specimens are placed face to face between filter papers and glass plates, and a load of 6·895 kN/m^2 (1 lb/in^2) applied at 50°C for 24 h. After the film has cooled to room temperature, the sheet must peel under a specified static load. A different approach is taken in ASTM D1893-61T which assesses blocking by measuring the force required per inch of sheet to draw a 6.35 mm ($\frac{1}{4}$ in) diameter rod perpendicular to its axis at a rate of 127 mm (5 in) per minute between the adhering films, thereby separating them.

9.10 SUMMARY

The mechanical properties of a film or laminate are important both to the packaging operation and to the end-use. The problems in the case of the packaging operation have been complicated by the fact that plastics films were late on the scene and packaging equipment was originally designed to run using stiffer materials with better dead-fold properties such as paper, cellulose film, aluminium foil or combinations of these.

The first part of a packaging line that makes its demands on the web is the reel unwind. Constant tension devices are often used for controlling tension during the unwind but the elongation or extensibility of the film is still an important factor. Too high a figure can lead to uneven stretch while too low a figure, if coupled with shock-brittleness might lead to film breakage when a sudden snatch occurs.

The frictional properties of a film are important when the film comes into close contact with metal surfaces such as 'ploughs', formers, guides, etc. which are used to preform tubes or to fold the wrapping material round the product. High slip (low coefficient of friction) is desirable but where low slip films are used then the metal surfaces which come into contact should be given a matt surface by sand blasting or some similar action.

Friction is also important when the film passes over free-running rollers. Too low a coefficient of friction would mean slippage instead of a positive drive of the rollers. Too high a coefficient of friction, on the other hand, could lead to wrap-around of the roller should there be a break.

Finally, stiffness is a factor in machine running whenever the film has to bridge a gap during its passage through the machine.

Mechanical properties of importance to the end-uses of a film include stiffness, a limp film being preferred for the packaging of woollen garments and similar goods.

If the film is to be used for making bags or sachets liable to rough treatment (especially when packaging heavy articles) then impact strength is important.

Tear propagation may be required to be high or low according to the particular end-use. Where tear tapes are to be incorporated then ease of tear propagation in one direction is desirable but for heavy duty sacks and similar applications, where puncturing during transit is possible, such propagation must be avoided.

The range of mechanical properties available is extremely wide when one considers possible combinations of plastics films, coupled with combinations using paper or aluminium foil. It is essential to be sure just what properties are essential at the design stage of any application. The dangers of under-performance are usually well understood but with costs becoming even higher, over-performance is equally to be avoided.

FURTHER READING

BROWN, R. P. (ed), *Handbook of Plastics Test Methods, 2nd edition,* George Godwin (1981)

10

Physical and Chemical Properties

10.1 OPTICAL PROPERTIES

10.1.1 LIGHT TRANSMISSION

Light transmission is measured by means of a photoelectric cell. The intensity of a light source is measured by the cell, both with and without the interposition of the film sample. The light transmission is the ratio of the light intensity measured with the film to that obtained without it, and is expressed as a percentage. The figure for light transmission takes no account of the quality of the light transmitted by the film and it is possible that a blurred image might be obtained even when looking through a film with a high light transmission value. Such a film might be suitable for a tight overwrap but would be undesirable when the film is acting essentially as a window.

10.1.2 'SEE-THROUGH' CLARITY

A better guide to the quality of an image when viewed through a film is given by the property known as 'see-through' clarity. This indicates the degree of distortion of an object when seen through the film. In one test the optical definition of a standard well-illuminated wire mesh grid viewed through the test film is

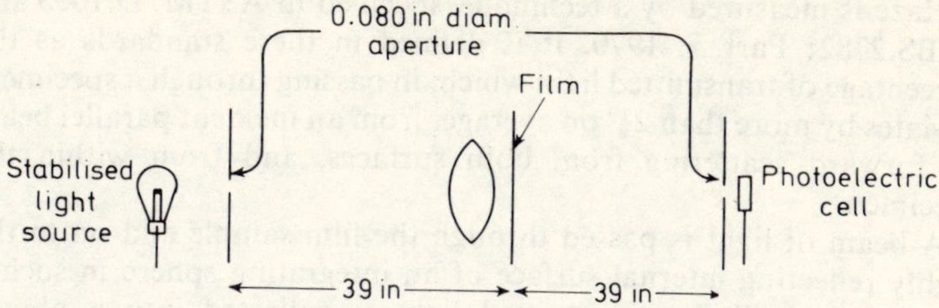

Figure 10.1. 'See-through' apparatus

compared with a set of eight standard photographs. These cover the range of clarities normally encountered in the material under investigation and the test sample is given a number corresponding to the standard it most closely resembles.

Another way of assessing 'see-through' clarity is based on the assumption that definition of an object is controlled by small-angle scatter. *Figure 10.1* shows the basis of the method.

A stabilised light source serves to illuminate a small (2.03 mm or 0.080 in diameter) orifice which then acts as the optical object. A large lens focuses this image on to a similar orifice placed in front of an electric cell. The distance between the orifices is 1.98 m (78 in) and the lens is placed midway between them. When a test sample of film is placed behind the lens, as shown, only light which is transmitted by the film with a deviation of less than 4 min will enter the photocell. The photocell reading, expressed as a percentage of the intensity of the incident beam, is the 'see-through' clarity. This figure is, of course, a measure not only of the distortion but also of the light transmission, and the 'see-through' clarity and the light transmission (determined as in the previous paragraph) must be compared in order to assess the effect of deviation alone.

10.1.3 HAZE

Haze can be taken as a measure of the 'milkiness' of the film and is usually of greater importance than 'see-through' clarity in the case of packaging films since these are usually closer to the contents of the pack. One cause of haze is surface imperfections of the film so that the amount of haze is not necessarily proportional to film thickness. The appearance of haze is caused by light being scattered by the surface imperfections or by inhomogeneities in the film. These latter can be caused by voids, large crystallites, incompletely dissolved additives or cross-linked material.

Haze is measured by a technique specified in ASTM. D.1003 and in BS.2782: Part 5: 1970. It is defined in these standards as the percentage of transmitted light which, in passing through a specimen, deviates by more than 2½° on average, from an incident parallel beam by forward scattering from both surfaces, and from within the specimen.

A beam of light is passed through the film sample and on to the highly reflecting internal surface of an integrating sphere in such a manner that all the transmitted light is collected into a photoelectric cell. Next, the sphere is moved through a small angle so that the path of the light beam now falls on an extension of the

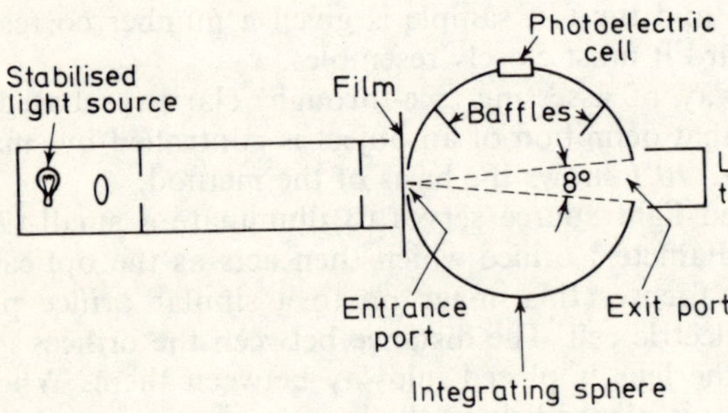

Figure 10.2. Hazemeter

sphere comprising a matt black surface that absorbs all the light transmitted straight forward along the path of the beam. Any light that is appreciably scattered in a forward direction is still gathered by the highly reflective internal surface of the sphere and into the photocell. The ratio of the second photocell reading to the first is a measure of the haze. The apparatus for the measurement of haze is shown diagrammatically in *Figure 10.2*.

10.1.4 GLOSS

Gloss is important to packaging films and is a measure of the ability of the film to reflect incident light specularly (i.e. as in a mirror—angle of incidence = angle of reflection). A high gloss will, therefore, produce a sharp image of any light source and will thus give rise to a pleasing sparkle on the film. Gloss is determined by a device that measures the percentage of the light, incident at an angle (usually 45°) to the surface of the film, that is reflected

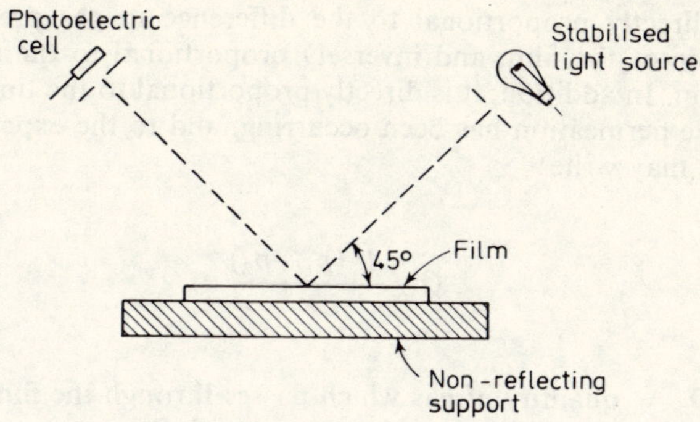

Figure 10.3. Glossmeter

at the same angle. The essentials of the standard test (BS 2782: Part 5: 1970) are shown in *Figure 10.3*.

The glossmeter is first calibrated by using an optically flat, highly polished, black glass plate of refractive index 1.52. This has a gloss value assigned to it of 53 units. A light trap, with matt black interior is first placed in position and the lamp switched on. The measuring device is then set at 0. The reference standard is next placed in position and the sensitivity control adjusted to give the correct reading for the standard. Finally, the test specimen is placed over the light trap and the new reading noted.

10.2 PERMEABILITY

In essence there are two mechanisms by which a gas or vapour can pass from one side of a plastics film to another. If the film is porous then the gas or vapour can flow through the holes. This effect is not usually present except in the case of very thin films. When porosity is not present then gases and vapours can still pass through the film by a process of solution (or absorption) and diffusion and this compound process is known as permeation. A full treatment of the permeability of gases and vapours through plastics is beyond the scope of this book. For this, and for further references, the reader is referred to an excellent monograph entitled, *Permeability of Plastics Films*, by B. J. Hennessey, J.A. Mead and T. C. Stening, and published by The Plastics Institute.

Suffice it to say that where permanent gases are concerned, and under conditions of constant temperature and a constant partial pressure differential, a steady state will be achieved after a certain time has elapsed. The quantity of gas, Q, passing through the film is directly proportional to the difference in gas pressure on either side of the film, and inversely proportional to the thickness of the film. In addition, it is directly proportional to the time during which the permeation has been occurring, and to the exposed area. Thus we may write:

$$Q \propto \frac{At\,(p_1 - p_2)}{x}$$

where Q = quantity of gas which passes through the film;
A = the surface area in contact with the gas;
t = time;
$(p_1 - p_2)$ = partial pressure differential;
x = thickness of plastic.

This expression can also be put in the form of an equation, thus:

$$Q = \frac{P\,At\,(p_1 - p_2)}{x}$$

where P is a constant for a specific combination of gas and plastic at a given temperature. The factor P is variously known as the permeability factor (or 'P-factor'), permeability coefficient or permeability constant. It should be emphasised that the above equation applies only to steady state conditions.

The position with regard to vapours, including water vapour, is less clear. For one thing, the steady state condition is reached more slowly, while there may also be chemical interactions between the permeant and the plastics film.

The practical importance of permeability hardly needs emphasising. One of the prime functions of a packaging film, very often, is to act as a barrier to gases and vapours. Biscuits, for example, need to be kept dry, while conversely, cigarettes and tobacco need to be protected from moisture loss. The position with regard to the permanent gases is similar. Fresh produce needs to be able to lose carbon dioxide and pick up oxygen while fatty foods may go rancid if oxygen is not kept out. Many foods are packed in a vacuum and a good barrier is essential if this condition is to be maintained. The measurement of permeability is, therefore, important and standard methods are available for both water vapour and the permanent gases.

10.2.1 WATER VAPOUR PERMEABILITY

Tests for water vapour permeability are described in BS.2782: Part 5: 1970. Methods 513A and 513B are carried out by sealing test specimens with wax over the mouth of metal dishes containing a desiccant.

The dishes are weighed initially and then placed in a temperature and humidity controlled cabinet. Weighings are carried out at regular intervals and the gains in weight are measured. In Method 513A, the temperature of the cabinet is maintained at 25°C±0·5°C and the relative humidity at 75%±2%. These conditions are taken as representative of 'temperate' climates. For 'tropical' conditions the tests are performed at 38°C±0·5°C and 90%±2% relative humidity.

It may often be found preferable to carry out water vapour permeability tests on completed packages rather than on dishes. Tests conditions are the same as above but the desiccant is contained in a heat sealed sachet instead of a metal dish. Weighings are carried out as before. In either case, the water vapour permeability is reported as $g/m^2/24h$.

Where a quick indication of the potential barrier properties of a film is required, the film can be clamped in an instrument where one side of the film is exposed to a high humidity atmosphere while the other side is in contact with dry air. As water vapour diffuses in, it is detected by an infra-red absorption cell or by a resistor, the value of which is affected by changes in relative humidity. The instrument then measures the time required for a particular change in relative humidity as detected by the cell or resistor.

10.2.2 GAS PERMEABILITY

Measurement of gas permeability is carried out under controlled conditions of pressure, as well as of temperature and relative humidity. In essence the usual tests consist in making the film a partition between a test cell and an evacuated manometer. The pressure across the film is usually one atmosphere. As the gas passes through the film sample the mercury in the capillary leg of the manometer is depressed. After a constant transmission rate is achieved, a plot of mercury height against time gives a straight line. The slope of this line can be used to calculate the gas transmission. Gas permeability measurements are described in ASTM D.1434 and BS.2782: Part 5: 1970.

10.2.3 ODOUR PERMEABILITY

There are no standard tests for the measurement of odour permeability although it is an important characteristic in many cases. One method which can be used to compare the efficiency of several films as odour barriers is to make up pouches with each film. The pouches can be filled with some odiferous material and then placed in separate clean glass bottles, sealed by crimping with aluminium foil. The minimum time for an odour to be apparent in the bottle can be measured and will give a rough ranking list of the test films. This ranking list may vary somewhat according to the type of odiferous material used so that the substance to be packed should be used as the test material if possible.

10.3 DENSITY

The density of a film sample is most satisfactorily measured by means of a density gradient column as described in ASTM D.1505-60T or BS.2782: Part 5: 1970. A density gradient column is a mixture of two fluids of different density, the proportions of which change uniformly from the top to the bottom of the column.

The liquids are chosen for their densities but care must also be taken that they do not have an effect on the film under test. Ethanol–water, and methanol–water mixtures have both been used successfully, especially for measuring the density of polyethylene films. The column is established initially as follows. The two liquids are de-gassed by boiling and then mixed in such proportions as to give compositions having the required density limits at top and bottom of the column. These are placed in separate flasks. The flask containing the lighter mixture has a capillary tube outlet which runs to the bottom of the column. This flask is then stirred and is connected to the flask containing the heavier mixture. The density in the first flask thus increases linearly with time and so a column with a linear change of density is produced. If the filling is carried out slowly (over about three hours) and the column is surrounded with a temperature controlled water jacket the column will be stable for 2–3 months.

A series of calibrated glass floats, covering the required density range is then placed in the column. Film samples are cut and then immersed for $\frac{1}{2}$ min in, say, ethanol or methanol, and then introduced into the column by means of tweezers, taking care not to introduce air bubbles. After 3 h measurement is made, against a scale, of the heights of the floating samples and the markers against their respective densities. The heights of the floats are then plotted against their respective densities and the best-fitting curve is drawn through the points. From this curve, the densities of the film samples can be read. Samples on which measurements have been completed are removed at intervals by means of a slowly moving wire gauze basket.

Another method of determining film densities is also given in BS.2782: Part 5: 1970. A series of liquids of different densities is made up in stoppered glass vessels, maintained at a constant temperature of 23°C. Film samples are placed in each of the vessels, care being taken that no air bubbles are attached to the film. Observations are made as to whether the sample floats or sinks and a note is made of the densest material in which the film sinks and the least dense in which it floats. The densities of these two liquids are then rapidly determined. The density of the film under test will then lie between these two values.

10.4 HEAT SEALABILITY

The heat sealability of a packaging film is one of the most important properties when considering its use on wrapping or bag making equipment and, of course, the integrity of the seal is also of tremendous importance to the ultimate package. The heat sealability of a film has to be considered in relation to many other factors, including the available pressure, the dwell time, temperature and the rate of heat transfer of the sealing bars. Any test of the heat seal qualities of a particular film must, therefore, simulate the conditions under which the film will be used as closely as possible.

Heat seals can be made by a number of different methods, apart from the straightforward application of heat to the layers of film. Seals can be made, for instance, by high frequency heating (as with PVC) or by ultrasonic welding. However the seal is made, the strength is determined by measuring the force required to pull apart the pieces of film which have been sealed together. The force can be applied in such a way as to cause the seal to fail in shear or in peel. If other factors, such as the temperature, dwell time, and pressure used in making the seal are equal, then peeling a seal will give a lower figure than shearing it. Seals are normally tested in peel, therefore, as this is the way they are likely to fail in practice.

Two tests are in common use, the dynamic and the static. In both tests a 25 mm (1·0 in) strip is cut through the heat seal. The dynamic test uses a sensitive tensile testing machine and the two free ends of the film strip are placed in the machine clamps. The force necessary to peel apart the two pieces of film is then measured. In the static test the strips are hung from a frame with one free end clamped and the other attached to a weight. The seals are examined at intervals for signs of failure. When recording the results of this test, the weight and the length of time the load was in operation are both noted.

Both of the above tests can, of course, be used for investigating the effect of changes in dwell time, temperature and pressure on heat sealing equipment if standard film samples are used.

10.5 DIMENSIONAL STABILITY

Dimensional stability is a desirable property in any film conversion process and particularly so in printing. Even small changes in film dimensions while passing through the printing process may lead to serious problems in print registration. In the finished package, too, dimensional stability is important. A reduction in dimensions of a film overwrap of the order of a few percent may be sufficient to crush a carton or cause rupture of the film. Conversely, a stretching of the film will lead to loosening of the wrap with creasing and wrinkling which, to say the least, are unsightly. Dimensional changes in a film may be caused by variations in either temperature or humidity. In some cases, a high humidity may cause a film to cockle.

Dimensional stability is normally tested by cutting film strips in both the machine and transverse directions, and then subjecting them to varying conditions and noting the percentage change in dimensions. Where possible, the test conditions should be closely related to the conditions likely to be encountered in conversion or end-use. If the precise conditions of end-use are not known then some arbitrary conditions can be used which will at least provide some measure of how the film will behave under changing conditions. For example, films which are sensitive to moisture may be exposed to a series of varying humidities at constant temperatures. The film strips should first be conditioned at some standard condition, say, 23°C and 50% relative humidity. After equilibrium has been reached the length of the film strips (or the distance between two previously scribed gauge markings) are measured. The strips can then be exposed to some high humidity, say 90%, for a specific period and the length remeasured. This could be followed by exposure to very low humidity (say 0–5%) followed by further measurement. Finally, the strips should be reconditioned at the standard conditions until equilibrium is again reached. If there is a permanent set, i.e. the strips do not return to their original length then this, too, is noted in addition to the changes at the various intermediate conditions.

An allied property, that of maximum shrinkage, is of interest in the case of heat shrinkable films. This can be determined by immersing marked film samples for 5 s in water (for temperatures up to 100°C) or in silicone 550 oil (for temperatures above 100°C). The temperature of testing is, of course, related to the shrink tunnel temperature in actual use.

10.6 WATER ABSORPTION

Water may be absorbed by a plastic film, either by direct contact or from the atmosphere. The effect varies considerably from film to film, ranging from negligible effect (as with the polyolefin films) to one of solution of the film (as with polyvinyl alcohol). Water absorption is measured by immersing a sample of film in water for a given time (usually 24 h) at a standard temperature and noting the change in weight.

10.7 EFFECT OF CHEMICALS

The effect of chemicals on a packaging film is obviously an important factor when assessing its suitability for containing a particular product. It might also, under certain circumstances, be important from an environmental point of view.

The normal test for chemical resistance involves subjecting a film sample to the chemical under test, usually by total immersion under specified conditions. Any change in appearance is noted at intervals as is the change in a particular property such as tensile strength.

It is not possible to generalise about the effect of specific chemicals so that tests should always be carried out where any doubt exists. Tests are particularly important in the case of mixtures even if information is available on the separate components.

One important phenomenon associated mainly with the polyolefins (although not polypropylene) is environmental stress cracking. This is caused by the action of stress in combination with particular chemical environments such as certain polar organic compounds. The chemicals are normally without action on the plastic in the absence of stress but the combination of stress and chemical environment can cause cracking in a very short time. The use of high molecular weight grades has been found to reduce the effect to a great extent.

Where a film laminate is to be used for a particular package this should be tested as a whole since one effect of chemical attack might well be delamination. Heat seals should also be investigated in this way as changes which may be unnoticeable in the film itself can sometimes vitiate the strength of the heat seal.

10.8 EFFECT OF LIGHT

Prolonged exposure to light may bring about many undesirable effects in certain films. In particular, light may catalyse certain other reactions such as oxidation. Brittleness, loss of clarity, colour changes and surface imperfections are some of the undesirable effects. If a film is intended for applications involving continued use in sunlight it should contain stabilisers such as ultra-violet absorbers.

Tests are usually carried out by exposing samples of the film to light of a specified wavelength, or combination of wavelengths, for a given time and noting the effect on various properties such as tensile strength or the various optical properties. The intended end-use is of importance when assessing the suitability of the film as the type of radiation may be very different. In addition to the effect of sunlight it is important in many applications to know the effect of shop or display cabinet lighting. The screening effect of glass is also important when packs are to be on display in shop windows.

In the case of coloured films, exposure to daylight or artificial light may cause undesirable colour changes.

10.9 EFFECT OF TEMPERATURE

10.9.1 HIGH TEMPERATURES

One consequence of a high temperature on a thermoplastic film may be to soften it and if high temperatures are likely to be encountered during use then a plastic with an appropriately high softening point must be chosen. For boil in the bag packs, for instance, films with a softening point above the boiling point of water are required. High temperatures may also lead to other undesirable changes, especially in the case of plasticised film, where loss of plasticiser can be accelerated leading to brittleness of the film.

10.9.2 LOW TEMPERATURE

Low temperature behaviour of film is also important in many applications, including the packaging of frozen foods. Some films, such as polypropylene and unplasticised PVC, suffer a loss in impact strength at temperatures around 0°C although oriented polypropylene is very much better in this respect. Where low temperature use is envisaged, it is preferable to carry out tests on the actual package since any stresses and strains in the film may influence the behaviour.

10.10 FLAMMABILITY

Plastics films are no more hazardous than most other wrapping materials but it may still be important to assess the way in which a film will behave when exposed to a flame. Tests depend on the material to be examined but one typical method consists in igniting a strip of film bent to the shape of an inverted 'U'. The distance over which the strip burns is an assessment of its flammability.

10.11 SUMMARY

Most of the properties dealt with in this chapter are concerned with the end-use behaviour of the film. The exception is Heat Sealability which is important both to end-users and to the packaging machine performance.

The optical properties are particularly important in these days of self-service selling. Light transmission tests give no indication of image distortion and are, therefore, not so important for packaging films. Light transmission characteristics can influence the choice of film in horticultural applications. Where the choice of a packaging film is concerned then 'see-through' clarity, haze and gloss may be more relevant, especially when sales appeal is critical, because of the influence of these properties on the visibility of the contents and on the 'sparkle' of the film.

No individual plastics film is a complete barrier to water vapour or to gases. Measurement of permeability is important, therefore, as an assessment of the protection likely to be given by a particular material. Complete impermeability is not always desirable, of course, and a knowledge of the exact permeability of the film is essential.

Good heat sealability usually means that packaging lines can be run at higher speeds. In addition, strong heat seals give packs that are less liable to leakage when roughly treated.

Finally, there are properties that can be grouped under the general heading of environmental resistance, i.e. resistance to the effects of light, chemicals and high or low temperature. These properties are relevant not only to packaging uses but also to building, agriculture and horticulture applications.

It should be re-emphasised that the performance requirements, in terms of physical and chemical properties, should be clearly stated so that the choice of film can avoid the extremes of failure in use, and over-packaging with consequent loss of profit.

FURTHER READING

BROWN, R. P. (Ed), *Handbook of Plastics Test Methods, 2nd edition*, George Godwin (1981)

11

Health Safety

by Dr L. L. Katan

11.1 INTRODUCTION

The majority of applications where health safety questions can arise relating to plastics films are indirect, i.e. possible effects on products in contact with them (e.g. packaging). There are three situations where such questions can arise directly from the film, and these are dealt with first.

11.2 DIRECT

11.2.1 INGESTION BY ANIMALS

It is widely believed that instances have occurred of damage caused to animals (cattle, horses, goats) due to ingestion of plastics films, either from packaging or wrapping (usually snack foods) or fertiliser bags. Damage is thought to have ranged from indigestion to death from blocking of the gastro-intestinal system or asphyxiation. In fact, the author has been unable to find any documented evidence of these cases, and whilst the hazard is perfectly possible, it seems not at all certain that it has actually occurred.

11.2.2 SUFFOCATION OF CHILDREN

This has occurred when young children played with used plastic bags, putting them over their heads. The hazard has largely been eliminated, by incorporation of holes into thin films used for making bags likely to come into the hands of children, or by using thicker film. Several countries have legislation. In the U.K., SI 1367 of 1974 specifies that bags with openings less than 190 mm and used for wrapping toys must be thicker than 38 microns. It also includes a standard warning to be printed on non-conforming bags which were permitted for an interim period (now over). Even where or when not mandatory, the warning is still often used and regarded as useful.

11.2.3 EDIBLE FILM

There are a few examples of edible film being used prior to the appearance of plastics, e.g. rice paper for wrapping nougat. A few attempts have been made to develop soluble edible plastics films for boil-in-the-bag applications; the idea being that the total package can be immersed in water, the film dissolves, the food cooks, is strained, and there is no need to remove the film—which is sometimes a messy business resulting in scalded fingers. There are two disadvantages. First, because the film is water soluble it cannot be used for wet foods. By the same token, the package is liable to be damaged by atmospheric humidity, rain, or even handling with wet hands unless it is further overwrapped. Second, some of the dirt on the outside of the package will be deposited on the food and consumed. This partly negates the main reason for using packaging—protection of food.

The polymers proposed for this application have included polyglycols (molecular weight *c.* 20 000) and ethylene vinyl alcohol copolymer.

Otherwise, the main health hazard interest associated with plastics films attaches to a number of end uses where toxic hazards can theoretically arise.

11.3 INDIRECT: TOXICITY-SENSITIVE END USES (TSEU)

These comprise:

(1) Food contact.
(2) Cosmetics and toiletries contact.
(3) Medicinals and drugs contact.
(4) Direct medical use.
(5) Children's toys.

The majority of applications in (1), (2) and (3) above, are in packaging; and, for brevity, we shall use 'packaging' to comprise all contact situations. In fact, the only other applications springing to mind are the covering of food or drugs manufacturing or processing equipment.

Food packaging is by far the biggest application, not only within TSEU, but for all plastics films. In the following, therefore, a full discussion will be given of food packaging; much of this also applies to the packaging of the other products.

The consumer is anxious about—and the plastics merchant responsible for—health safety of the packaged food. Many other aspects of food quality are also affected by packaging, and it is essential to evaluate the sum total of interactive effects to assess the acceptance of the food by the consumer. However, any of the relevant aspects, e.g. nutritive value, are common to all packaged food and are best assessed by the food manufacturer or merchant. In this monograph, we concentrate on the two particular aspects of most concern for plastics films, namely, health safety dealt with in this chapter; and organolepsis, dealt with in the next chapter.

In almost all instances of plastics films food packaging, contact with the food is more or less continuous. There are a few instances where plastics/food contact is intermittent, e.g. throw-away tableware which may be pressed from sheet, plastics paper napkins, etc. Whilst the following applies essentially to packaging, these cases of intermittent contact can be included making allowance for the hazard (if any) being lower because of the short contact time.

11.4 BASE LINES FOR EVALUATION

In nature, growing foods are protected from their environment by the wide range of defence mechanisms developed by evolution and natural selection. As soon as the food is garnered, not only do these defence mechanisms cease to operate, but decay usually sets in. For example, fruit not only ceases to be protected by skin and hence becomes prone to attack by micro-organisms, but also specific decay mechanisms start to operate. Even superficially inert materials, such as salt or sugar, are liable to be influenced by atmospheric humidity: the vast majority of foodstuffs are much more sensitive. The changes that take place are complex, and include not only straight interactions between foodstuffs and the environment but also interactions between different phases within the food itself.

The above general argument applies also to processed food, with extra modifications. On the one hand, processed foods are often formulated (e.g. with additives) or processed (e.g. dehydrated) so as to enhance storage life and reduce decay; on the other hand, the mixture of different foods, or comminution, can lead to further reactions.

Thus, foods almost invariably change with time—usually for the worse—and it is hence unrealistic to compare foods which have been in contact with plastics films for a certain time with the freshly garnered or prepared commodities. Comparisons should be made on either of two bases: (a) an equivalent lifetime in the absence of any package, or (b) an equivalent lifetime in an alternative environment to the plastic. However, especially when considering regulations, the base line is taken as a hypothetical zero risk situation. This is actually a meaningless concept, since all human activity involves some risk, however small.

11.5 COST/RISK/BENEFIT

A better approach is to use a cost/risk/benefit analysis, in which a reasonable balance is achieved. The basic risk equation (for all hazards) is

$$\text{Risk} = f^1\,(\text{exposure}) \times (\text{damage per unit exposure}) \quad [\text{eqn } 1]$$

If one sets risk = 0, and accepts that exposure is finite, then damage per unit exposure must be zero, which is virtually impossible to prove for any hazard, including packaging.

Equation 1 does not include cost, i.e. the cost of reducing hazard (for plastics film testing, quality control, etc.). Including this in the equation gives:

$$\text{Risk/unit cost} = f^1\,(\text{exposure}) \times f^2\,(\text{damage per unit exposure, reduced by unit cost}) \quad [\text{eqn } 2]$$

If risk is now set at zero, and exposure is finite, then cost becomes infinite.

Neither equation takes benefit into account; and the ultimate equation, which should be (but seldom is) used would be in terms of risk/unit cost/benefit (dividing both sides of eqn 2 by benefit). Of course, the factors in the equation are difficult if not impossible to quantify in common units, but *Table 11.1* shows the situation qualitatively, from which the imbalance for packaging is obvious.

Also, to appreciate some numerical values, data are given in *Table 11.2* for some risks.

Table 11.1. COST/RISK/BENEFIT

	Risk	*Benefit*	*Cost of testing*
Drugs	+++	+++	+++
Food additives	++	++	++
Food packaging	+	++	+++

+++ high
++ medium
+ low

Table 11.2. COMPARATIVE RISKS QUANTIFIED

	Average annual casualties, E.E.C. (1974–1978)	
	Deaths	Injuries
Road transport*	52 440	1 552 190
Food poisoning† (bacterial)	200	54 000
Food packaging	Nil	Few hundred‡

* Statistical office of the E.E.C.
† Estimate based on data for England and Wales, and some others.
‡ Broken glass.

In this chapter we give the scientific basis for evaluation of hazard or risk from film packaging, stressing that this should be an input to the overall social safety evaluation, and not an end in itself.

11.6 INTERACTIONS

To evaluate the scientific basis of health safety, it is necessary to set up a standardised model system. This reduces any food packaging situation (or indeed any food contact situation) to its elements. A typical element can be considered as a barrier between food and its environment; any hazard arises from interactions between components of all three. The element should be representative of the package as a whole. In reality, of course, packages are not uniform, and more than one element type may be involved—notably closures. Also, there may be scale effects. Nevertheless, taking appropriate precautions, the model shown in *Figure 11.1* can be used to integrate up to the whole package.

The model comprises three phases: environment/film/food. Each phase has numerous components; for the present purposes, it is

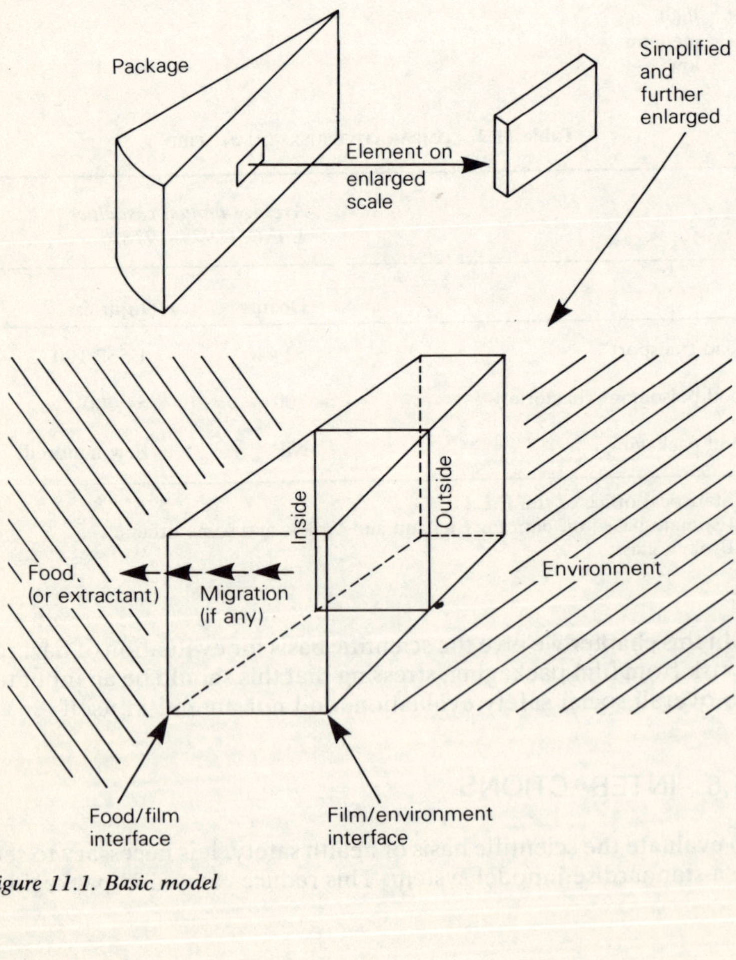

Figure 11.1. Basic model

Foodstuff	Plastics	Environment
Nonvolatile components	Base polymer	Vapour
Volatile components	Volatile components	Odorous components
	Additives	Micro-organisms
	Residual reactants	Macro-organisms
		Radiation

Figure 11.2. Components for interactions

convenient to define these as in *Figure 11.2* according to the following criteria.

Food: Volatile components are those which, under the physico-chemical conditions pertaining at the time, can undergo mass transport to a substantial extent; non-volatile components are the remainder.

Plastics film: polymer is defined as the high molecular weight polymeric components. Volatile components are defined in exactly the same way as for foodstuffs above. Additives are the non-polymeric components added subsequently to the manufacture of the original polymer. They include processing agents such as heat stabilisers, and end-use improvers such as UV stabilisers, anti-static agents, etc. Residual reactants are those traces of raw materials which did not react to form polymer in the original manufacturing process, and were not removed by subsequent purification. These include unreacted monomers (e.g. styrene in polystyrene, caprolactam in nylon-6, VCM in PVC, etc.), but traces of solvents and unchanged catalysts (if any) would also be included. For thermosets (e.g. polyurethane) the category would include residual basic components of the basic formulation from which the thermoset has been made.

Decomposition products arising at any stage (e.g. acetaldehyde from PETP) can be classified as volatile components, or residual reactants according to convenience; in the vector analysis used later, the distinction is not important.

Environment: vapour includes all non-odorous components which can diffuse into or through the plastic itself. The most important materials concerned are oxygen, water vapour and carbon dioxide; although in certain situations other materials may be significant (e.g. chlorine from sterilisation). Odorous components are those which are capable of changing the taste or smell properties of the food or plastic. Micro-organisms, macro-organisms and radiation are defined more fully below as vectors.

Since there are 11 component categories in all, each of which can interact with two other phases, there are 22 possible interactions. Some of these are of purely technological significance, or of no importance; nine may be relevant to health safety, and are listed in *Table 11.3* with their possible consequences.

Table 11.3. INTERACTIONS RELEVANT TO HEALTH

Number	*Component*	*From*	*To*
1	Non-volatile	Foodstuff	Plastics film
2	Volatile	Foodstuff	Environment
3	Polymer	Plastics film	Foodstuff
4	Volatile	Plastics film	Foodstuff
5	Additives	Plastics film	Foodstuff
6	Vapour	Environment	Foodstuff
7	Micro-organisms	Environment	Foodstuff
8	Macro-organisms	Environment	Foodstuff
9	Radiation	Environment	Foodstuff

One can study each interaction in turn, and sometimes for specific problems this is useful. For a general study, however, it is more convenient to classify interactions according to the major physical factor, or vector, concerned.

The vectors are:

(1) Macro-organisms.
(2) Micro-organisms.
(3) Gas and vapour.
(4) Radiation.
(5) Migration.

These relate to interactions as follows:

Interaction	*Vector*
1	5
2	3

3	5
4	5
5	5
6	3
7	2
8	1
9	4

11.7 MACRO-ORGANISMS

These are animals which are visible to the naked eye, or under a low power lens. Most of those with which we are concerned in the present context would be usually described as pests and vermin.

Plastics are not very attractive as food to animals, and any attack is likely to be in pursuit of the food itself, or emergence from the package.

Considering the spectrum of macro-organisms, from large animals which are usually accidental consumers, through small predatory mammals, to insects, plastics films as a whole have rather better resistance than would be expected, certainly better than cellulosic products.

The larger animals, which are capable of consuming packaged articles, will not be deterred by plastics films. Sheep and goats will consume plastics bags of food, and cows will sometimes munch empty fertiliser sacks. In some cases, the incentive may be hunger, but in others curiosity or mere chance. As already mentioned in section 2.1, such incidents might cause hazard due to asphyxia following obstruction of breathing passages, or obstruction of the alimentary tract. In fact, there do not seem to have been any cases worse than indigestion.

However, children and human babies are severely at risk when handling plastic bags large enough to cover their heads or breathing passages. As mentioned in section 11.2.2, it is essential to ensure that holes are perforated in any bag likely to come into the hands of young children or babies.

All these effects are mechanical and functional, not toxic or organic.

The smaller animals, e.g. rodents, are the classical pests and vermin whose activities seriously affect man's supply of food. It is well known that a determined rat can penetrate virtually any barrier; plastics films are no exception in not presenting a complete defence. However, plastics films do present a substantial defence—closer to that of metal and glass than paper. The problem is probably most severe with bulk storage, i.e. sacks or film used to cover food in bulk. By the same token, the opportunities offered to plastics are also the most interesting. Bulk-stored foods, e.g. rice, grain, root vegetables, are difficult to confine in rodent-free buildings, although this is indeed the ideal solution where possible. In the absence of defences, severe attack is very possible: it has been estimated that 25–30% of the world's supply of cereals is lost by pest infestation, the majority of this to rodents. In addition to the direct loss, pests and vermin leave behind a trail of their own waste matter and secondary infections which, in turn, spoil substantial quantities of food over and above those actually consumed.

For bulk storage, metal or glass containers are usually impracticable, and paper not very effective (including against weather). Plastics film (in sack or larger form) now offers the farmer and food distributor, for the first time, a realistic and good defence against rodents without having to protect the whole building. As already mentioned, no material offers total protection; but the barrier effects of plastics films are quite high. General guidelines are as follows:

Better barrier	*Worse barrier*
Thicker film	Thinner film
Smooth surface	Rough surface
Coherent film	Pinholes
Hard and abrasion-resistant	Soft or readily abraded
e.g.:	e.g.:
Polyamides	Polystyrene
Polycarbonates	LDPE
HDPE	Plasticised PVC
PP	PMMA
PETP	

For smaller, or retail packaging, insects present a major threat. All plastics constitute a good barrier to insects except those insects with mouthparts equipped for cutting and boring. *Table 11.4* shows that there are significant differences between different plastics films. However, in all instances the interposition of a plastics film between insect and food reduces the rate of penetration substantially.

An important factor in insect penetration is package geometry. For maximum penetration, an insect must press down with its legs in order to apply pressure with its mouthparts. Hence, attack is at a maximum where there are corners or crevices with two surfaces

Table 11.4. RESISTANCE TO INSECTS

Test material		*Test insect*			
Type	Thickness (mm)	Beetle*	Beetle†	Beetle‡	Grub§
Rigid PVC, white unoriented	0·06	++	++	++	++
Polyester	0·05	+	++	++	++
Rigid PVC, white, oriented	0·06	+	++	++	+
PP	0·03	+	++	++	+
PS	0·03	+	+	++	+
PAN	0·05	+	+	++	n.t.
Rigid PVC, uncoloured unoriented	0·06	+	+	++	+
Plasticised PVC	0·06	+	+	++	+
LDPE	0·05	–	–	+	+
Regenerated cellulose, uncoated¶	0·05	–	+	+	–
Regenerated cellulose, (PVDC coated)¶	0·05	–	–	+	+
Cellulose acetate¶	0·05	–	–	+	+
Regenerated cellulose, nitrocellulose coated¶	0·05	–	–	+	–

* Rhizopertha dominica
† Stegobium paniceum
‡ Tribolium castaneum
§ Plodia interpunctilla (larva)
¶ For comparison

++ Fully resistant, no penetration
+ Partly resistant; up to 50% penetration
– Not resistant; more than 50% penetration
n.t. Not tested

(Adapted from H.-U. Schmidt, *Süsswaren,* 1979, 5, 34–39.)

close together, especially at an angle optimum to the insect species (see *Figure 11.3*).

It should also be noted that some instars (a metamorphic form at some stage of life cycle) of some insects are extremely small, e.g. linear dimension 0·05 mm. Such instars can penetrate film packages through large pinholes, or imperfect closures. They can then grow in the food to a very much larger size before either boring their way out or dying in it. This stresses the point that, if insect infestation is found in packaged food, investigation may need to

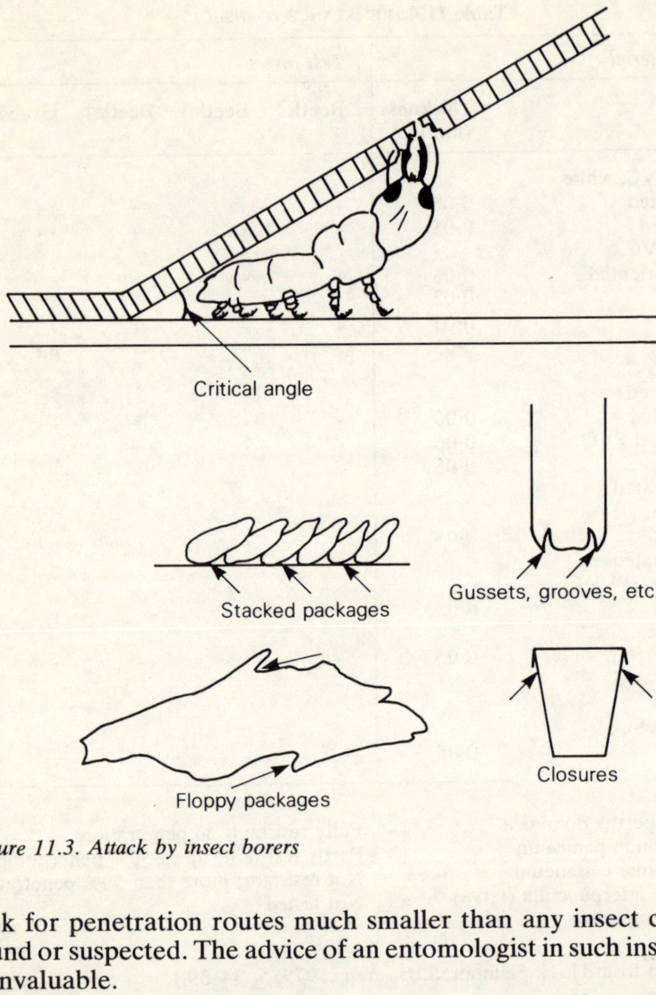

Figure 11.3. Attack by insect borers

look for penetration routes much smaller than any insect carcass found or suspected. The advice of an entomologist in such instances is invaluable.

Summing up, although plastics films can not offer complete defence, they are sufficiently good when used at the appropriate quality and in the correct manner to be regarded as very practical barriers to pests and vermin.

11.8 MICRO-ORGANISMS

In nature, most things destined to become foods have very good life-related defence mechanisms—skins, normal metabolism and waste disposal, antibodies, etc. As soon as the normal processes are interrupted, by death or harvesting, as described in section 11.4, not only do the defence mechanisms cease to function but decay is usually accelerated by modified (and often localised) metabolic conditions and specific decay mechanisms. Food is then prone to attack by micro-organisms. Of course, some purely chemical changes (notably oxidation or hydrolysis) also occur; but many of these, even, are mediated by or part of a reaction chain including micro-organisms.

Since food is usually wet and by definition nutritious, it is bound to be a satisfactory substrate for a wide range of micro-organisms. These include bacteria, fungi, slimes and moulds.

The two most dangerous bacterial types in this context are *Salmonella* and *Clostridia.* Both flourish over a fairly wide range of temperatures, and are a major cause of illness and death due to food poisoning. Mycotoxins—poisons produced by fungi, moulds or slimes—can also be very dangerous; they include aflatoxin, produced from *Aspergillus Flavus,* one of the most deadly poisons known. It can grow on barley, corn, rice, peanuts, soya, wheat and other starchy plants. In industrialised countries, however, fungi, although unpleasant, do not present a common toxic hazard. One reason is that, if stored for any length of time, food is *preserved* from micro-biological, principally bacterial, attack. There are two fundamentally different ways of doing this. The first—bacteriostatic—provides conditions in which bacteria, although present, cannot multiply. Freezing and drying are the prime examples of this. In the second—bacteriocidal—the vast majority of the bacteria are killed. This may be done by heating, or incorporation of chemicals, such as sugar, salt or preservatives.

Packaged food in the first group is *not* sterile: microbiological change can continue irrespective of packaging if the bacteriostatic conditions cease to exist. This is quite a common situation (e.g. holding of frozen foods in warm rooms) and it is not unknown for the consequences to be blamed on the plastic film. However, plastics films may sometimes promote conditions favourable to the growth of micro-organisms, usually by allowing increase of relative humidity, or providing anaerobic conditions suitable for anaerobic bacteria which would be killed by oxygen.

To summarise, attack by micro-organisms is one of the most important single factors in food storage. All plastics are very good barriers to micro-organisms, hence where micro-biological attack does occur in food packed in, or in contact with, a continuous layer of film, the outbreaks can usually be traced to bacteria or spores already present in the foodstuff, or to leakage through inadequate closures. Only very occasionally does penetration take place through pinholes or other imperfections (e.g. cracks) in the film. Thus, as a general rule use of plastics film vastly increases the safety in use of food.

11.9 GASES AND VAPOURS

Because all plastics films are, to a greater or lesser extent, permeable to gases and vapours, some transport from the environment to the food, or loss from the food to the film or environment, will occur whenever there is a pressure or concentration gradient across the film.

The two most important vapours in the first category are oxygen and water vapour. Most of the effects of oxygen are to accelerate or initiate decay processes, the effects of which are to promote staleness and deterioration in nutritive value. However undesirable, these effects are seldom toxic unless associated with biological change—which would only apply to non-sterile foods.

The ingress of water vapour can lead to quality loss, especially of dried foods. However, the main hazard arises from providing a medium in which micro-organisms can flourish—the critical level for many foods is about 13% water by weight. Again, this only applies to non-sterile foods.

Ingress of other gases may sometimes also be significant; for example, carbon dioxide, chlorine (from sterilisation) and odorous micro-contaminants. In the reverse direction (transport from food to plastic or environment) loss of water from moist foods may affect their quality, and hence indirectly have an effect on health; but the most important potential health hazard is the loss of volatile preservatives, when these are present. Ethylene oxide is not only volatile, but reasonably soluble in many plastics. Another example is sulphur dioxide, although this is now usually present in a relatively non-volatile form as sulphite or other compound.

Increasingly, food is being stored under controlled atmospheres, e.g. nitrogen, carbon dioxide, ethylene, oxygen/nitrogen mixtures or even carbon monoxide, etc. These improve keeping qualities, and may even afford preservation. The controlled atmosphere may be produced at the packing stage, when it is important to ensure minimum change, which predicates film of minimum permeability. Alternatively, however, film may be chosen (usually a laminate) which has a pattern of permeabilities to different gases leading directly to the gas composition required. Evidently, the second method is only applicable to oxygen/nitrogen/carbon dioxide mixtures.

See Appendix A for data on film permeability.

The process of vapour transmission is remarkably complex, comprising the following sequential transport or chemical stages for ingress.

Vapour diffuses through the film, the rate being controlled by pressure difference between the partial pressures of the vapour in the environment and in the package (head space). Vapour in the head space is then absorbed by the food, the rate depending on surface area and concentration in food and partial pressure. In the food, a concentration gradient is set up depending on both diffusion and chemical reactions. There are usually several of these, some being purely chemical, some micro-biological. They are always mediated by temperature, and sometimes by radiation (notably UV). They are often classified as fast and slow. Each predominates at a different time range after storage.

Apart from the vapour concentration in the environment, all other concentrations can change with time; hence there is a series of unsteady state transports or chemical reactions. Clearly the overall situation is very complex, and hardly susceptible to a general mathematical model.

Some semi-mathematical analyses have been produced, and some equipment is available, with mathematical models to be used in conjunction with them. These are useful for screening tests, preliminary studies, and perhaps quality control. For development, however, there is really no substitute for field trials under realistic conditions.

11.10 RADIATION

Radiation may be applied deliberately to food, film or a filled package, for sterilisation. Its use for this purpose is quite rare, and largely confined to storage for military use. (It is used more for packaging of pharmaceutical products—see section 11.14.) Care must be exercised on two counts. First, legislative constraints apply to the limits of radiation that may be used in connection with given foods. Second, intense radiation of this kind leads to degradation of many plastics, especially polyolefins (by chain scission, cross-linking, oxidation, etc.) and the products from this are very likely to give rise to odour.

Adventitious radiation is largely UV (and a certain amount of infra-red) from daylight or fluorescent lamps. The effects on food may be significant; for example, exposure of milk to North European sunlight for three hours will reduce the Vitamin C content by 50% and largely destroy the riboflavin content. These effects, and similar ones on other foods, all relate to nutrition as opposed to toxicity, and hence the effects on health are seldom serious and never acute; in fact, UV radiation has some beneficial sterilising effects on pathogens, if present, as has the high intensity radiation mentioned above.

However, a transparent film is often required for visibility of the food at point of sale. Although not directly proportional (transmission is different at different radiation frequencies) such film is also likely to have significant transmission for other radiation, notably UV. Hence a balance between conflicting requirements may need to be struck, unless special methods are adopted as described below.

Where the greatest barrier to radiation *is* required, this is best achieved not by selecting a particular plastic, but by pigmentation of any plastic. Over 90% of *all* radiation transmission can be eliminated in *any* plastic by the pigment levels used to achieve normal colouring, i.e. 0·1–0·2% pigment. Where radiation must be reduced to an even lower level, carbon black can be used up to about 2%, where radiation transmission is virtually zero for all plastics films even at low thicknessses (e.g. 50 μm).

Some reduction in UV transmission can also be achieved by incorporation of UV absorbers. There is a variety of these, e.g. benzophenones, only a limited number of which are approved for food contact. Some pigments have recently been developed, mostly based on small particle size hydrated iron oxide, which are transparent to visible light, but relatively opaque to UV. These may overcome the problem of reducing UV transmission while retaining transparency mentioned above.

Of course, radiation can also be reduced or eliminated by thick coatings, printing inks, or combination with opaque components in laminates, e.g. paper.

11.11 MIGRATION

11.11.1 DEFINITION

Migration is mass transfer (transport) between plastics and food. It can operate in two ways, from plastic to food, which is the normal meaning, or from food to plastic, which may be termed negative migration.

11.11.2 NEGATIVE MIGRATION

This is interaction 1 in Section 11.6. Its main direct result is psychological, but it can have effect on the nutritional quality of food if certain components of the food are lost to a significant extent. The main influence is loss of preservative, but some cases have occurred leading to nutritional or organoleptic changes, e.g. extraction of fat component of milk into polyolefins. If colourant is extracted from food, the effect on the food is usually not significant, but the consequential discoloration of the film (staining) is likely to be unattractive.

11.11.3 IMPORTANCE OF MIGRATION

There is no documented case of any proven health hazard arising from migration to food from plastics film (or indeed any plastic). Nevertheless, the vector is of great importance for the following reasons.

(1) All the other vectors can occur in the absence of packaging; only migration is caused by it.
(2) The migration, however undesirable, from some other packaging materials, especially metals, where it is usually due to chemical reaction, has been known for some time and become accepted within specified limits. (In fact, this is no longer fully valid—lead from can seals will soon be unacceptable; and attention is being paid to migration from ceramics, glass and cellulosics.)
(3) Plastics are the newest group of packaging materials, and hence regarded with some apprehension by consumers and the authorities. This is greatly accentuated by the discovery, early in the 1970s, of vinyl chloride monomer in significant quantities in food packaged in PVC (it was first noticed in alcoholic beverages packed in small bottles for airline use). This coincided with the revelation that the monomer is a carcinogen if inhaled.
(4) Plastics are certainly very complex, and contain numerous chemical species. Theoretically migration can occur of a large number of different components. This applies to other package materials, and, to a lesser extent, attention is being focused on these also.
(5) Most legislation or regulations cover migration and organolepsis, seldom the other vectors.

11.11.4 GENERAL MODEL FOR ONE-SIDED MIGRATION

The basic model shown in *Figure 11.1* can be quantitatively amplified to show concentration of plastic components as shown in *Figure 11.4*. The component can be any chemical species in the plastic; the following are the main ones of interest.

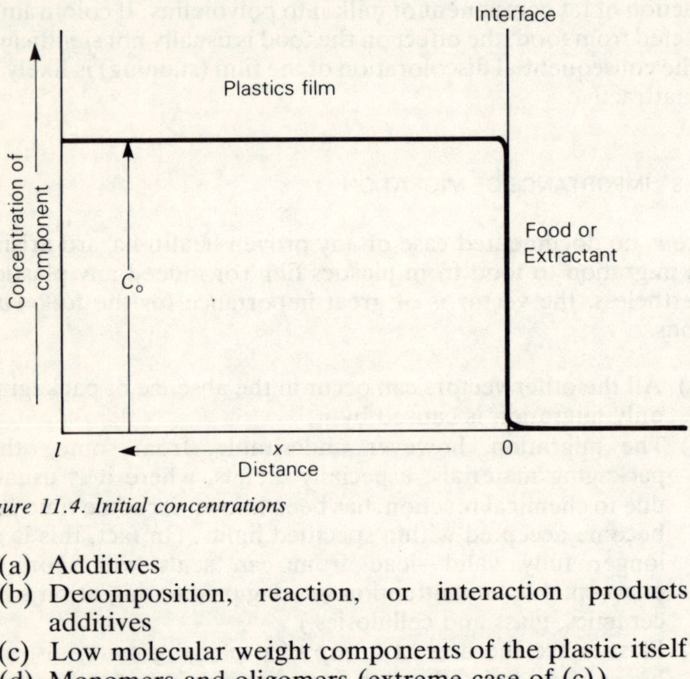

Figure 11.4. Initial concentrations

(a) Additives
(b) Decomposition, reaction, or interaction products of additives
(c) Low molecular weight components of the plastic itself
(d) Monomers and oligomers (extreme case of (c))
(e) Unreacted reagents (see Section 11.6)
(f) Catalysts and processing residues

If migration occurs, the concentration will fall in the plastic, the general situation being shown in *Figure 11.5*.

If

A = area of contact with food
x = distance from interface
l = thickness of film
c = local concentration of component
t = exposure time
M_m = mass of component remaining in film.

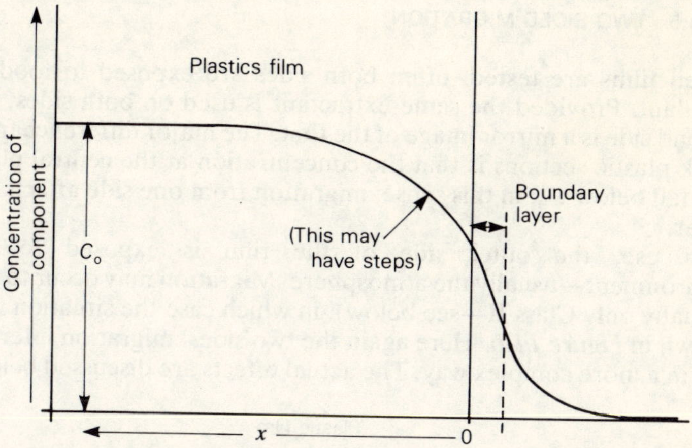

Figure 11.5. Concentrations: generalised

$$M_m = A\int_{x=0}^{x=l} C_{x,t}\,dx \qquad \text{[eqn 3]}$$

Hence, if M_t is the mass migrating, and M_0 is the total original mass of component in film, then

$$M_t \pm M_0 - M_m \qquad \text{[eqn 4]}$$

But, if c=original concentration of component,

$$M_0 = Acl$$

And if R_t is the proportion migrated, then, from eqns 3 and 4

$$R_t = M_t/M_0$$
$$= 1 - \frac{1}{Cl}\int_{x=0}^{x=l} c_{x,t}\,dx. \qquad \text{[eqn 5]}$$

Since c_x varies with time, the general integration of this equation is not straightforward. It is also complicated by the fact that there are several different mechanisms by which migration can occur. These are discussed below.

11.11.5 TWO-SIDED MIGRATION

When films are tested, often both sides are exposed to food or simulant. Provided the same extractant is used on both sides, the second side is a mirror image of the first. The major difference from thick plastic sections is that the concentration at the central plane can fall below C_0; in this sense, migration from one side affects the other.

In use, the outer side of the film is exposed to the environment—usually the atmosphere. Migration may occur to this (usually only Class II—see below), in which case the situation is as shown in *Figure 11.6.* Here again the two sides' migration interact, but in a more complex way. The actual effects are discussed below.

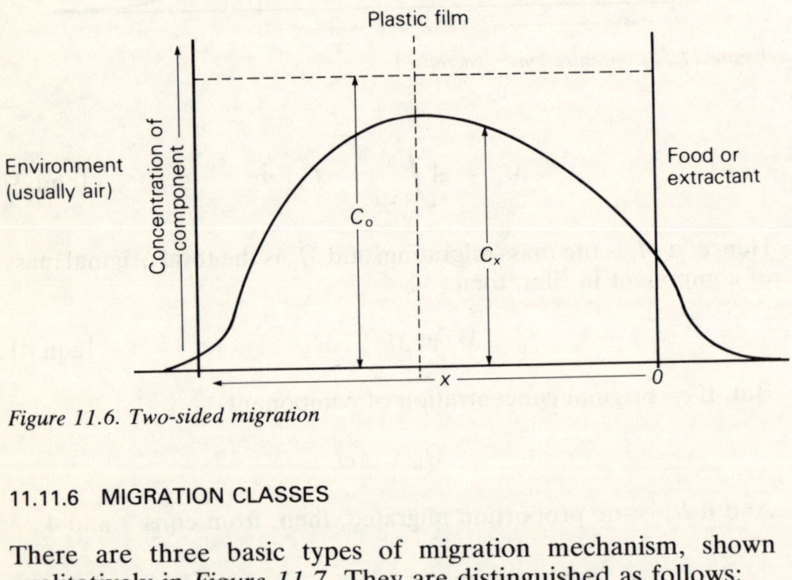

Figure 11.6. Two-sided migration

11.11.6 MIGRATION CLASSES

There are three basic types of migration mechanism, shown qualitatively in *Figure 11.7.* They are distinguished as follows:

Class I: non-migrating
Class II: spontaneously migrating
Class III: leaching.

Class I

In eqn 5, $M_t = 0$ and $R_t = 0$. The class includes high molecular weight polymer components contacting most foods, and some inorganic residues or pigments contacting neutral foods, and a few

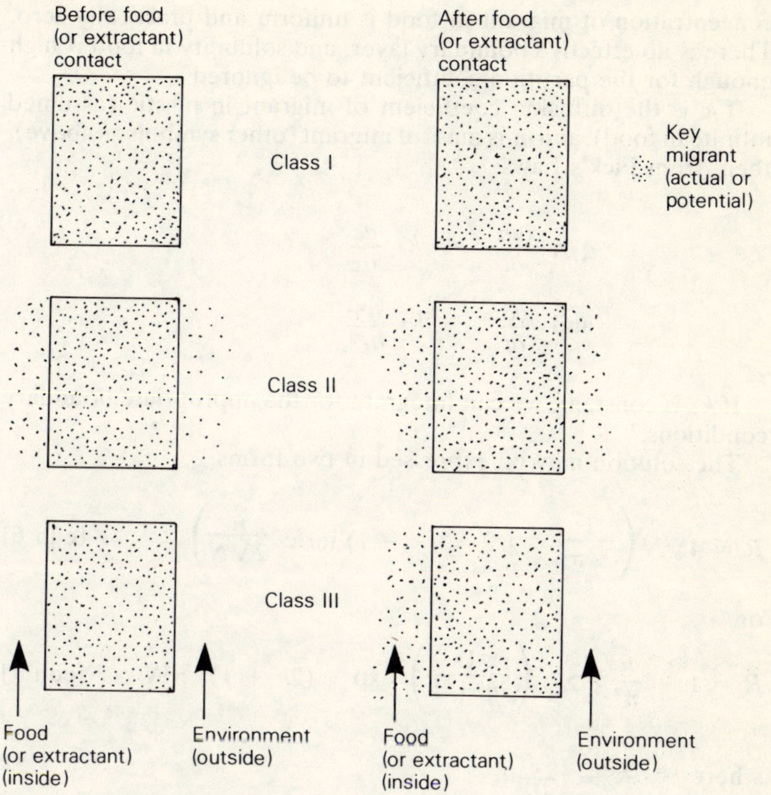

Figure 11.7. Migration classes

inert (relative to plastics) foods, e.g. dry sugar and salt—although abrasion may be a factor here.

The only problem is to demonstrate that a system falls into this class.

Class II

(a) Migration occurs in the absence of food contact (see *Figure 11.7*), i.e. the migrant diffuses out, into the environment and the food, as shown in *Figure 11.6*. Equation 5 can be integrated using the appropriate diffusion regime.

The simplest case assumes that the mass ratio of food to plastics is sufficiently large, and mobility or stirring adequate, to ensure that concentration of migrant in food is uniform and practically zero. There is no effective boundary layer, and solubility in food is high enough for the partition coefficient to be ignored.

If k is the diffusion coefficient of migrant in plastic (assumed infinite in food), and m is mass of migrant (other symbols as above), then, from Fick's Laws:

$$1/A \frac{\partial m}{\partial t} = -k_1 \frac{\partial c}{\partial x}$$

$$\text{and} \quad \frac{\partial c}{\partial t} = k_1 \frac{\partial^2 c}{\partial x^2}.$$

If k_1 is constant, we can integrate for the appropriate boundary conditions.

The solution may be expressed in two forms:

$$R = 4\sqrt{N}\left(\frac{1}{\sqrt{\pi}} + 2 \sum_{n=0}^{n=\infty} (-1) \operatorname{ierfc} \frac{n}{2\sqrt{N}}\right) \quad \text{[eqn 6]}$$

or

$$R = 1 - \frac{8}{\pi^2} \sum_{n=0}^{n=\infty} \left(\frac{1}{2n+1}\right)^2 \exp - (2n+1)^2 \pi^2 N \quad \text{[eqn 7]}$$

where $N = \dfrac{k_1 t}{l^2}$

N is a dimensionless number, analogous to thermal diffusivity. It is useful in arriving at generalised correlations, reducing time and thickness to a standardized form.

The solution of the above equations shows the behaviour of R, illustrated in *Figure 11.8.* It will be noted that in the early stages

$$R \propto \sqrt{N}$$

$$\propto \frac{1}{l}\sqrt{k_1 t}.$$

Or, for constant l (i.e. a specific film or container)

$$R \propto \sqrt{t}.$$

Hence migration is relatively rapid up to $N = 0{\cdot}1$, where $R > 50\%$, and proportional to $\sqrt{t}$.

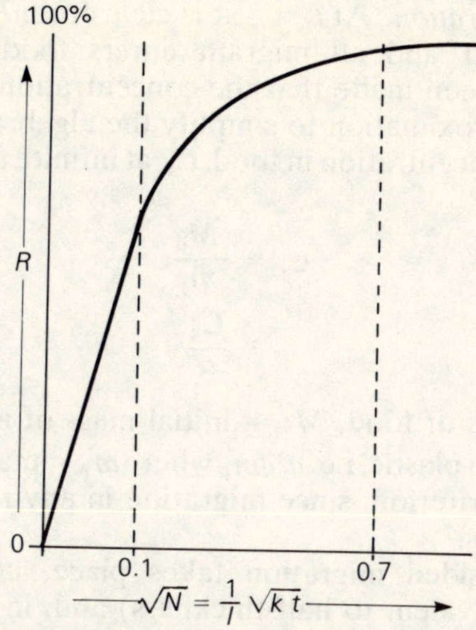

Figure 11.8. Class II parameters

(b) *Two-sided migration*. If we now introduce migration to the environment, the situation is similar to that of two-sided extraction (Section 11.11.5). Provided the same conditions apply, i.e. the surface concentration is zero, then the situation is as shown in *Figure 11.9*. Integration of eqns 6 and 7 is rather more complex, but the general behaviour of *R* is similar.

(c) *Limiting cases*. Two limiting cases are very important in practice: indeed, these probably cover the majority of real-life situations.

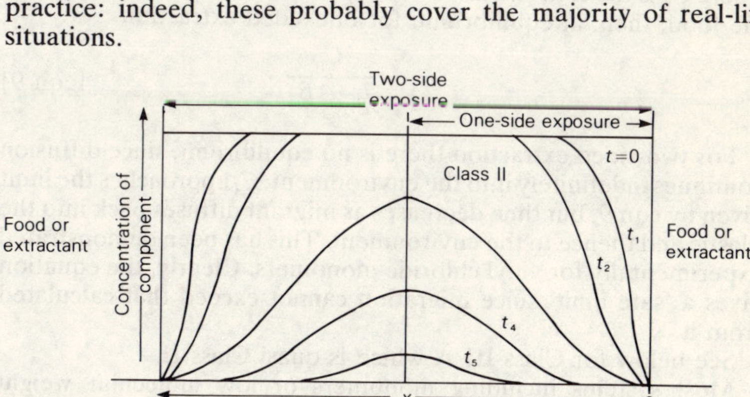

Figure 11.9

(i) *Complete migration.* At $t = \infty$ it is clear from *Figures 11.7* and *11.8* that $R = 1$ and all migrant enters food. Although the assumption has been made that the concentration in food is zero, this is a first approximation to simplify the algebra. For one-sided migration, the concentration in food, C_f, at infinite time is given by

$$C_f = \frac{M_0}{m_f}$$

$$= \frac{C_p}{\alpha} \qquad \text{[eqn 8]}$$

where m_f = mass of food, M_0 = initial mass of migrant and α = mass ratio food to plastic, i.e. m_f/m_p where m_p = mass of plastic. This is a 'safe limit' criterion, since migration in any real situation can never exceed this.

Usually, two-sided migration takes place, however, roughly halving M_0 (equivalent to half-thickness) and, in this case

$$C_f = \frac{C_p}{2\alpha}$$

(ii) *Migration limited by partition coefficient.* If solubility in food is limited, or relatively high in the plastic, the partition coefficient may become limiting. If the partition coefficient, β, is defined as

$$\beta = \frac{C_p}{C_f}$$

where C_p is the concentration in the plastic, and C_f concentration in the food, then, at equilibrium, for one-sided extraction

$$C_f = \frac{C_p}{\alpha + \beta} \qquad \text{[eqn 9]}$$

For two-sided extraction there is no equilibrium, since diffusion continues indefinitely into the environment. C_f approaches the limit given by eqn 9, but then decreases as migrant diffuses back into the plastic and thence to the environment. This has been demonstrated experimentally for vinyl chloride monomers. Clearly, the equation gives a 'safe limit' since migration cannot exceed that calculated from it.

See below for Class III A which is quasi Class II.

Most systems including monomers or low molecular weight oligomers are in Class II, as well as some relatively volatile additives. The criterion for inclusion is a self diffusion coefficient in polymer (in the absence of extractant) greater than about 10^{-12} sec. cm^{-2}.

Class III

(a) Migration only occurs if the plastic is in contact with food or other extractant (see *Figure 11.7*). It is obvious that there must be some physical or chemical action which changes the transport mechanism of the migrant: this can be in two ways.

(b) *Class III A.* If the migrant has a relatively high diffusion coefficient in the plastic, but is not volatile, it falls into Class III. However, as soon as contact is established, the surface layer of migrant is dissolved, and the concentration falls to a low level. Diffusion can then continue exactly as for Class II, and the above analysis for Class II applies.

The most important systems in Class III A are those including antistatics or slip additives contacting food in which the additives are soluble.

(c) *Class III B.* This class is the most difficult in terms of mathematical and scientific analysis, and has only recently become understood. It is also one of the most important, including as it does most additives in most plastics contacting most foods.

There are two stages, as shown in *Figure 11.7*. First, the food, or one or more components of it, penetrates the plastic to a certain depth—say j. In this penetrated layer the plastic matrix is substantially changed to the point where mobility of the component within it is increased greatly. The component then diffuses out through this layer into the food.

Diffusion in polymers is seldom simple, and the general mathematics of Class III B have not been, and perhaps never will be, fully established. The simplest case, however, where Fick's Laws apply, is susceptible to analysis as follows:

For the first, penetration, stage, it can be shown that penetration follows the pattern shown in *Figure 11.10*. The depth of the j-layer is proportional to $\sqrt{t}$ (note similarity to Class II in reverse direction). In the second stage, the pattern depends on diffusion coefficient, which varies with distance depending on the first stage. Explicit mathematical calculation of this is very difficult, but the general picture is as shown in *Figure 11.11*.

Following the Principle of Safe Limits, 100% migration from the j-layer would be assumed. Migration must be equal to, or less than, this.

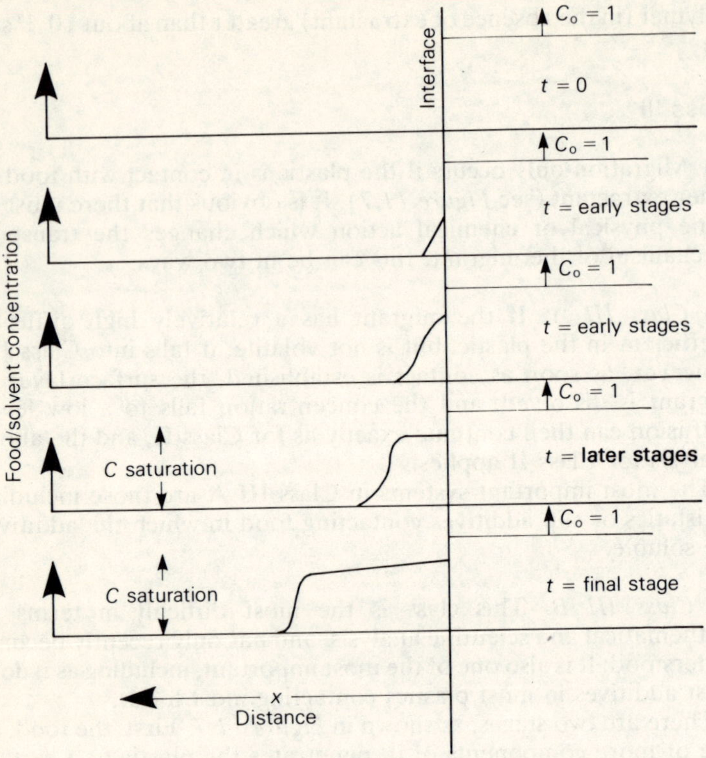

Figure 11.10. Class IIIB: solvent/food concentrations

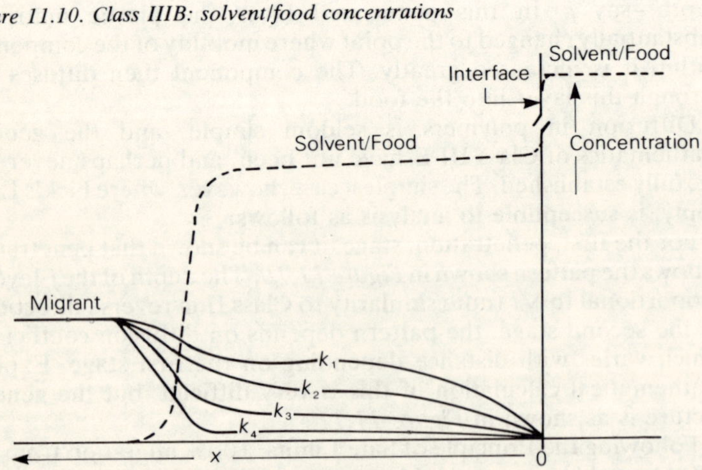

Figure 11.11. Class IIIB: overall concentrations

In experimental work, it has been found that a Fickian diffusion regime does not operate fully, especially over long time periods. In many cases, penetration does take place as described above, roughly proportional to $\sqrt{t}$ for a time (order of weeks at ambient temperature in many systems), but then stops. It is believed that this is due to polymer inhomogeneity; penetration takes place only through domains which are compatible with a food component. These are the amorphous, or low molecular weight areas. Eventually the solvent (food component) is faced by a 'wall' of high molecular weight, crystalline, impenetrable, domains. There are sometimes other binding mechanisms, however, including chemical reaction between additives and polymer.

For a bulk plastic (semi-infinite) migration is limited either by time (if in the time-dependent regime) or by limiting penetration, but for films further limitations exist since the *j*-layer may exceed the thickness of the film. These are illustrated in *Figure 11.12*. It will be seen that in the first case (no *j*-layer, or relatively thick film), limiting migration is proportional to the *j*-layer, but in the second case, migration may be total—which is the same as Class II.

Which case exists is easily determined by two-sided extraction experiments since migration reaches a limit when the *j*-layers for both sides overlap—see *Figure 11.12*. Again, this is similar to Class II.

(d) *Practical implications*. It will be seen from the above that a first, fail-safe, assumption for the film is 100% migration. If this is acceptable on toxicity grounds (see below) no more need be done, irrespective of whether the system is in Class II or III.

If not acceptable, it is necessary to determine the depth of the

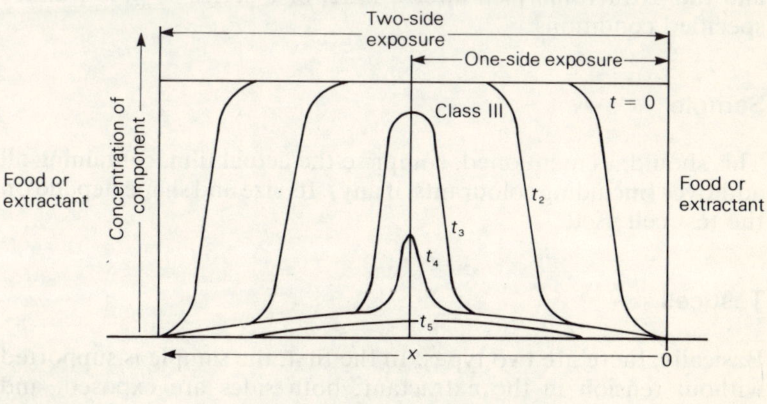

j-layer for the particular system, and contact time (shelf life from initial packaging). If assumption of 100% migration from the *j*-layer is acceptable on toxicity grounds, no more need be done.

Again, if this is not acceptable, then the actual migration has to be determined. This can be done either using the food itself, or a food simulant (see section 11.11.7) under the actual conditions of use or under experimental conditions that can be related to actual conditions on theoretical grounds.

11.11.7 TESTING

From the previous section it will be seen that fail-safe estimates can be made of migration without actual testing. However, if these are not acceptable from the toxicity point of view, migration testing has to be carried out.

It has been taken for granted in the past that tests on bulk (i.e. moulded) samples could be applied to film, using appropriate parameters, e.g. migration per unit area. In fact such tests are prone to two types of errors. The first comprises effects of scale and geometry (see section 11.11.8 below). The second arises from the major differences in morphology, surface and visco-elastic strain between samples of identical raw material processed in different ways. Such differences are not trivial: if migration of a particular additive per unit area to a standard solvent is measured on film and also on cast or pressed samples, factors of five between the results are not unusual. Hence it is best to carry out tests on the film itself.

The ideal procedure is to use actual foods for test, but for the reasons given it is usually difficult to do this and extractants are used. The sample is exposed to the extractant (one or both sides) and the extract analysed after a time, or a series of times, under specified conditions.

Sample

This should, as mentioned, comprise the actual film, containing all additives (including colourants, if any). Its size and shape depend on the test cell used.

Test cell

Basically, there are two types. In the first, the sample is supported without tension in the extractant; both sides are exposed, and corrections made (simplest to halve, but this is not always valid as described on p. 142—Two-sided migration, and shown in *Figure 11.9*).

In the second type, the film is gripped between cylinders of stainless steel or glass. Extractant is placed above or below the film, which is under tension. There are some standard cells, e.g. ASTM. *Figure 11.13* shows the Eurocell, developed at the E.E.C. laboratory, Petten.

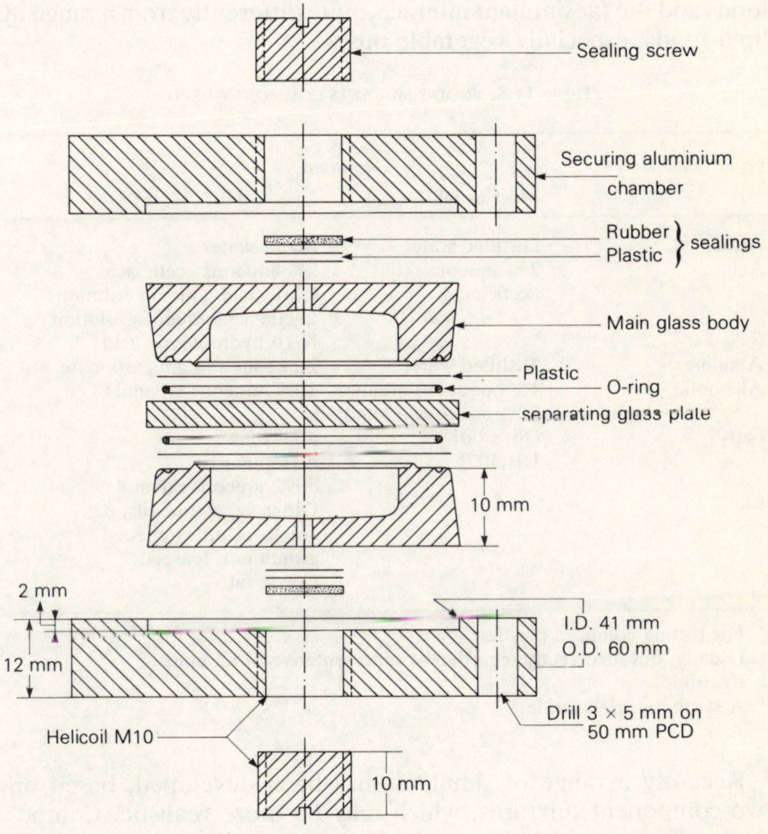

Figure 11.13. Eurocell

Extractant

As mentioned, this should ideally be the food to be packaged, and sometimes this can be used. However, severe problems usually arise, namely decomposition of the food, difficulty in the analysis, non-homogeneous distribution of migrant and the usual need to ensure that the film is suitable for a wide range of foods. Therefore, so-called food simulants are used instead. These are liquids which are convenient for analysis and mimic the action of food. Those commonly used are shown in *Table 11.5.* The extent to which these liquids do simulate real foods is dubious. For Class II, solubility of migrant is a major factor, whereas for Class III, absorption into the plastic is controlling. Acetic acid behaves quite differently from acid foods and the fat simulant interacts quite differently from a range of lipid foods, especially vegetable oils.

Table 11.5. FOOD SIMULANTS COMMONLY USED

Food type	*Simulant*	
	Most usual	Less usual
Aqueous	Distilled water	Mains water
Acidic	3% aqueous acetic acid	2% aqueous acetic acid Citric acid aqueous solution Lactic acid aqueous solution N/10 hydrochloric acid*
Alkaline	Distilled water	Aqueous sodium carbonate
Alcoholic – low	15% aqueous ethanol	10% aqueous ethanol
– high	50% aqueous ethanol	
Fatty†	Olive oil‡ HB 307§	n-Hexane n-Heptane 50% aqueous ethanol Other vegetable oils, e.g. arachis, sunflower seed, groundnut, teaseed. Cocoa fat

* For testing coloured plastics.
† Usually, but incorrectly regarded as representative of all lipids.
‡ Rectified.
§ A synthetic triglyceride.

Recently a range of simulants has been developed, based on two-component mixtures, which may be more realistic. Components of these include tetrahydrofuran, methanol, water and chloroform.

Temperature. Tests should be carried out at normal use temperature, the following being typical:

For sterilisation: 115°C

For boil-in-the-bag: 100°C
For tropical storage: 38°C
For temperate storage: 23°C
For normal refrigeration: 4 or 5°C.

Frequently, 40°C is used (or even 60°C) in what are assumed to be accelerated tests equivalent to migration at 23°C for a longer period. At best this is a poor approximation, at worst very misleading results are obtained because of changes in polymer or extractant morphology.

Time. Ideally the test should be carried out for the maximum storage time (shelf life) envisaged. This may not be practicable, and extrapolation from shorter times is necessary. This is totally valid if migration approaches equilibrium, and usually can give adequate order-of-magnitude results if testing time is not too short. Ten days is a commonly used test period.

Although not universally valid (especially where migration reaches a limit) a good approximation can often be made by assuming (see pp. 143–147—Class II and Class III) migration to be proportional to $\sqrt{t}$. Ten-day tests can then be extrapolated as follows:

Storage period	*Migration*
10 days	X
90 days	3X
1000 days	10X

Since the deviations from a $\sqrt{t}$ regime are always negative with increasing time, assumption of migration at 1000 days (*c*. 3 years) to be 10 times migration at 10 days is safe.

Correlations of this kind are often built into regulatory systems, FDA being a recent convert.

11.11.8 DIMENSIONS AND UNITS

To define fully a migration test, it is necessary (for a given plastic and food) to quote ten variables as follows:

Intensive: temperature, time, density of plastic, density of food.

Extensive: thickness of film, thickness of food, contact area, mass of film, mass of food, and geometry.

Various combinations of these are often used, e.g. initial concentration in film, concentration in food, etc. Information (on scale) is lost, and extra variables must be quoted to restore the total. Some limiting cases exist, e.g. when partition coefficient is controlling (see p. 146—Class II); or where migration is so low (in relation to volume of food) that depth of food has negligible effect.

Some other combined variables often used, e.g.
α as defined on p. 146—Class II
γ = mass of extractant/area
Note that some of these have dimensions, for example, γ can be expressed as g/cm^{-2}. If, using densities, masses are converted to volume γ still has dimensions, e.g. cm or cm^3/dm^2 (= 0·1 mm).

11.11.9 SAFETY EVALUATION

From the above, an estimate can be made of migration from the sample of film under appropriate conditions. It is necessary to derive from this an estimate of human daily intake, and for this purpose, assumptions must be made of the consumption of food packed in the film as well as the ratio of food to film.

There are various ways of doing this, of which the simplest is as follows:

If:

> D is migration per unit area, mg/dm^2
> A is typical ratio film area:food mass, dm$_2$/kg.
> and B is typical food consumption, kg/person/day
> then consumption of migrant is ABD mg/day.

If F mg/day is the consumption of migrant from other sources in diet (if any), then the total daily consumption, $ABD + F$ must not exceed the acceptable daily intake *(ADI)* of the migrant, i.e.

$$ABD + F \leqslant ADI$$

$$\text{or } ABD \leqslant ADI - F.$$

For general plastics packaging, A is usually taken to be in the order of 6 dm^2/kg, but for flat packages, e.g. sachets, a more appropriate figure is around 20 to 30 dm^2/kg.

B obviously depends on the food. For minor items, e.g. coffee, spices, etc., it is perhaps a few grammes, but for some liquid foods, e.g. salad oil, milk, wine, etc., it may reach 1 or 2 kg/day. A typical figure, frequently assumed, is 1 kg/day.

Note that D is not a general constant for the plastic with respect to A; but should be the actual figure for the film concerned.

ADI is derived from toxicity studies. These are based on animal experiments extrapolated to humans using various assumptions and a safety factor. It usually errs on the safe side by a very substantial margin, and certainly cannot be regarded as more accurate than within an order (or even 2 or 3 orders) of magnitude. This being so, there is no justification to seek greater accuracy than this in estimated migration.

11.11.10 EXPRESSION OF SAFE LIMIT

The above discussion represents a rather idealised situation, in which the *ADI* for a migrant is established, and the migration known for actual film sample in contact with actual food concerned under realistic conditions. Often this situation is not at all realised. Either in codes of practice, legislation or suppliers' evaluations, approximations or estimates have to be made for one or more of the conditions.

In any event, it is not possible to control manufacture on the above basis, but the results must be converted into a more practical form. There are two ways in which this is usually done in regulations—

(a) limiting concentration in food, i.e. *ADI* divided by *B* (mg/kg or ppm).

(b) limiting concentration in film, i.e. C_0.

The former is more directly related to health safety, but requires more investigation by the manufacturer; the second calls for little or no assessment by him, but more work by the authorities.

11.11.11 ADULTERATION

Definition: Adulteration is the presence in food of significant quantities of harmless non-food contaminants, not originally present in the food. (There are other definitions in legislation, many of which imply deliberate addition.)

Presence of such contamination does not (by definition) constitute a health hazard, but is to the detriment of the consumer since he is purchasing useless material at food cost. Essentially this is fraud, and has been regulated by legislation since antiquity. Common examples include water added to milk, vegetables or butter, and chalk or other fillers added to bread. Small amounts may be unavoidable, but larger amounts are restrained by legislation either specific or under general food laws.

Although the health safety evaluation, and migration methodology given above are generally applicable, in practice the migration concerned is at a very low level. Usually C_p is below 1% and C_f below 1ppm (these figures are not correlated; they merely indicate order of magnitude).

A few plastic components are present in some instances at higher levels, notably fillers, plasticisers and, of course, low molecular weight components of the polymer itself.

Most inorganic fillers are rather inert and migration very small (though not necessarily negligible). A few, e.g. chalk and starch, have potential for substantial migration, and leach into aqueous extractants, and these, together with the others mentioned, can give rise to C_f levels well above 1 ppm—more often under test conditions than in real life. Also, some additives, especially (but not only) plasticisers, flow promoters and internal lubricants, can so change the polymer matrix that diffusion is enhanced of other components, also leading to the same results (i.e. transfer from Class III to Class II). Also, of course, these additives are prone to substantial migration themselves.

Major examples of the above (i.e. where C_f can be well above 1 ppm) are:

LDPE of high melt flow index or low density or both, contacting lipid extractants.
PS (and especially high impact PS) contacting lipid and alcoholic extractants.
Plasticised PVC contacting lipid extractants.
Toughened PP (copolymers) contacting lipid extractants.
Relatively highly polar polyamides (e.g. nylon-6) contacting aqueous and especially acid extractants.
More paraffinic polyamides (e.g. nylon-11) contacting lipid or alcoholic extractants.
Alkaline or soluble fillers contacting aqueous or acid extractants.

For these, C_f can reach levels in the range 10–100 ppm or even more (depending, of course, on many other factors, especially α and β). The question then arises as to whether, even if there is no toxic hazard, these levels constitute adulteration.

The criteria for acceptable levels of true adulteration are social and political, finding expression in legal form. They vary from country to country, and for different foods, but are normally in the percent order of magnitude.

Unfortunately, there is some confusion, both in legislation and social understanding, between adulteration and health hazard. Since toxicity is always dose related, there are a few instances where the health hazard level is in the adulteration region, but this is rare; usually health hazard limits lie in the ppm region, and adulteration in the percent—a factor of thousand to a million.

This confusion has led to the concept of limiting overall or global migration.

11.11.12 OVERALL OR GLOBAL MIGRATION

Much legislation, including that proposed for E.E.C. Directives, incorporates an overall or global limit on migration. This may be defined either in terms of concentration in food, where 50 to 60 ppm is typical, or migration per unit area, where 10 mg/dm^2 is typical. (These limits are often regarded as equivalent, but in fact are only so where $\alpha = 5$ or 6 dm^2/kg which applies to a 1-litre cubical package; they are substantially different for film packaging, where α is usually over 20 dm^2/kg.)

The arguments for the justification of these limits are:

(1) Protection from toxic hazard. The totality of toxic migrants cannot exceed the limit. This argument has little validity, since toxic effects can be significant at levels very much below those set for global migration.

(2) Protection from adulteration. This again is hardly valid, since the limits are well below those normally accepted for adulteration (see section 11.11.11).

(3) Reduction in analytical testing. It is only necessary to test an extractant for migrants presenting a health hazard at levels below the global migration limit. This is also not valid, since, if migration testing for health safety is to be meaningful, it is necessary to test for all possible migrants (whether believed to be present or not), and the elimination of testing for a few relatively innocuous components is a negligible saving (in fact, much exceeded by the global migration test itself).

Apart from basic objections to the global migration test, it has been found in practice to present surprisingly high analytical difficulties, especially for lipids, where major corrections have to be made for extractant penetrating and absorbed by the plastic.

Nevertheless, the test is regarded—perhaps for emotional reasons—as a contribution to consumer protection or an impression of this. As we go to press, it remains a major item in the legislation of many states and is a cornerstone of proposed E.E.C. legislation.

It should also be noted that specification of global migration is usually attached to the bulk plastic, which may well give higher values than the film itself, as explained in sections 11.11.5 and 11.11.6. There may be provision in regulations for measurements on the film to be used, instead of the bulk plastic, and where this is so advantage should be taken of this.

11.12 SPECIAL SITUATIONS

11.12.1 THIN FILMS

It was already made clear on p. 146—Class II) that, for thin films, a major proportion of a plastic component may migrate. This should be borne in mind, especially for films less than 50 μm thick. The consequences for health hazard evaluation have been dealt with above, but the effects may be significant in technological terms. (Most of these cases fall into Class III.)

If processing aids, residual reactants, etc., are lost, the indirect effects are usually of no importance; but if end-use additives are lost the effects may be significant. This applies particularly to antistatics (loss of which may encourage dust deposition), and UV absorbers, the loss of which may affect food quality.

11.12.2 COMPOSITES, LAMINATES

Composite films comprise two or more plies, laminated together. The plies may be different plastics, variants on the same plastic, metals (usually aluminium, but others are now being studied) and cellulosics (regenerated cellulose or paper). With increasing emphasis on economy, because of the higher cost of plastics, such composites are increasingly popular. Usually each ply is chosen for a specific purpose, such as vapour barrier (e.g. Saran), melt adhesion (e.g. LDPE), colour, printability (e.g. paper) and rigidity. An important example is the retortable pouch, in which an aluminium ply providing sufficient rigidity and complete vapour barrier is laminated to a PP ply, giving the whole integrity at sterilising temperature of 115°C.

With reference to health safety, it is usually adequate to consider the interaction between the food or other product packaged, and the inner ply (that contacting the product). There are the following exceptions.

(a) If a non-contacting ply contains a volatile component, this can diffuse through the other ply or plies into the packaged product, and must be taken into account.

(b) If the laminate is folded, e.g. at edges, or cut, or in any other way contacts the packaged product end on, then there is direct contact—admittedly over very small area—with the outer plies.

(c) If adhesives are used to bind the laminate, these are very liable to contain residual solvents or reactants, which can behave as under (a) above. Since some of the possible components of adhesives are rather toxic, this is important, and has caused substantial delay in approval being forthcoming for retortable pouches in some countries.

(d) Fibrous components (see section 11.12.5 below).

11.12.3 SEALANTS

Sealants between plies can be considered as adhesives, covered in section 11.12.2 above. Sometimes they are used at edges or joins, and may contact the packaged product directly. The health hazard arising from this (if any) has to be considered on an ad hoc basis.

11.12.4 FOAMED OR EXPANDED FILM

The use of foamed or expanded thin film is increasing rather slowly, and the main uses lie in areas which do not involve health safety, such as wallpaper. However, their use for food wrapping is likely to increase, both for technological and economic reasons. Thicker films, usually called sheet, are vacuum-formed for packaging of meat, eggs, etc. The general methodology given above is equally relevant to all these with the addition that possible diffusion of the blowing agent and its decomposition products into the food must be taken into account. Where the gaseous phase is carbon dioxide, air or nitrogen, no problem arises. However, some hydrocarbons (which may be used in polystyrene) or some labile nitrogen compound, e.g. azodicarbonamide, cannot be assumed to be harmless.

11.12.5 FIBRES

Polypropylene stretched film, usually in the form of tape, is used for the manufacture of sacks; and these are used for transport of bulk foodstuffs, e.g. grain, root vegetables, salt and cocoa beans. The above methodology can be applied to these but the possibility of substantially increased surface area contact due to fibrillation must be taken into account. Also, although not a direct consequence of the plastic nature of the materials, the irregular, large and rough surfaces of the fibres present an ideal substrate for the accumulation of other materials which function as a breeding ground for micro-organisms, particularly fungi, with consequent health hazard.

Composite film may now include a fibrous ply, usually paper. The fibres in this may penetrate the other plies—especially that contacting the packaged product—and have two effects. First, the fibrous component may contribute to migration in its own right. Second, it may absorb the product by a wicking action, not only conducting food into non-contact plies or layers, but also distorting the morphology of the contact ply, thus changing its characteristics with regard to migration.

General methodology of such systems is still under study: the only advice that can be given at the moment is to carry out tests as close to realistic conditions as possible.

As already mentioned, in the majority of examples the relevant fibres are cellulosic from paper, but there is a small amount of polypropylene and other fibres now being used in teabags.

11.12.6 BOIL-IN-THE-BAG

Boil-in-the-bag pouches are susceptible to the methodology given above in section 11.7, carrying out the tests at 100°C for the limited time (maximum 30 min) likely to be used.

11.12.7 OVENABLE FILM

Some films, notably PETP, are used to cover ready meals (usually deep frozen) which are subsequently oven heated up to 200–250°C. There is only limited contact of film with food; nevertheless, some testing is required to ensure that migration either of the original plastic components, or decomposition or degradation products, present no health hazard.

With increased usage of micro-wave ovens, similar consideration has to be given to films used in these. In general the duty is less severe, since the film itself is not greatly heated by the radiation, its temperature rising largely by contact with the hot food.

At higher temperatures, quite significant decomposition may

occur, with evolution of carbon monoxide, formaldehyde and acrolein from all plastics, hydrogen chloride from PVC, styrene from polystyrene; and hydrogen cyanide, ammonia or nitrous fumes from polyamides or acrylonitrile-containing polymers; acetic acid and vinyl acetate from PVA (most of these are unlikely to be used). Even if no health hazard exists, some tainting is possible (i.e. organoleptic change—see Chapter 12).

11.13 COSMETICS AND TOILETRIES PACKAGING

The methodology, procedures and discussion in sections 11.11 and 11.12, which are oriented to food packaging, apply with some changes, mostly of emphasis, to packaging of other TSEU. For cosmetics and toiletries packaging, the changes are as follows, relating to the five vectors described in section 11.6.

(1) Macro-organisms. Although there are a few instances where animals (or even humans!) will eat some cosmetics and toiletries, by and large there is no attraction and hence this vector is of little or no importance.
(2) Micro-organisms. This is also not very important. Cosmetics and toiletries are usually intrinsically antiseptic, or contain antiseptic additives, and hence are not susceptible to infection, irrespective of packaging. There are a few exceptions, e.g. soap in humid or tropical conditions, but even here the deterioration, although serious from a marketing point of view, seldom presents a health hazard.
(3) Vapour transmission. This is of great importance from a quality point of view, e.g. loss of perfume, change of moisture content, etc., but again there is seldom if ever a health hazard.
(4) Radiation. A few effects, e.g. change of colour, can be significant in terms of quality and consumer appeal, and activity of ingredients may be modified. Again it seldom, if ever, constitutes a health hazard.
(5) Migration. This can be important in terms of quality due to negative migration (loss of perfume, colour, active ingredient) but this seldom leads to health hazards. Positive migration can, however, be significant for health safety.

Procedures are the same as for food (see section 11.11) with the major difference that toxicity testing is not usually by ingestion, but on contact with skin or other body surface, or inhalation for aerosols.

11.14 MEDICINALS AND DRUGS PACKAGING

By definition, this group of materials is physiologically active, and any change may have a health hazard, if only by reduction in activity. Also, because these materials are prescribed for patients whose condition is already weakened, even greater attention to health hazard may be justified.

Differences from food, in terms of vectors, are as follows:

(1) Macro-organisms. Same as for cosmetics and toiletries (section 11.13).

(2) Micro-organisms. This can be very important. Many medicaments are carefully sterilised, and maintenance of sterile conditions can be vital. Although the general principles in section 11.8 apply, even greater care must be taken, and stringent quality control systems applied.

(3) Vapour transmission. This may be of importance in leading to other changes, e.g. providing conditions for micro-organism growth. Comments are similar to those in section 11.9, again with, if anything, more care and stringent quality control.

(4) Radiation. This may affect activity of ingredients sensitive to it. Similar comments to those in section 11.10 apply, but again more care has to be taken, since the effects may be of greater importance to health rather than to nutrition or quality. Films can be coloured, loaded with UV absorbers or covered with opaque coatings or UV absorbers, lacquers, etc., to reduce radiation transmission to acceptable levels.

(5) Migration. Methodology is the same as in section 11.11, with the following qualifications or changes of emphasis.

(a) Negative migration can lead not merely to quality change, but loss or reduction of active ingredients (possibly selectively). This can (occasionally) be highly significant for health hazard.

(b) For medicines taken by the mouth, similar procedures apply as in section 11.11, except that (i) the quantity ingested can be estimated more accurately and is usually much less than for food, (ii) the consumer is not an average healthy person.

(c) For medicaments applied other than by the mouth, e.g. skin or by subcutaneous injections, toxicity testing must be by the appropriate route. Effects may be quite different, and in some instances (notably injection) much more severe.

In addition, there may be specific regulations applicable (see section 11.17).

11.15 LAW AND REGULATIONS: FOOD

It will be clear from the above that devising a system for regulating plastics films in contact with food, designed to safeguard public health, is a complex scientific problem. Also, since food is such a basic need, legislation on it goes back to antiquity, and is complicated by social needs and norms. The complicated weft of scientific analysis is woven into the warps of existing legislative systems, local customs and social habits. Hence it is not really surprising that the resultant world pattern is somewhat confused and contradictory, although there is an increasing tendency for legislation in different countries to converge.

Both packaging and plastics films have had an explosive growth in the last few decades, and hence relevant regulatory systems have found it difficult to keep up with progress. They are therefore continually under review and subject to change. Indeed, any detailed statement of the world situation made at any time is bound to be out of date very soon. Consequently, in the following, attention is focused more on basic principles, likely developments and harmonised legislation, than on current details of regulations in any particular area.

As mentioned, food laws go back to antiquity. In the course of history, individual authorities—in modern times governments—have developed their food laws in accordance with their philosophies of legislation and jurisprudence. The consequent variations in systems are characterised by different emphasis placed on three basic types of law: Religious, Common and Statute.

11.15.1 TYPES OF LAW

Religious Law

This normally deals with food itself. Islam has certain prohibited foods; orthodox Judaism has both prohibited foods and specifically encouraged foods. Hinduism also has prohibited foods. There are many other specialised rules. In some cases, the rules can be extended from the food to cover also articles in contact with it, and consequently, plastics film could be concerned.

The main problem is that components of non-permitted foods could be absorbed into the plastic, and hence by migration 'contaminate' permitted food consumed from the same object on subsequent occasions. This is extremely unlikely to occur with film, since it is normally only used once.

There is also a possibility that 'forbidden' components may be present in an item contacting the food, including plastics film, and these can enter the food by migration. An example is unsaturated fatty acid amide slip agents, made from pig fat; whether the amide is covered by the prohibition on pig products is a knotty theological problem. Obviously, in this field, each problem has to be considered individually.

It could be asserted that, since plastics are totally synthetic, they are 'neutral' in relation to any religious code and hence should be regarded as unexceptional. This has been successfully argued on occasion, and there do not seem to be instances in modern times of sanctions by secular authority on a religious basis. It would therefore seem safe for a plastics film merchant not to take this problem into account unless it is raised by a specific consumer; in which case the above arguments apply. In any event, any unfavourable reaction could be regarded as consumer resistance rather than infringement of the law.

Common Law

Common Law is formalised tradition. It is a body of law deriving from socially accepted norms or customs, interpreted and specified by the judiciary. It leans heavily on precedent.

Increasingly, Statute Law (see below) is superseding or codifying sections of Common Law. Even in countries such as the U.K. which are regarded as Common Law countries it is now probably correct to regard the main body of law as Statute, with Common Law (and some other systems, e.g. Equity) filling in. This leads to an important fact—in Common Law countries, nothing can be regarded as outside the civil law (see below). If there is no Statute, Common Law is always available even if there are no direct precedents.

Common Law may be criminal or civil (see p. 168—Criminal and civil law), and although not apparently as specific as Statute Law, is equally enforceable.

The basic philosophy of Common Law relevant in the present connection is payment for damage. That is to say, an eye for an eye and a tooth for a tooth, with substitution of money's worth for eyes and teeth where these are in short supply. The way this should operate in connection with packaging food in plastics films is, on the face of it, very simple. If the merchant of a plastics film sachet containing frozen peas were to sell such quality that the unwitting customer, having consumed his peas, were to lose his eyes or teeth the customer could successfully sue the supplier for the current market value of the appropriate number of eyes and/or teeth; in addition, he could also sue for the money equivalent of pain and suffering, loss of earnings when laid up (consequential damage), etc. Unfortunately, there are two severe practical problems.

The first is the difficulty of separating effects due to some fault in food from any effects due to the packaging. The second is that the hazard—if any—due to packaging is both slight and subtle and (with some important but rare exceptions) it is difficult to determine, at least in a short period, whether it has happened at all. Consequently, there has been a tendency even in strong Common Law countries for it to be reinforced by Statute Law in the area of food and packaging.

Another area of Common Law, which is highly relevant, is the law of contract (much of it also covered by Statute Law). This enters where the plastics film merchant guarantees, warrants, or otherwise specifies that a plastic film is suitable for use in contact with food. Provided he has received consideration for the film to be sold, he is then bound by the contract to provide plastics film of this appropriate quality—and would be liable for damages if the plastics were not of the correct quality. In practice, the effects of this are similar to the basic rule of Common Law on damage, but, where a contractual relationship exists, the matter is more clearly defined. The responsibilities of the merchant are less open to argument, and the damages may extend to consequential matters (e.g. loss of reputation) beyond the more specific damages mentioned above. Also, contract law is more relevant between companies (e.g. raw material supplier and converter, or converter and packager) than between retailer and consumer.

Contract law is also important in countries which do not have specific legislation, but do have codes of practice. The latter, whilst not having themselves the force of law, can be effectively brought within its scope by enshrining them in a contract.

Under Common Law, responsibility for safety rests entirely with the merchant; the Government acts, effectively, as a kind of neutral referee between the merchant and the customer. Over the years, of course, Common Law has developed a large body of rules based on precedent, many of which are similar to the principles of Statute Law described below.

Statute Law

This is a body of legislation promulgated by the State, usually Parliament, but sometimes by the head of state (president, monarch) or by government ministers under authority of other legislation. It is quite specific, and attaches considerable responsibility to the government or administering authority. In the field of food packaging, Statute Laws often do little more than extend the backing of government to the Common Law situation. That is to say, such laws often merely state that packaging materials should present no hazards to public health. Whereas, however, under Common Law the penalty would probably be under civil law, under Statute Law infringement can be criminal.

Legislation on food contact and packaging is often part of food law. Whereas all law is more or less complex—especially to the layman—food law is particularly difficult because there is no generally accepted definition of what is meant by food, what is its purpose and what is the purpose of the law in relation to it. Disparate interests, such as health, hygiene, trading, commerce and equity, are usually involved (see full discussion in the treatise by Bigwood and Gerard, referred to at the end of this chapter). An important point for the present purpose is that there are *two* principles underlying all modern regulatory systems. These are (a) quality and (b) honesty.

The former, *quality,* includes the *essential* needs of the consumer for health safety, for freedom from toxicity in any shape or form. It also includes less essential requirements, such as nutritional value, acceptable organoleptic properties, etc.

The second principle, *honesty in trading,* covers a *desirable* consumer need, and is related to the general law covering commodities offered for sale and trade. The relevant aspects are those concerning false weights and measures (metrology), misleading advertising and presentation (ornamentation). In general merchandise, these aspects are concerned with commerce and trading, but when dealing with food are also related to quality; sometimes the borderline becomes blurred. Depending on the countries concerned, the above principles may be given more or less emphasis; and also may be dealt with either in separate or combined laws. There is a strong tendency, especially in Europe and the U.S.A., to transfer legislation dealing with honesty in trade from the fabric of Common Law to consumer protection acts.

It is important to distinguish between the two principles underlying law, since matters such as the size and ornamentation of a package, which are relevant to honesty in trade, are under the control of the converter or packager; whereas health hazard effects are largely (not entirely) under the control of the raw material supplier. Also, as already mentioned, it is important to distinguish between health hazard and adulteration; the former lies in the field of quality, the latter should be regarded as fraudulent trading.

Much legislation in the present field concentrates on migration of plastic components into food. From the point of view of Statute Law, such migrants can be considered in three ways:

(1) They can be regarded as unintentional food additives; and hence subject to the laws dealing with additives in food themselves.
(2) They can be regarded as extraneous, or foreign matter in food. This puts them on a par with dirt from whatever origin, and tends to associate them with adulteration.
(3) They can be dealt with as a special case.

All three methods are currently to be found in use; and it is wise for the plastics film merchant to be alert to the methods implicit in the legislation with which he is concerned. Method (1) is scientifically sound, but presents formidable problems in terms of implementation. Method (2) is historically favoured, but is now dying out because of its obvious lack of assessment of hazard to health. This leaves method (3).

Under this, a wide variety of procedures can be adopted. In fact, the majority of elements of legislation which do not utilise methods (1) or (2), adopt either (a) composition, or (b) specification systems. The scientific aspects of these have been described in section 11.11.10. Legally, under the *composition* rationale, constraints are placed on the composition of the plastics film contacting the food. Such constraints may take a variety of forms, but are usually positive lists of permitted components (from base polymer to additives), or negative lists of forbidden components. Nothing is said about the migrants themselves; but experimentation carried out prior to framing of the legislation is assumed to have provided sufficient data for correlation to be established.

The *specification* method lays down certain criteria (usually analytical limitations in food or simulants) which must be met; but does not directly place any restriction on the actual composition of the plastic itself.

Both methods are often to be found within the same overall piece of legislation.

Licensing type systems

To overcome some of the problems mentioned above, several countries have introduced 'frame' laws. Such laws specify that food packaging must be suitable for the purpose and present no health hazard, the interpretation of this requirement to be made by government officials usually operating under the ministries of health or social welfare. Sometimes precise criteria are laid down, e.g. positive lists, in which case the system is practically equivalent to straightforward Statute Law, but sometimes it is left to ad hoc decisions based on individual cases. Where this is so, it is essential for the plastics film merchant to get clearance from the authorities before marketing.

Criminal and civil law

In criminal law, the adversaries are the state and the defendant, with penalties ranging from social service or fines to imprisonment or (rarely today) physical punishment.

In civil law, the adversaries are two persons (a person may be an individual, or a corporate body, e.g. a company). The plaintiff takes the initiative against the defendant. Penalties are most commonly financial, although specific performance or other remedies are available. Enforcement is also by the state, but punishment (e.g. imprisonment) only arises from contempt of court if the defendant does not accept the remedy imposed.

Statute Law is usually criminal law, Common Law may be either. Plastics or packaging legislation nowadays may fall under either.

11.15.2 INTERNATIONAL TRADE

Because of the variations already mentioned between the legislative or regulatory systems applying in different countries, problems may arise in international trade. The plastics raw material may be manufactured in country 1, utilise additives from country 2 (or more), be converted into film in country 3, further processed (e.g. printed and shaped) in country 4, for use with food ultimately sold in country 5. Whose laws apply?

There are minor aspects which may have an influence here and there; for example, one of the additives may be subject to control by industrial hygiene legislation, in which case the laws of country 2 would be relevant (or country 3 if the additive were used in a conversion process). A very few countries prohibit the processing for export of material which would not be legally saleable in that country; in such exceptional cases, the laws of country 3 apply. However, in the majority of cases and as a general rule, the laws of country 5 are controlling. That is to say, it is the legislation applicable to the ultimate customer, either of the food or the film in contact with it, which is controlling.

11.15.3 INDIVIDUAL COUNTRIES

For full details, reference must be made to appropriate textbooks, or to official published national legislation. The following gives a summary of the relevant legislative systems in the U.K. and some other countries. The U.S.A. and West Germany are included particularly, because their regulations are often accepted by countries having no detailed statute legislation, and compliance with them is also often required by customers seeking guarantees.

In many countries, food packaging and health safety legislation is currently under revision, not least in the member states of the E.E.C. (see below). Eventually it is likely that a consensus will be achieved, with most countries' legislation on a similar basis. However, even when this arrives, there will be differences, especially in administration and policing, based on different systems of jurisprudence.

United Kingdom

The U.K. is the archetypical Common Law country, and any adverse effects of food packaging can be covered by this as explained on p. 164—Common Law. However, partly as component of a general trend, and partly because of the requirement to conform with E.E.C. Directives, there is an increasing amount of Statute Law.

The earliest relevant legislation was the Food and Drugs Act (current version, 1955). This gives wide powers to the Minister to issue regulations concerning food quality in all its aspects, and it is possible to bring packaging (or any other food contacting situation) into the orbit of the Act by regarding packaging as a source of adventitious food additive (see p. 165—Statute Law). That is to say, if a packaging material were to transmit foreign substances to food, it thereby became contaminated, adulterated, tainted or unsafe for consumption, the food itself would infringe some aspects of the Act and the packaging held to blame. This approach is logical, since in the interests of public health it is only what is consumed that really matters. In practice, however, it is not at all easy to determine what adventitious food additives *have* derived from packaging, and the approach does not take into account any other major aspect of packaging, notably changes in microbiological contamination (almost invariably a reduction).

The Act is implemented in detail by the issue of regulations (prepared by the Minister, and usually approved by Parliament), and several of these have been issued. These cover anti-oxidants, colouring matters, mineral oils, preservatives, etc., in food (not packaging). All these give positive lists of permitted food additives. With very few exceptions (notably BHT which is a permitted anti-oxidant at certain levels in various foods, and mineral oil) none of the regulations deals with additives likely to be encountered in plastics film. Theoretically, therefore, it could be held that all plastics additives which can be regarded as covered by these regulations are illegal; but the point has never been raised in court except where a genuine health hazard has been thought to exist (e.g. lead from metal cans).

Recommendations for detailed implementation of the Act are made by the Food Standards Committee and the Food Additives and Contaminant Committee (FACC). Recognising the possible need for more detailed legislation on packaging, the FACC was commissioned to prepare an advisory report. This was published in November 1969, but its recommendations were not very definite, and are now considered to be superseded by the imminence of E.E.C. legislation. The British Plastic Federation (BPF), in conjunction with the British Industrial Biological Research Association (BIBRA) has published a Code of Practice for plastics for food contact application. This provides specifications for base plastics, and lists of permitted additives (both under chemical and trade names) with concentration and other limitations where relevant. This code, enforced by contract or warranty if necessary, but even used as 'best possible advice', would usually satisfy any requirements under Common Law or the Food and Drugs Act, but not the more recent legislation described below.

Some regulations have now been issued, or are on the way, covering packaging and plastics in specific detail. The first was Statutory Instrument (SI) 1927 of 1978 'Food and Drugs: The Materials and Articles in Contact with Food Regulations, 1978' made under the European Communities Act, 1972. It applies to 'materials and articles which are in their finished state and are intended to come into contact with food or which are in contact with food and are intended for that purpose'. These must be 'manufactured in accordance with good manufacturing practice, that is to say in such a way that under normal or foreseeable conditions of use they do not transfer their constituents to food with which they are, or are likely to be, in contact, in quantities which could—

(1) endanger human health or
(2) bring about a deterioration in the organoleptic characteristics of such food or an unacceptable change in its nature, substance or quality.'

Materials and articles not conforming must not be sold, imported or used (Clause 4).

There must be suitable indication in terms of labelling or description that retail packaging is suitable for food use (Clauses 5, 6 and 7). Clause 8 extends the regulation to materials and articles offered as prizes.

Clauses 9 and 10 cover enforcement. The authorised officers are empowered to enter manufacturing premises; but information obtained must be kept confidential (Clause 11).

Enforcement is by food and drugs authorities in England and Wales, regional or island councils in Scotland, or port health authorities for imports (Clauses 2 and 9).

Clauses 14 and 15 deal with implementation and responsibility. In certain circumstances, an individual may be guilty of an offence as well as the corporate body of which he is a member.

Penalties range up to a fine of £400.

Defences are available based on 'due diligence' (Clause 19) or warranties (Clause 20).

The above is a summary of the salient points of the SI. It is essential for any company or individual concerned to refer to the original, or a legal adviser for full details. However, it is clear that the SI does not substantially change the technical requirements for plastics films contacting food from the status ante quo under the Food and Drugs Act and Common Law. The novel features are as follows:

Clarification of implementation.
Introduction of organolepsis as a formal requirement.
Legal requirements for labelling and/or description.
Powers of entry into manufacturing premises.
Fixed maximum penalties (these will probably be increased).
Clarification of defences based on due diligence or warranty.

It is important to note that the SI applies only to migration and not to other vectors.

Other regulations are following, implementing other E.E.C. Directives. SI 1838 of 1980 implements the E.E.C. Directives on

VCM limits, analytical methods for VCM in PVC, and food contact symbol (see p. 176—Existing Directives.)

France

France is a strong Statute Law country, with the Code Napoléon reinforcing its Roman Law tradition. The basic relevant legislation stems from 1905 and specifies in essence that nothing may be, intentionally or otherwise, transmitted to food unless it has been specifically authorised. The latest basic version is the Ministerial Decree 73-138 of 12 February 1973. The positive list of additives permitted in plastics is administered by the Service de la Répression des Fraudes. Ministerial Decree 73-138 included the list current at the time; amendments are issued as they arise, and up-to-date collections at intervals in the Journal Officiel (latest being 1227 of 1980). The individual circulars and collections are published ad hoc: there is no classification.

The law also includes the proviso that packaging materials (and their components) are permitted in such cases where it can be assured that 'not a trace' of the material or its components will migrate into food. Thus, if the plastics supplier can be sure that this applies to the film, no attention need be paid to the positive list.

As E.E.C. Directives are promulgated, the law will be adjusted to conform with them (see p. 176—E.E.C.).

Federal Republic of Germany

The F.R.G. (West Germany) has a basic frame law, the earliest versions of which date back to before the Second World War. The latest version, of 15 August 1974 (Bundesgesundheitsblatt 1974, No. 95,p.1945), provides in essence that extraneous (foreign) matter must not migrate from utensils (which are taken to include packaging) into a food or on to its surface. It also provides that the extraneous matter must be not only non-injurious and non-tainting, but must also be technologically unavoidable. Advice on the implementation of the law is given by the Federal Health Office (Bundesgesundheitsamt—BGA) in the form of Recommendations. Whilst these have no formal legal validity, they are recognised by the courts and tend to function as *de facto* legislation. Nevertheless, it is possible to ignore the Recommendations, provided the supplier is satisfied that his material conforms with the requirements of the law, and accepts responsibility for any consequences.

As in all E.E.C. countries, changes are under way. It is likely that new legislation will be introduced, making the BGA Recommendations statutory requirements. Also, some Ministerial Decrees have been issued, beginning with 26 October 1979 on VCM (Bundesgesundsheitsblatt Part 1, 1979, No. 64, p.1773) and more will be, implementing E.E.C. Directives (see p. 176—E.E.C.).

Netherlands

The Netherlands is also a Statute Law country, and has legislation covering general requirements for the quality of food. Under the authority of this legislation, Ministerial Decrees of 1 October and 21 December 1979, which came into force 1 October 1980 (Netherlands state journal of 25 January 1980, No. 18) have been promulgated implementing the E.E.C. Directives (see p. 176—E.E.C.). However, they also include schedules with quality specifications for all plastics likely to be used, a positive list of all additives likely to be used, specifications for overall or global migration, and analytical procedures to determine the migration of additives to food or a simulant.

A Decree of 15 August 1979 provides that products used for maintenance and cleaning in food factories must be labelled adequately. It also covers packaging of additives or other plastic ingredients, which must be labelled with composition data.

Italy

Italy is a fully Statute Law country, most relevant legislation being enacted by Ministerial Decree. The first decree still relevant was that of 19 January 1963 which defined plastics migrants as 'indirect' food additives, and included a positive list of permitted components as well as specifications of safe plastics. It also included an overall (global) migration limit. It has been amended from time to time to update the specification, and bring it in line with E.E.C. Directives (see p. 176—E.E.C.).

Compliance is mandatory.

Belgium

Belgium is also a Statute Law country, and has legislation which prescribes general requirements for quality in food. In accordance with this, Royal Decrees are issued covering detailed implementation, and these now cover food contact and plastics packaging.

Probably the first decree still relevant is that of 25 August 1976, modified by those of 29 July 1977 and 30 January 1979. They are similar to the Netherlands legislation (see above) though somewhat less detailed. Further decrees are following implementing newer E.E.C. Directives, the latest being 30 June 1981.

Scandinavia

Each Scandinavian country has different legislation, but in principle all have a frame law under which plastics compositions for food contact are considered as a whole. Consequently, although not essential, it is wise for a plastics merchant to obtain clearance from the authorities for his product. This may be in terms of specific additives, but is usually more in terms of the finished grade: the approval is normally sought by the raw material manufacturer.

In addition, Denmark has a small positive list of materials which may be used without further query (prepared by the National Health Service). It also, of course, now has to conform with E.E.C. Directives. That on VCM in PVC was implemented by an Order of the Ministry of Environment dated 8 November 1979; similar legislative instruments will doubtless follow.

Sweden has a basic Food Decree of 1971, SFS 1971-807, paragraph 7 of which covers food packaging in general terms; amended in 1973 by SNFS 1973-334 (in Swedish code of Statutes, published 30 August 1979). Some detailed items, such as VCM limits, are also covered by the Ordinance concerning foreign substances in food of 1978, SLV FS 1978-34.

In Sweden, increasingly, authority is attached to organisations including trades unions and consumer representative bodies as well as government. In particular, the Product Control Board makes recommendations to the Ministry of Agriculture which are then implemented by legislation, or is authorised to implement legislation itself. The two relevant examples are the Ordinance containing stipulations regarding certain household goods of the Product Control Board, 12 September 1975, and the ban (with some exceptions) on the use of cadmium and its compounds implemented by the Control Board with effect from July 1982. This will, *inter alia,* prohibit the use of cadmium pigments in plastics film.

U.S.A.

Although the U.S.A. inherited a Common Law philosophy from English law, it is, in the present context, a Statute Law country, the relevant legislation being enacted and implemented at Federal and State level. The basic Federal Law is the Food, Drugs and Cosmetics Act of 1938 which was amended in 1958 to include food additives. This brought packaging materials within the scope of the Food and Drugs Administration (FDA) of the Department of Health, Education and Welfare. The FDA regulations include an enormous positive list giving specifications of base polymers and additives. Usage of plastics and their components is permitted in terms of type of foodstuff (nine categories), temperature, applicational type (e.g. film, moulding, or 'polymeric composition' which allows any application), etc. A material may therefore be approved for use in plastics films either under the general plastics sections, or specifically as film. Different limitations may apply to specific cases, and relaxation of regulations for general use are often made for thin films, since the migration may be acceptable at the 100% level (see section 11.12.1). The FDA consider evidence submitted to them, and may take action in cases of infringement. They do not issue certificates (i.e. they are not a licensing authority).

A major complication for the FDA is the Delaney Amendment. This absolutely prohibits use at any concentration in food of an additive deemed or found after appropriate tests 'to induce cancer in man or animal'. Following this, the FDA sought to ban the use of PVC and ACN polymers in contact with food. (The relevant monomers are proved or highly suspected to be carcinogens, based on animal tests.) These systems are Class II (see p. 143—Class II) and hence some monomer, however little and whether detectable or not, could be assumed to migrate into food. Industry appealed through the courts, one ground being that, at very low levels (below 1 ppm in polymer), the monomer is bound in some way and it does not migrate at all (i.e. $\beta = \infty$). The courts were not satisfied that this contention was valid, but nor were they satisfied with the FDA contention. They also thought the FDA could judge a level to be *de minimis,* or acceptable on risk/benefit grounds where the risk was exceptionally low. As we go to press, the question has been referred back to the FDA for reconsideration.

In addition, the FDA are undertaking a fundamental reappraisal of their food packaging or contact regulations; this includes a major research study. The objective is to produce simpler and more effective regulations.

11.15.4 E.E.C.

It can be argued that different regulations on food packaging material in member states in the E.E.C. constitute a non-tariff barrier to trade. If this is accepted, it is valid, under the Treaty of Rome (Articles 100, 101 and 102) for Directives to be promulgated which harmonise ('approximate') such legislation.

Directives are not instant Community law—they are instructions to governments of member states to bring their own legislation into line within a certain period. In general they must be agreed by all member states represented at the Council of Ministers.

There is a detailed process of consultation, study and preparation leading up to the promulgation of a Directive. Central to this is the E.E.C. Commission (originally Directorate General VI, and more recently III, is concerned with food contact legislation). Although the resultant legislation is intended to represent a consensus, the views of the Commission carry considerable weight, and these tend strongly towards Roman Law.

At the time of writing, five Directives are in existence, and others are in various stages of preparation.

Existing directives

(a) *General Packaging and Food Contact (Materials and Articles).* Directive 76/893 (of 23 November 1976) often known as the General Packaging, or Materials and Articles, Directive covers most food and water contact, i.e. packaging, plant (including pipes), food processing and handling machinery, and domestic potable water systems. It excludes covers and coatings, potable water distribution systems and antiques. It also excludes direct contact with the human body, but includes (in the preamble only) buccal contact (contact simultaneously with the mouth and with foodstuff).

Health safety and organolepsis are covered in Article 2—the requirements are those given above under U.K. legislation (p. 169).

Composition requirements are covered by Article 3 which refers to the issue of further Directives (see below) which 'may' include overall (global) and specific migration limitations, and buccal contact. Labelling of finished articles is covered by Article 7. Specified phrases (in English 'for food use') must be used, or a symbol (see below), but these are not required if the article is obviously a food container.

This Directive has been implemented in all member States. Those Statute Law countries which already had legislation (Belgium, Italy,

France, Luxembourg) have ensured that their legislation was in conformity, or have adjusted it accordingly. Those which did not (Netherlands) have introduced such legislation. Common Law countries (Denmark, Ireland and U.K.) have introduced suitable new legislation. In some cases (e.g. Belgium, Netherlands) the new legislation goes well beyond the requirement of conformity with the E.E.C. Directive.

(b) *VCM.* Directive 78/142 (of 30 January 1978) was the first detailed Directive made following the basis provided by 76/893. It covers the VCM content of PVC or vinyl copolymers contacting food limits are 1 ppm (mg/kg) in polymer, or 10 ppb (0·01 mg/kg) in food (strictly it requires a not detectable level using a method with a detection limit of 10 ppb).

This Directive could not be fully implemented until analytical methods had been agreed. The method for analysis in plastic was issued as Commission Directive 80/766 on 8 July 1980. It uses headspace gas chromatography with N,N-dimethyl acetamide as solvent. The method for analysis in food took longer to agree, the Directive (81/432) being issued on 29 April 1981. By publication date, both should be in force in all member States.

(c) *Symbol.* The general Directive 76/893 requires an indication that the material or article is suitable for food contact use. This can be done in various ways, depending on the stage of production/marketing chain. For example, raw material may be accredited by documentation. At the retail stage, member states have the option not to insist on marking where articles are 'by their nature clearly intended to come into the contact with foodstuffs'.

Another option at this stage, or an alternative for others, is the use of a symbol. This had been laid down in Directive 80/590 of 9 June 1980 and is shown in *Figure 11.14*.

Proposed directives

The above comprises E.E.C. legislation in force at the time of going to press. The following are at various stages of development.

(a) *Overall migration.* As mentioned above (section 11.11.12) it is widely believed that a limitation on overall or global migration has certain advantages. A Directive has been proposed for plastics on these lines. The latest published version (OJ of 16 June 1978, C141/4) proposed limits of 60 mg/kg in food or simulant, or 10

Figure 11.14. Symbol

mg/dm^2. Further drafts have been issued with very substantial modifications. The majority of these concern implementation—it is proposed that the Directive be not mandatory until other aspects (e.g. positive lists, classification of foods, etc.) are introduced.

The limitation based on area (10 mg/dm^2) applies to '(a) containers or articles which are comparable to containers or can be filled, with a capacity of less than 200 ml or more than 10 litres.

'(b) sheet, film or other articles which can not be filled and in respect of which it is not possible to calculate the relation between the surface of the material or articles and the quantity of foodstuff in contact therewith'.*

It is clear, therefore, that in general this second limitation will be the one applying to plastic films. As discussed in section 11.11.12, it is clear that it corresponds to a higher concentration in food than 60 ppm.*

(b) *Positive lists.* A European list of permitted components is seen as a desirable ultimate objective. However, formidable problems arise in preparing such a list. Some are due to paucity of scientific knowledge (composition and migration as well as toxicity) but equally troublesome are political and industrial conflicts of interest.

The Commission has therefore established priorities, top of the list being monomers. These are currently being studied by the E.E.C. Scientific Committee for Food.

(c) *Regenerated cellulose.* Although regenerated cellulose as such

*Withdrawn in October 1982, published version

is outside the scope of this book, the E.E.C. proposals for a Directive on regenerated cellulose film are of interest (1) because of its use in composites with plastic, and (2) the approach used for regenerated cellulose film could be applied to plastics film, separating it from bulk plastics packaging and hence possibly removing some of the restrictions or burdens mentioned above.

The proposed Directive has been under discussion for a considerable time, and reached the stage of a final proposal in 1981 (Official Journal C235/3 of 15 September 1981). It includes two positive lists, one for uncoated, and one for coated film. To a considerable extent the restrictions or specifications for permitted components are dealt with pragmatically, by concentration limits for individual items, limits for groups of items, limits on coating thickness, purity specifications, etc. There are no requirements for overall migration limits, and no specific migration limits. The principle of 'lowest technological level' has been used in assessing some restrictions.

Peripheral legislation

There are many other Directives which may impinge on plastics film—i.e. they have a peripheral effect, or have a small component related to packaging.

(a) *Packaged products*. There are many Directives on individual foods. If migration reaches a level where the migrant can be regarded as a food component, the package as a whole might be held to be in breach of the legislation deriving from the Directive. This is, perhaps, rather unlikely—see p. 176—Existing Directives.

There are also several Directives on other products which may be packed in plastics films, e.g. 67/548 on classification, packaging and labelling of dangerous products (amended several times), 73/173 on solvents (amended by 80/781), 73/404 on detergents, 76/116 on fertilisers, 77/728 on paints, varnishes, inks and adhesives, and 78/631 on pesticides. These all have small sections on packaging, which may be relevant to plastics film—the usual requirement concerns labelling and security of closures, but it is essential that any packager concerned refers to the implementation of the legislation in the relevant country.

(b) *Packaging*. There are several Directives covering all packaging. The aspects covered are indicated weights and volumes, metrology and standard sizes. Some of the Directives cover all packaging, some food only. One Directive, 79/112, deals specifically with labelling, presentation and advertising of foodstuffs for sale to consumers.

(c) *Consumer protection.* There is an increasing number of consumer protection Directives (as well, of course, as national legislation) including a proposal on product liability. Plastics films are covered in the same way as other consumer products.

Increasingly, consumer complaints arise concerning packaging which is superficially misleading, especially as to contents. For example, a yoghurt or cosmetics pot with a relatively large hollow base can give the impression that there is more in the pot than there actually is. To some extent this is now being alleviated by requirements for quantity indication in standard units, as mentioned above, and it is less likely to affect packaging in film than in rigid containers.

In the U.K. the subject is covered by the Packaging Code for retail products, monitored by the Packaging Council.

(d) *Environmental.* There is an increasing number of these Directives also. They cover such matters as water and air pollution, manufacturing practice (including occupational hygiene), etc., and mostly affect manufacturing or processing. However, some of them (e.g. on ground water pollution) may have some relevance to waste disposal of end product.

A Directive has been proposed, and a draft issued, on returnable beverage containers. This would impose restrictions (possibly including tax) on 1-way containers, and encourage returnables. It would discriminate against plastics films, since these are invariably non-returnable. At the time of going to press, opposition to the proposed Directive is mounting, but it cannot be guaranteed that it will not be implemented, possibly modified.

11.16 LAW AND REGULATIONS: COSMETICS AND TOILETRIES

Most of the general information in sections 11.15.1 and 11.15.2 is equally relevant to cosmetics and toiletries packaging.

11.16.1 INDIVIDUAL COUNTRIES

In Common Law countries, any proven hazard arising from

packaging cosmetics and toiletries would be covered by Common Law; and in Statute Law countries (as well as, now, increasingly in Common Law countries) legislation on consumer protection might be relevant. See section 11.13 for health safety criteria.

Several countries do have regulations on cosmetics *per se.* It is theoretically possible for these to be infringed by migration from plastics film of components not permitted as cosmetic ingredients. In practice, this does not seem to have occurred; but, if it were to happen, it would be necessary to seek inclusion of the migrant on the cosmetics positive list (this is exactly analogous to the situation for food, see p. 179—Peripheral legislation).

It is possible that some countries (e.g. Sweden) will introduce a licensing system. If so, these are likely to be for the total packaged product, in which case the film would be covered.

11.16.2 E.E.C.

No direct E.E.C. legislation on packaging of cosmetics and toiletries is currently proposed, but Directive 76/768 (amended by 79/661) covers limitations on the products themselves. The remarks in section 11.16.1 apply.

Also, most of the peripheral legislation described on p. 179—Peripheral legislation is relevant.

11.17 LAW AND REGULATIONS: MEDICINES, DRUGS AND PHARMACEUTICALS

Stringent legislation or regulations (e.g. pharmacopaeia) exist in all countries, and it is essential to conform with these.

No E.E.C. legislation is currently proposed.

FURTHER READING

BIGWOOD, E. J. and GERARD, A. *Fundamental Principles and Objectives of a Comparable Food Law,* Vol. 4, S. Karger, Basel and New York, 1972.

BRISTON, J. H. and KATAN, L. L., *Plastics in Contact with Food,* Food Trade Press, London, 1974.

CROSBY, N. T., *Food Packaging Materials,* Applied Science Publishers, London, 1981.

HIGHLAND, H. A., 'Insect Resistance of Food Packages—A Review'. *J. Food Processing and Preservation,* 1978, 2, 123–130.

LEFAUX, R., *Les Matières Plastiques dans l'Industrie Alimentaire,* Publications Techniques Associées (CFE), Paris, 1972 (English version of earlier edition, Iliffe, London, 1968).

12

Organolepsis

by Dr L. L. Katan

12.1 INTRODUCTION

In choosing a food item, a consumer usually decides in principle on a type—e.g. meat or poultry for protein; potatoes, rice, or bread for carbohydrate; some favoured vegetables; fruit, etc. When choosing the actual product to purchase within the type, however, although stated nutritive value or content may have an influence, the major factors are related to perception through the physical senses. These factors are, therefore, of vital importance in marketing food; hence the influence on them of packaging film is, likewise, vitally important.

There are five major physical senses: sight, hearing, touch, taste and smell, which may be involved in the sensory perception of food. Together with the several others (electric stimulation or shock, radiation, etc.) these are called organoleptic effects, and the totality is organolepsis.

Note: some writers confine the term to the senses of smell and taste only.

There are essentially two types of recognition by the brain of organoleptic patterns. The first type constitutes the instinctive or hereditary pattern, essentially defence mechanisms, which find certain primitive colours, tastes, smells, etc. (e.g. natural colour of vegetables, simple food tastes) attractive, and others, usually associated with putrefaction and decay, repulsive. These patterns generate favourable or unfavourable reactions which at high levels may be quite strong such as salivation, catching of the breath or nausea.

The second type of appreciation depends on patterns established by personal experience since birth. Where these patterns do not conform with instinctive patterns, they are called 'acquired taste'. There is an enormous variation in personal idiosyncratic behaviour in this context, but there is also a wide measure of conformity depending on social factors.

The practical significance of this is that, whereas most of the physical properties of films discussed elsewhere in this book can be determined objectively, and related to performance in a mechanistic way, organolepsis and organoleptic effects are much more subjective and can not be dealt with by physical methods in isolation. A new discipline has evolved to deal with it.

12.2 PSYCHOPHYSICS

As suggested above, in a marketing economy, products are purchased not only for their functions, but also for sensory reasons. It is nowadays difficult to think of artefacts where sensory impressions are completely irrelevant; these are usually where the article is not directly open to the senses before purchase, e.g. engines, bedsprings. Otherwise, the importance of sensory impression varies from the significant in regard to highly technical items such as buildings, bridges, electric cables, through major importance for all consumer durables, to overriding importance for furniture, food and clothing, where sensory properties are the major sales feature.

For the several thousand years in which some sort of market economy has existed, evaluation of sensory properties was largely non-deliberate, and measured by *de facto* acceptability or sales. One ancient Roman chairmaker would sell more chairs than another; Wedgwood beat his competition; fresh fish sold better than stale fish; and more recently, in plastics, some picnicware would be preferred to other.

Within the last few decades, however, the subject has been put on a more quantitative footing. Aesthetic design still remains more or less subjective, but organoleptic effects, taken in isolation, are increasingly being measured and placed on quantitative scales. The new science of psychophysics has been born; it deals with the correlation between consumer sensory response, which is psychological, and corresponding physical characteristics.

The 'consumer' is most frequently human, but other animals are also important for some products, notably animal food. It has been found that there is a general correlation between a stimulus and its perceived sensory effect. The basic law of psychophysics is

$$F = kI^n$$

where F = intensity of perceived sensory effect (see below for meaning of this).

I = quantitative stimulus (e.g. physical, chemical, electrical).

k = constant for a specific stimulus (e.g. specific chemical, particular light wavelength).

n = constant for the type of stimulus (e.g. taste, smell, colour).

The highest value of n (3·5) applies to electric shock (perhaps a defence mechanism). For the senses considered in this chapter typical values are as follows.

Sight. Very variable, depending on specific stimulus. Brightness, for example, varies from 0·3 to 0·5, whilst lightness is about 1·2 (reflectance of grey paper). Also, there are major differences between colourants—especially pigments and dyes.

Hearing. 0·5 to 0·6 for loudness.

Touch. Variable, depending on stimulus. For hardness it is *c.* 0·8, for roughness *c.* 1·5, for vibration 0·6 to 1·0 depending on frequency. For a static force on the skin it reaches 1·1.

Taste. Depends greatly on stimulus. For saccharin it is 0·8, for sucrose and salt 1·3.

Smell. 0·5 to 0·6 (but many exceptions—also influenced by fatigue: see also p. 204, Threshold and fatigue).

12.2.1 HEDONISTIC SCALES

For many subjects now covered by psychophysics, scales have been developed, some of which (e.g. brightness of stars, hardness of minerals, loudness of sound and even intelligence) are of considerable antiquity. A classification scale system has now been established, and is being standardised (e.g. BS 5929 Part 1).

All the scales correlate F (perceived quality) with I (stimulus). There are four types of scale, in increasing quantification order.

(1) *Nominal, or classification.* This is a qualitative scale, sorting into predetermined categories. For plastics film, the ratings would be tainting and non-tainting, acceptable or non-acceptable.

(2) *Ordinal or ranking.* This places samples in order. If numbers are used, they are merely to indicate order, and have no quantitative significance. For plastics film the designations would be arranged from best to worst.

(3) *Interval or rating.* In this, a numerical scale is established, in which the intervals are equal. Although expressed numerically, the descriptions of the numbers are hedonistic, e.g. for plastics film, severe (foul), bad, moderate, slight, 'zero' odour (this being arbitrary).

(4) *Ratio or scoring.* This is similar to (3), except that the intervals are quantified.

All the scales, but particularly (3) and (4), facilitate correlation between the physical stimulus (for plastics films concentration of colourant, microcontaminant, etc.) with perceived sensory effect (colour, texture, taint, etc.). Although all scales, especially (4), are likely to be used for the packaged product (especially food), for the film itself and its effect it is usually quite sufficient to use the first scale, namely classification into acceptable and non-acceptable. This may require use of the other scales to establish the criterion of acceptability. This has always been a major marketing factor, and hence highly subjective since the relevant consumers constitute a restricted population (see subsequent discussion in this chapter, especially section 12.7). Knowing their market, producers have been able to assess and correlate it with criteria established by their testing panels.

However, in future it will also be a legal requirement (see Chapter 11, p.171) which presents serious difficulties. Courts can, of course, attempt to set up their own criteria, but these will still be dependent both on subjective views and on testing panels. They can use methodology conforming to established standards, but will put the standard-making bodies in difficulty if the level of acceptability, as opposed to the method, is called for as a standard.

12.3 DIRECT AND INDIRECT EFFECTS

There are two main ways in which organoleptic effects relate to plastic films: direct and indirect. *Direct* refers to the properties of the film itself; *indirect* refers to the effects of the film on the packaged product.

Direct organoleptic properties of the film are significant for three reasons:

(1) Where related to the actual application, e.g. colour in agriculture and horticulture, building or furniture; colour, touch and possibly hearing for paper substitutes.
(2) Marketing appeal to processor or packager.
(3) Fear by packager of indirect effects, especially for food packaging.

The last two are important in marketing: they are based on the subjective belief that indirect effects will occur. The major example is the fear that film with a detectable odour will impart that odour to the packaged product (food, cosmetic or toiletry); which is very possible but not necessarily so.

Indirect effects on the product in contact with the film are also significant; and in the packaging of sensitive products are in fact more important.

Both types are covered in this chapter under the headings of each organoleptic sense.

12.4 SIGHT

There are two main divisions of this property—colour; and surface gloss and drape. The latter is covered in Chapter 10.

Colour is a huge subject (see Further Reading) and its main importance in plastics is for decoration (paints) or moulded articles. The following covers only those aspects relevant to plastics films.

12.4.1 DIRECT EFFECTS

These are significant where the colour of the film itself is a marketing feature. As mentioned above, the main areas are thick films, especially in agriculture and horticulture, building, decorative laminates, paper substitutes, tapes and films.

The majority are hedonistic, but there are a few which are more technological. The main examples of the latter are black films for UV protection, or light elimination; grey, blue or brown for light change or reduction (e.g. aircraft windows), and identification (e.g. product identification—sacks).

Another major application is brand or retailer identification (many stores use special colours for their carrier bags).

12.4.2 INDIRECT EFFECTS

These are essentially confined to packaging applications. Major effects can be caused by bleeding or blooming, in both of which colourant is directly transferred to product.

Nowadays, these phenomena are rare, but if they do occur the effects are very serious. More common are the subtle effects on sensitive products—food, pharmaceuticals and medical products and drugs. Migration may occur from film to product and either colour it directly, or change its properties in other ways. The latter can be due to colourless impurities in the colourant (see Chapter 11).

Coloured film affects the transmission of light and other radiation; and this in turn can have effects, especially on food and medicinal products (see Chapter 11).

12.4.3 ASSESSMENT OF COLOUR

Visible light is electromagnetic radiation visible to the human eye. Its fundamental property is, of course, wavelength (inversely proportional to frequency) and the visible band lies between 380 and 750 nm. As the wavelength increases, the perceived colour changes through the spectrum violet–indigo–blue–green–yellow–orange–red. Somewhat paradoxically, the wavelengths immediately below 30 nm are called ultra-violet (UV) and those above red, infra-red light (which transmit heat). A complete mixture of all the wavelengths, such as is transmitted by the sun, constitutes white light.

When a beam of light impinges on a plastics film, several things happen. First, some of the light can be transmitted, i.e. pass through largely unchanged. If this occurs with virtually all of the light, the film would be transparent and not coloured.

Some of the light may be reflected. This depends on surface gloss, angle of incidence, as well as colouring. However, the reflected light corresponds to the colour of the colourant.

Third, the light may be absorbed. The remaining light transmitted is therefore deficient in the wavelengths that have been absorbed, and hence its colour has been changed. This is the main way in which colourants operate; by removing certain wavelengths from wide-spectrum light (usually white); the perceived colour is the sum of the remainder.

Finally, some light may be scattered. This is the main method by which inorganic pigments effect colouring; dyes are largely light-absorbing.

The overall effect of a colourant is, therefore, complex; and whilst the colour of a film will be similar to that of the colourants used to make it, the differences are sufficiently large for it to be always essential to check colour on the finished compounded film.

It will be clear from the above that the light emanating from a plastic film is a complex mixture of wavelengths; moreover, it is present at different intensities. The human eye has sensors for three colours: red, green and cyan (blue), none of which is a single wavelength. The brain analyses the totality of the stimuli, and builds up an image from this.

To describe this situation, various systems have been developed. Most of them use 'colour co-ordinates' in terms of three variables. These comprise *hue,* which is the actual colour in subjective terms (e.g. pink, brown, as well as the pure spectral colours); *value,* which describes lightness or darkness; and *chroma,* which describes the purity or saturation of the colour.

For practical purposes, the simplest way of describing a coloured plastic is to inspect a plaque (chip), bearing in mind the following. First, the perceived colour will change with thickness; as the film gets thinner, transmission increases, whilst scattering and dispersion decrease. Second, and more important, the perceived colour will depend on the incident light. Standard light sources should be used, although diffuse daylight is quite good. Light of limited wavelength (e.g. fluorescent) can give very misleading results. Some colourant formulations are more prone than others to give different colours when viewed with various incident light sources. The phenomenon is known as *metamerism,* and a colourant which gives the same perceived colour to plastics under different light sources is said to be *non-metameric*. No colourant formulation is perfectly non-metameric for all light sources, but some approach close to it for the two main light sources likely to be encountered, namely tungsten filament and daylight.

Finally, of course, the base plastic in which the colourant is compounded must be the same as that which is to be used for the final film.

12.4.4 MEASUREMENT

A wide variety of instruments is now available for measuring various aspects of colouring. These can assist greatly in formulation, preliminary testing and quality control. However, the human eye is extremely sensitive, and the processes of the eye and brain, based on the three stimuli received by it, so complex that no instrument can totally match them. Hence the human eye remains the ultimate arbiter.

12.4.5 MANUFACTURE—COMPOUNDING

Plastics are coloured by compounding into them pigments or dyes. The former are more or less insoluble in plastic, and operate largely by scattering and reflection. Most are inorganic, notably cadmium sulphoselenides (yellows, reds, browns), titanium dioxide (white) and carbon black. A few are inorganic or organic-inorganic, notably the phthalocyanines (greens and blues). Dyes are soluble or partly soluble, and operate largely by absorption. They are complex organics, frequently containing nitrogen.

Compounding is usually done in two stages. First the colourant (mixture of pigment and/or dyes), sometimes wetted or emulsified in a carrier, is thoroughly mixed with the polymer (in granular or powder form). This mixture is then extruded, usually in a special compounding extruder.

The ultimate colour is significantly affected by the compounding process. Not only is the dispersion (at various aggregate levels) controlled by compounding, but extrusion temperatures can affect colour. It is important to choose a colourant that can tolerate the extrusion temperature needed for the polymer—a colourant for LDPE may not be suitable for PP.

12.4.6 FILM SURFACE

As already mentioned, film surface can affect colour. In addition, and more directly, surface gloss can add lustre to a film or product packaged in it. This applies to most retail packaging film, but for some other applications, e.g. photographic film or toy components, a dull or matt appearance is required with minimum gloss. See Chapter 10 for more detail on gloss, including methods of measurement.

12.4.7 OPACITY

For many applications, e.g. packaging of textiles and some foods, good clarity is essential; but for some applications the reverse, high-opacity, is required. These include mulching film, refuse bags and packaging of products sensitive to UV and visible radiation. Opacity is the inverse of transmission, and is measured in the same way.

12.4.8 ORNAMENTATION

For the majority of retail packaging applications, and some others, film is printed. There are many processes, from silk screen to offset lithography; and the process can be done on reel, or at a more finished stage. Details are given in Chapter 14; as far as organolepsis is concerned, the following aspects are relevant.

Components of printing media (usually solvents or carriers) may be odorous and hence impart odour either to the film or to the product packaged in it. Ink manufacturers are aware of the problem, and can usually supply formulations which do not give rise to tainting if the ink is correctly applied. However, if the print formulation is *not* correctly applied (e.g. too thickly, or with too much solvent), or insufficiently dried, baked or cured, problems can arise. This is particularly true for UV-cured inks, since little heat is available for driving off residual volatiles. Also, it is possible for over-baking to give odoriferous breakdown products.

12.5 HEARING

12.5.1 DIRECT EFFECTS

Sound effects are not of major importance, and no generalisations can be made; only specific examples given. Probably the most important favourable one is the crackle of film for retail textile packaging (e.g. shirts). Unfavourable examples include the noise from flapping building film.

In most instances, the only way to assess the property is on an ad hoc basis, but in general the more crystalline films (notably OPP) give the most noise, and the most amorphous (e.g. conventional LDPE) the least.

12.5.2 INDIRECT EFFECTS

For some foods, notably breakfast cereals ('Pop-n crackle'), fresh salad vegetables, and fun foods (popcorn, etc.) hearing is an important sensory property. It is also associated with texture (see below).

Although film packaging can affect the hearing properties of food in several ways, by far the main effect is due to water vapour transmission. Most of the relevant foods have maximum crackle when dry or at their optimum low water content; consequently, for these, very good water vapour barrier properties are needed. This almost always predicates the use of metal foil laminates, usually aluminium, although for fast moving products occasionally a high barrier plastic, e.g. PVDC coating may be acceptable.

12.6 TOUCH

12.6.1 DIRECT

Some consumer appeal or repugnance can be associated with surface movements or roughness. These are, of course, also major factors in processing (slip, block, etc.) and are often associated with antistatic properties. Often the same properties (e.g. non-stick) are required for both; but where contradictory properties are required, an optimum has to be sought between adequate processing and customer appeal.

Touch (as well as, of course, other properties) is substantially modified by fillers (chalk, talc, starch, etc.). Since these are being increasingly used for technological or cost reasons, the effects of touch (as well as, of course, visual) should not be overlooked.

12.6.2 INDIRECT

The indirect effects of touch are best regarded as a component of texture (see section 12.7.7 below).

12.7 TASTE AND SMELL

It is clear from the above that several sensory effects are appreciated together by the brain. This is especially so for the two senses of taste and smell, and not infrequently it is these that are called organoleptic (although, as made clear earlier, the term is applicable to all senses).

For the direct effects of film itself, only smell is relevant; this is also true for the indirect effects on a few packaged products, notably perfumes, cosmetics and toiletries. For a very few foods (sugar, salt) only taste is relevant, but for no food is only smell relevant. Apart from these instances, taste and smell act jointly and are usually tested together; methods for the two are very similar. In this section, therefore, taste and smell will be discussed together.

12.7.1 DIRECT EFFECTS

Several plastics, notably polyolefins, have characteristic smells. The effect is smaller for films than larger artefacts, because the larger area/mass ratio encourages loss of volatile microcontaminants. Nevertheless, the effects cannot be overlooked. Either the packager or the consumer may be deterred by smell, either for direct subjective reasons, or because he believes it may contaminate the packaged product. This fear is now enhanced by legislative requirements (see Chapter 11, p. 171).

12.7.2 INDIRECT EFFECTS

The majority of problems arise from consequential effects on sensitive products—mainly food, but also perfumes, cosmetics and toiletries, medical products; very rarely other products, such as textiles or toys.

The following discussion is oriented to indirect effects, but much of it, including all the methodology, is relevant to direct effects also.

12.7.3 BASICS OF TASTE AND SMELL

Taste is sensed by organelles on the tongue known as taste buds. They vary in sensitivity to different flavours, groups on different parts of the tongue responding either to a larger extent or wholly to different sensations. With certain exceptions, such as salt, sugar and acid, any foodstuff stimulates a range of taste effects which are signalled to the brain essentially as a spectrum. This spectrum or pattern is not analysed by the brain into its basic components (although these may well be recognised), but is essentially recognised by comparison with or relation to previously established reference patterns.

Smell is detected by the olfactory nervous system, which is located partly in the nose, and partly in the cavities at the rear of the throat. It is also signalled to the brain in the form of a spectrum or pattern. The sensitivity of the olfactory nerves is quite astounding; the peach moth, for example, can detect the aroma of its mate at a distance of several miles. In this, as in many other cases, it is evident that detection occurs at levels below those of chemical analysis.

Plastics films contacting food are not usually required to contribute to taste or smell of the food; on the contrary, it is usually required that they should *not* do so. If the taste/smell properties of the food are changed in any way, the result is almost invariably considered unfavourable; if the change is sufficiently unpleasant, the result is called 'off odour', and 'off flavour', or 'tainting'. For convenience, the term 'tainting' will be used to cover all cases of deterioration in smell/taste of the foodstuff due to the plastics film contacting it.

There are essentially two types of recognition by the brain of smell/taste patterns. The first type constitutes the instinctive or hereditary patterns, essentially defence mechanisms, which find certain primitive tastes and smells (e.g. flower perfumes, simple food tastes) attractive; and others, usually associated with putrefaction and decay (e.g. hydrogen sulphide), repulsive. These patterns generate Pavlovian or reflex reactions such as salivation, catching of the breath, or nausea. The unfavourable reactions in this class would be the epitomes of tainting.

The second type of appreciation depends on patterns established by personal experience since birth. Where these patterns contradict instinctive patterns, they are called 'acquired tastes'. There is an enormous variation in personal idiosyncratic behaviour in this context, but there is also a wide measure of conformity depending on social factors.

Historically, tainting was descriptive of food which had decomposed and was hence unsafe to consume; more recently it came to mean the *acquisition* of an unpleasant flavour from an external source. Most recently, however, tainting (and off-flavour) have become accepted as describing all cases where consumers show a significant reaction (usually unfavourable) to the smell/taste properties of the food; this leads to the definition given above. It is important to note that, whereas the earlier meanings of tainting implied knowledge of the cause (putrefaction of the food, or physical transfer of an odour from outside the food), modern technology is more modest.* Tainting is, indeed, often attributable to these reasons but there are sufficient others to make exclusion of these misleading.

So far, the discussion has been exclusively concerned with tainting of the foodstuff and this is obviously most important from the point of view of the consumer. However, although a tainted plastics film will usually pass on its taint to the foodstuff, this is not invariably so. Conversely, it is quite common for a plastics film with no detectable taint to impart taint to the foodstuff (this may be by loss of some component, or other interaction with the environment). But even though a tainted plastics film does not impart taint to the

*But not legislation which is confined to migration (see pp. 171 and 176).

foodstuff, it would present an unattractive product for the plastics merchant. Consequently, when studying and evaluating tainting, it is necessary to consider both tainting of the plastics film itself, and tainting of the foodstuff in contact with it. The norms for the foodstuff itself have been given above; the norm for the plastics film itself is is virtually zero taste, and a bland or slight smell.

To complete a statement of the problem, it must be added that (unlike in the study of toxicology) trials with experimental animals serve little purpose: correlation with human beings is negligible. It therefore follows that all studies must be on human populations related to actual specific situations. Fortunately, this is permissible, since no question of toxic or health hazard arises.* However, trials in this field present formidable problems since the human response, even within a closely defined sociological group, is very variable. Thus any trial on tainting must be with a carefully defined human population, on a specific (or at least standard) food contacting a specific plastics film under specific conditions of storage (temperature, time, aeration, etc.). The following inferences flow from the above:

(1) Although very accurate analysis can contribute towards the detection of already known patterns, there are no analytical procedures yet available which can function fully as analogues for organolepsis. All testing must be, basically, with human populations or panels.

(2) There are not, and in the foreseeable future cannot be, fully objective standards for organolepic properties and, consequently, for the effects of plastics films on these, i.e. tainting.

(3) No *a priori* rules can be laid down for the presence or otherwise of tainting, although guide lines may be established based on experience.

12.7.4 CAUSES OF TAINTING

Tainting has a similar mechanistic rationale to toxic hazard, in that it arises from interactions between the foodstuff, the plastics film, and the environment. The basic model for the situation has been described in Chapter 11 (q.v.); and *Table 11.3* lists the interactions which may affect health. *Table 12.1* gives the interactions which may also lead to tainting.

It should be noted that only two of these (Nos. 4 and 5) involve

*This must be previously confirmed

Table 12.1. INTERACTIONS RELEVANT TO TAINTING

Number (As in Chapter 11)	*Component*	*From*	*To*
2	Volatile	Foodstuff	Environment
4, 5	Volatiles and additives	Plastics film	Foodstuff
6	Vapour	Environment	Foodstuff
7	Micro-organisms	Environment	Foodstuff
8	Macro-organisms	Environment	Foodstuff
9	Radiation	Environment	Foodstuff

separate, prior, tainting of the film itself; the remainder are not directly connected with this.

We shall now consider each interaction in turn.

Loss of volatile material from food to environment

There are two distinct types of this interaction:

(1) Loss of flavouring ingredient;

(2) Loss of preservative.

(1) Where the first of these types occurs, it leads to loss of flavour, or to flatness. Although quite significant from a marketing point of view, it constitutes low levels of tainting. Theoretically the diffusion is calculable, but since the amounts concerned are usually quantitatively minute, the only practical method of evaluation is empirical.

Qualitatively, foodstuffs that depend on highly volatile components, e.g. flavours in potato crisps, biscuits or boiled sweets, oxygen in water, carbon dioxide in beverages, etc., are particularly susceptible.

(2) The main volatile preservatives in use are sulphur dioxide and ethylene oxide (there are others, e.g. diphenyl, but their use is highly specialised). Loss of the preservatives in significant amounts may lead to deterioration in the food, and hence to tainting. As mentioned in Chapter 11, however, the potential toxic or dietic effects are usually sufficiently serious to outweigh the effects of tainting.

Migration of volatiles, additives, and volatile residual reactants from plastic to food

This is the most important interaction that may lead to tainting. Its importance depends not so much in its frequency of occurrence, but on the facts that:

(1) it is peculiar to the plastics film, and independent of the environment (except insofar as temperature and other intensive properties of the environment control the physical condition of the film); and

(2) the consumer is currently especially sensitive to flavour changes deriving from plastics films because of their unfamiliarity compared with more conventional materials.

With rare exceptions, all high polymers* are tasteless and odourless; thus the majority component of all commercial plastics films (which fall into this category) will not give rise to tainting of any foodstuff. This is a remarkable generalisation that can only be made for glass and non-corroding metals, but not for many other packaging materials.

Volatiles liable to diffuse from the plastic to the food divide into (1) those residual from the manufacturing process (hence also including residual reactants); (2) those formed during the conversion process; (3) additives used to impart certain processing or end-use advantages.

(1) *Volatiles from the plastics manufacturing process.* The biggest single group of compounds in this category comprises monomers and their oligomers with molecular weights up to around 1000. In the polyolefins these components are paraffins or olefins but in PVC and PVDC, chlorine compounds also occur. Polyamides contain monomers, oligomers, and other nitrogen compounds. At the higher molecular weight end of the spectrum, the paraffinic component predominates, hence tainting is 'oily' or 'paraffinic' in nature. Nearer to the monomer, tainting will correspond identifiably with the monomer, e.g. ethylene for polyethylene, styrene for polystyrene, etc.

A variety of other materials may also be present, deriving from the polymerisation process. The most common amongst those likely to cause tainting are oxidation products—aldehydes, ketones, and carboxylic or hydroxy acids—especially if these contain unsatura-

* This refers to the truly high molecular weight components—say above 1000. Sometimes plastics as a whole are described as high molecular weight (or macromolecular); this refers to the average and not each species.

tion elsewhere in the molecule. Another major group of impurities comprises residues of catalysts or polymerisation processing aids used with thermoplastics. Generally speaking all these materials finish up at very low levels, and either in the form of simple gases which disappear from the product on processing, or as metallic oxides or salts which do not migrate (and are hence unable to taint by taste) and are odourless. Important exceptions are organic nitrogen blowing agents in expanded polystyrene, solvents, and organic catalysts.

Thermosets rarely occur in film form, but are occasionally used as coatings. The residues may include the original reactants (phenol and cresol are the most likely to cause tainting but others such as ethylene oxide, low molecular weight epoxies, toluene di-ioscyanate may also be present at very low levels), impurities in them (e.g. epichlorhydrin), catalysts (e.g. diamines), accelerators, inhibitors or hardeners.

(2) *Degradation products formed during conversion.* These are similar to those mentioned above arising from polymerisation. Most plastics decompose slightly on heating. In a few cases, notably polystyrene and nylon-6, the main reaction is depolymerisation, and the product is monomer or oligomer. In the majority of cases the products are not those which would be obvious.

Polyolefins give a range of oxidation products, including formaldehyde and acrolein. Polypropylene is particularly susceptible to decomposition, giving under severe conditions (which should not occur in normal processing) quite significant amounts of aldehydes, ketones and acids down to acetic and formic. Polyethylenes tend also to cross-link thus reducing decomposition, but the smell of overheated polyethylene is well known, deriving from oxygenated compounds and tarry residues. (Polyethylene made by the high pressure process may also be tainted by decomposed compressor lubricating oil.)

PVC is extremely sensitive to overheating, and can easily decompose to the point of charring. Polystyrene is relatively stable, but, even so, usually evolves 0·01–0·02% styrene monomer on extrusion. Nylon-6 tends towards its equilibrium of 9·0% monomer, but this is readily removed by water washing, and does not normally cause tainting. Polyethylene terephthalate (PETP) and related thermoplastic polyesters decompose to a variety of products, of which by far the most important with regard to tainting is acetaldehyde. A careful time/temperature profile must be maintained during extrusion to keep the acetaldehyde below the tainting level. (The cola industry has set up a specification of

3 μg/litre, but for other foods, typical threshold levels range up to around 60 ppb.)

(3) *Additives.* Additives include such a wide range of chemical types that no generalisations are valid beyond those flowing simply from the formulae of the compounds concerned. That is to say: sulphur containing compounds (e.g. many antioxidants) may give rise to mercaptans; phenolic compounds (antioxidants and ultra-violet stabilisers) may produce quinones and hence their characteristic odours; amine derivatives (antioxidants and antistatics) may produce fishy odours. In general, each additive must be considered, together with its probable spectrum of thermal and oxidative degradation products as well as products formed by interaction either with the polymer itself or other additives.

Vapour from environment to food

The importance of this interaction increases as the thickness of the plastics barrier decreases. Consequently the problem is negligible with moulded objects and is greatest for plastics films, especially thin films.

Together with film thickness, the other major factor is the affinity of the foodstuff for the migrant odour. Thus lipid foods, notably butter, margarine, cheese and chocolate are prone to tainting by hydrocarbons such as kerosene or gasoline, or other lyophilic materials, notably rancid oils and fats, mercaptans, and amines from fish. Aqueous foods, such as fruit juices, milk, cut fruits, are more prone to tainting by hydrophilic materials including ammonia from faulty refrigerators, sulphur compounds from atmospheric pollution, and phenolic compounds from a variety of sources (e.g. overheating electrical insulation and wood preservatives).

To all such tainting influences, plastics films present a significant but not total barrier, because of their vapour permeabilities.

Although not strictly part of the environment, it is convenient to include in this category tainting by materials on the external surface of the film itself. Such tainting is almost entirely confined to the components of printing inks or to external packaging materials (overpacking). In principle, the mechanism is the same as given above for general tainting from the environment, with the qualification that, because of the high concentration and close proximity of the potential tainting material, the problem—where it occurs—is likely to be particularly serious.

External packaging seldom presents a problem, but occasionally strong smelling wood or cardboard is encountered.

Micro-organisms to food

In general, in the absence of any protection, this is a major source of tainting of foodstuffs. There is no way in which plastics films can directly promote such tainting; on the contrary, their function as a cover or packaging is normally designed, inter alia, to prevent ingress of micro-organisms. However, (a) micro-organisms may enter through inadequate closures, or directly into open systems; and, (b) micro-organisms may already be present in foodstuffs which have not been sterilised. As discussed in Chapter 11, this interaction is likely to be accompanied by toxic degradation; where this occurs, it is of far greater importance and, in overcoming it, the relatively less important tainting effect may automatically also be remedied.

Macro-organisms to food

Theoretically, the effects are similar to those caused by micro-organisms. However, other effects, e.g. spoilage, loss of quality, toxicity, etc., are so much more important that little study of tainting from this cause is called for.

Radiation from environment to foodstuff

As already discussed in connection with toxicity (Chapter 11, q.v.) unpigmented plastics films can transmit sufficient radiation to affect foodstuff quality. Where organoleptic changes occur, they are almost invariably associated with other major changes in food quality. Tainting effects are relatively minor, leading to bland and stale taints. The main example is the destruction by UV radiation of riboflavin in milk stored out-of-doors; this leads to a so-called 'chalky' taint. Destruction of Vitamin C, also by ultra-violet radiation, can lead to an insipid taste in milk or other products, e.g. fruit juices.

12.7.5 ASSESSMENT

As already discussed, a plastics film may have an odour (rarely, also, a taste), which may, or may not, be imparted to the food. Irrespective of its so doing, such film would be unattractive for sale.

Contrariwise, food may become tainted in contact with film which has no detectable taint in isolation. Thus in assessing the

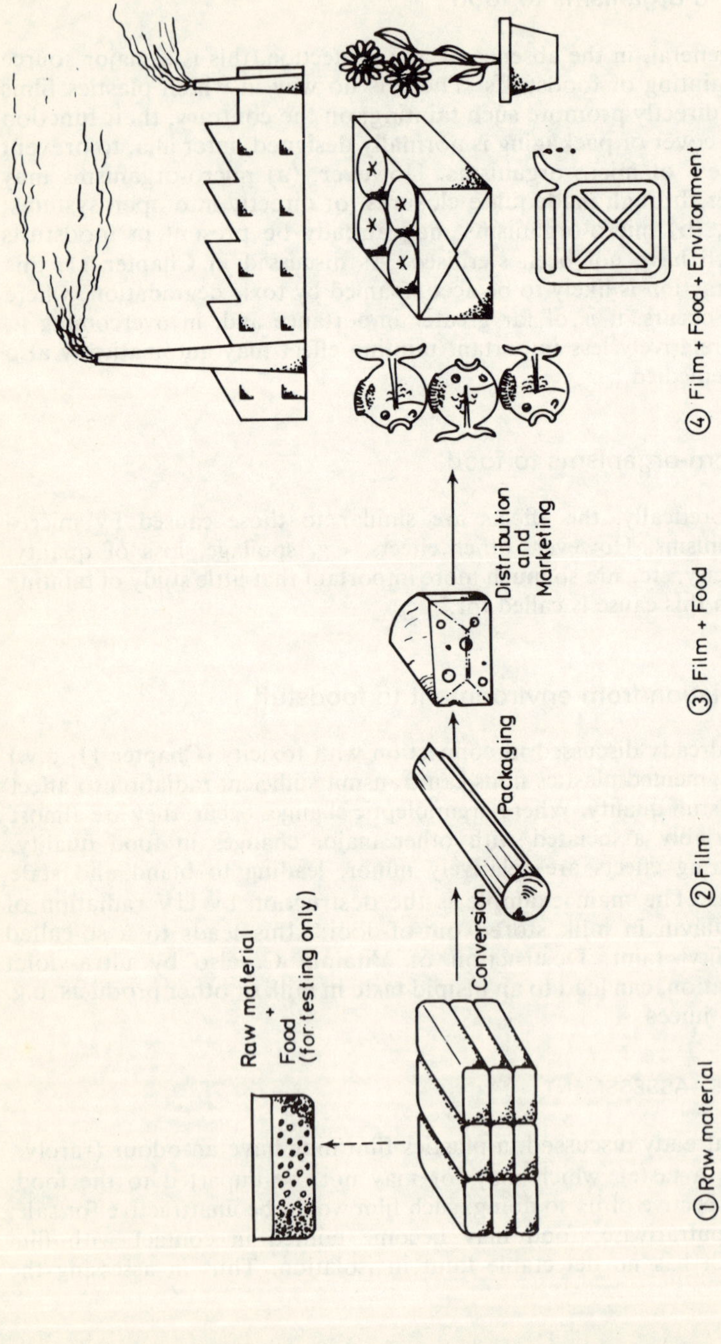

Figure 12.1. *Sources of tainting*

origins of tainting it is necessary to examine the whole supply chain, as shown in *Figure 12.1*. It will be seen that tainting may arise at any one of the four stages:

(1) Raw material (i.e. plastic) alone.

(2) Finished film alone.

(3) Finished film plus food.

(4) Finished film plus food in contact with environment.

Samples

(a) *Plastics*. The choice of samples to test will depend on whether the objective of the assessment is

(1) to confirm that a film/foodstuff/environment situation is satisfactory; or

(2) where tainting *has* occurred, to determine its cause.

In the former case, it is usually sufficient to examine a number of samples at Stage 3, i.e. the finished film package containing the food.

In the latter case, samples should be examined at Stages 4, 3, 2, in that order. Since tainting is already known to take place, results from Stage 4 will already be available. Using a testing panel (see below: possibly an expert testing panel in this case) it may be possible to identify the cause of the tainting from examination of the samples. Also, as mentioned above, it may be possible to locate it from an actual study of storage conditions.

If, however, the cause cannot be isolated at Stage 4, tests on Stages 3 and 2 must then be done.

If these tests do not lead to the cause of tainting, tests must then be continued at Stage 1, i.e. on the raw material alone. For the reasons mentioned above, this test itself must be in two parts, namely: on the raw material in isolation, and on the raw material in contact with the food (not necessarily the actual food; standard foods can be used as described below).

If tainting is found at Stage 2 but not Stage 1, then it must be due to the film manufacturing process.

For studies of Stages 2, 3 and 4, part of the film as used is most appropriate. A minimum of 10 dm² surface is appropriate. For studying Stage 1 (raw material) samples should be pressed into the form of sheet or film using the minimum of heat and taking great care to avoid all contamination. If this cannot be done, granules are preferable to powder because of the likelihood otherwise of substantial contamination by absorption.

Stages 2, 3 and 4 of sampling should be carried out using standard statistical methods for quality control. Individual samples should then be taken from different parts of the package and randomised. Samples available at Stage 1 will normally already be homogeneous.

(b) *Food.* As discussed above, tainting is a somewhat subjective sensation, and often quite specific to a given food. Where possible, therefore, practical trials should also be carried out using the food concerned. However, where there are difficulties in doing so, e.g. because of the unstable nature of the food, or where model trials are to be carried out, a selection of relatively susceptible, representative, and stable foods can be used. These usually comprise water; butter (fresh, unsalted, and taint free); chocolate; and sugar.

Panels

As already discussed, no mechanical equipment yet exists which can be used reliably for odour/taste testing. Also, although animals can occasionally be used for special cases, they are not suitable for testing of plastics tainting. Consequently, human groups must be used; and the possibilities are as follows:

(1) Whole population;
(2) Statistical sample of whole population;
(3) Random panel;
(4) Selected panel;
(5) Expert panel;
(6) Expert individual.

Methods (1) and (2) constitute Market Research; (3) is not recommended. Methods (5) and (6) are used in special situations, particularly where it is hoped to identify a specific component. The recommended method for assessment of tainting by plastic films is (4), selected panels. A typical basis for the selection is given in British Standard 3755 'The test panel should consist of individuals who normally give consistent levels of odour and taint assessment, and before they are appointed to the Panel, some form of test should be carried out to ascertain whether the individual concerned is suitable. In general, the individuals of a test panel should agree with the average of the entire panel and consistent disagreement is a basis for replacement. Any members with respiratory infection should be omitted, since sensitivity is thereby impaired'.

Test methodology

(a) *Identification of stimuli.* It will be clear from the above that tainting tests for plastics films may have one or both of two immediate objectives, namely (i) to determine whether a tainting problem exists, and (ii) to identify the source of tainting.

For the latter it is necessary for the panel to give an indication of the nature of the tainting; and though not *prima facie* essential for the former objective, it is still desirable as assisting in scientific control. Hence, in selecting individuals for a panel, their sensory reactions are checked against identification of specified stimuli.

For tastc, thc usual sensory effects are: sweet; sour; salt; bitter (standard solutions for tasting are available for each). For smell (as well as other taste effects) a 'vocabulary' or 'thesaurus' is usually established including terminology such as aromatic, green, almond, musty, etc. as well as smell or taste, effects directly related to chemical stimuli, such as ammonia, styrene, phenolic, aldehyde, etc.

(b) *Panel selection.* The basic principles have been outlined above; to these must be added that the individuals for the panel must be chosen from the same sociological group as the ultimate consumers. For statistical sampling, the panel should be as large as possible but in practice, of course the size has to be restricted because expense, difficulty in assembly, absenteeism, etc., increase rapidly with size. The British Standard suggests 2, 3, 4, 6, or 12; but 12 is really the minimum, and 15 or 18 the preferred number. This implies a group of 25–30 people on call, so that 15 or 16 will be available at all times.

It is best if the candidates for the panel can be nominated—this avoids the inherent bias of volunteers. It is almost equally satisfactory, however, to select at random people from a much larger group of volunteers (say 250).

The fifty or so preliminary candidates should be screened for their ability to detect and identify the four basic flavours already mentioned. The procedure is as follows.

A series of groups of test samples should be prepared, each group containing ten samples, each of which has double the concentration of stimulus material of the previous sample. Thus, numbering the samples in a group 1–10, concentrations will be in the ratio 1, 2, 4 . . .512 (this is because, in the basic law of psychophysics—scc scction 12.2—the relation between sensory effect and stimulus is logarithmic). To some extent, the test samples can be tailored to the specific end-use. There are several Standards for these, e.g. ISO 3972:1979. For the present purpose, an appropriate set of tests would include four aqueous solutes largely or entirely detected by taste (e.g. common salt, sucrose, caffeine) and four vapours largely or entirely detected by smell (acetone, styrene monomer, ammonia, etc.).

The solutions are tasted in the case of liquids, or smelt in the case of the vapours, in ascending concentration until a sample is (a) detected, and (b) identified. The result is recorded by sample number, which automatically measures the response on a logarithmic scale. Several criteria can be adopted to select panels. The simplest is merely to select those candidates with the lowest total scores; these would be those who are on average most sensitive. However, as already explained, the required choice is of a *selected* panel and not an *expert* panel. Consequently, it is desired to choose those candidates nearest to the panel median, and to reject those deviating substantially. The following procedure is usual:

(1) Reject all candidates whose identification is poor, i.e. those whose detection and identification samples are further apart than adjacent.

(2) Reject those individuals whose score for any one single test lies outside the 90% confidence level for all the candidates' results for that particular test.

(3) Reject candidates whose total scores lie outside the 95% confidence level for the corresponding scores of the candidates as a whole.

(4) Accept those that remain as the Panel.

(c) *Threshold and fatigue.* It is obvious (and implicit in the method of panel selection already described) that there is, for any given sensory effect, a threshold below which the effect cannot be detected. That is to say, a certain minimum concentration for taste, or minimum concentration for smell, is required. This may be very low indeed, but it must have some finite value. This is the *threshold*, and varies not only from material to material but also from individual to individual. In the same way that some sounds can be detected against the background noise, and others can not, the threshold value for a given material may be much higher when it is present in a mixture.

An even greater change in threshold takes place with time. This is because the olfactory mechanisms become rapidly fatigued, and cease to detect the effect. The process can be looked at as one of rapid adaptation to an environment, or as the more physical process of saturation of the parts of the body in immediate contact with the stimulus. Whatever the actual mechanism, the effect is very large and very rapid: in most cases, the threshold increases by orders of magnitude above 100, and within a few seconds.

Figure 12.2 illustrates the effect in terms of time, showing a

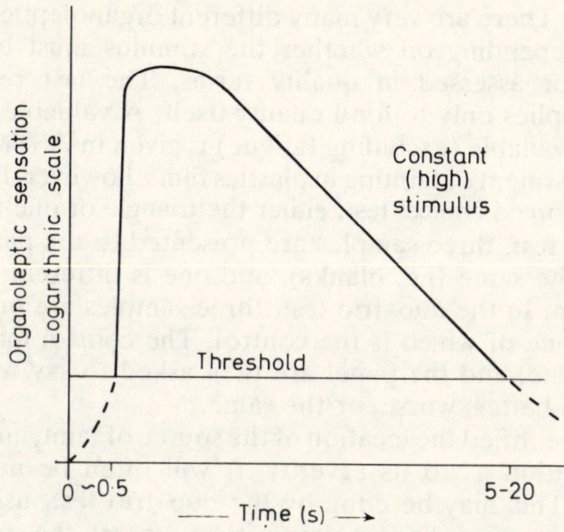

Figure 12.2

typical sensory reaction to a given (relatively high) stimulus. This figure also shows the induction period, which is typical, and analogous to other animal reactions. It follows that individual sample testing, whether by trained or untrained personnel, must be carried out in short sniffs, or tastes. Sufficient time must be allowed between each test for the tester to recover his initial threshold value.

It also follows that, if the taint is close to the average threshold value, many testers will pronounce there to be zero taint. After a short while of testing, all testers will confirm this. Since, however, the ultimate consumer is probably sensing the taint in a situation where fatigue to the taint cannot have taken place, his will be the most sensitive response.

The phenomenon of fatigue can, on the other hand, be used as an effective method of identification. The procedure is as follows. Let us suppose that a taint sample contains compounds A, B, C, . . ., X, . . ., etc.; and that the taint is in fact due to compound X. The panel are first exposed to a high concentration of compound A, such that they are all fatigued to it. Reverting to the original sample, they will find no change. Carrying out the same procedure fatigued with B, C, etc., they will still find no change. However, when they have been fatigued with compound X, they will report either that the taint has disappeared, or that the odour or taste is dramatically different. In either case, the taint will have been identified as due to compound X.

(d) *Testing*. There are very many different organoleptic test forms available, depending on whether the stimulus must be located, identified, or assessed in quality terms. The last requirement normally applies only to food quality itself. A valuable resumé of test forms available (excluding fatigue) is given in ASTM STP434. For the assessment of tainting in plastics films, however, it is usual to use only a 'forced choice' test, either the triangle or due-trio test. In the triangle test, three samples are presented to the panel, two of which are the same (i.e. blanks), and one is different, i.e. under investigation. In the duo–trio test, three samples are presented to the panel, one of which is the control. The control may have an odour or taste, and the panel are then asked to say whether the samples are better, worse, or the same.

Having identified the location of the source of taint, and obtained some indication as to its severity, it will often be necessary to identify it. This may be done by the duo–trio test, using control samples containing known impurities; or by the fatigue test described above. These methods are often successful, but they require a shrewd idea concerning the possible cause of the taint.

Where this is not known or suspected, the most sophisticated analytical methods have to be used to detect micro-contaminants present, possibly at concentration levels down to 10^{-12} or 10^{-13}. Any of the most sensitive techniques, from mass spectrometry to HPLC and fluorescence, may need to be used. The only generalisation that can be made is that preliminary concentration may be helpful. If this is carried out by a procedure involving heating or volatilisation, great care must be exercised to ensure that the odoriferous agent is not lost, since it is most likely to be volatile and possibly unstable.

The one technique that can be always recommended is head space concentration, taking care to ensure as near 100% recovery as possible. (This can be checked using a volatile internal standard.)

(e) *Combined panel/analysis test*. In a procedure unique to odour analysis, the sensory effects noted by a panel are correlated with stimuli identified from analysis. The sample in suitable form (i.e. solvent extract, head space concentrate, etc.) is passed through a chromatograph (usually HPLC, but other types can be used or even other analytical methods providing the sample is separated into components which appear at different times or places). The panel members sniff the effluent from the instrument, and their sensory reaction related to the concurrent instrumental readings, e.g. peaks on chromatograph, fractions on mass spectrometer, readings on FID, fluorescence, etc.; as shown in *Figure 12.3*.

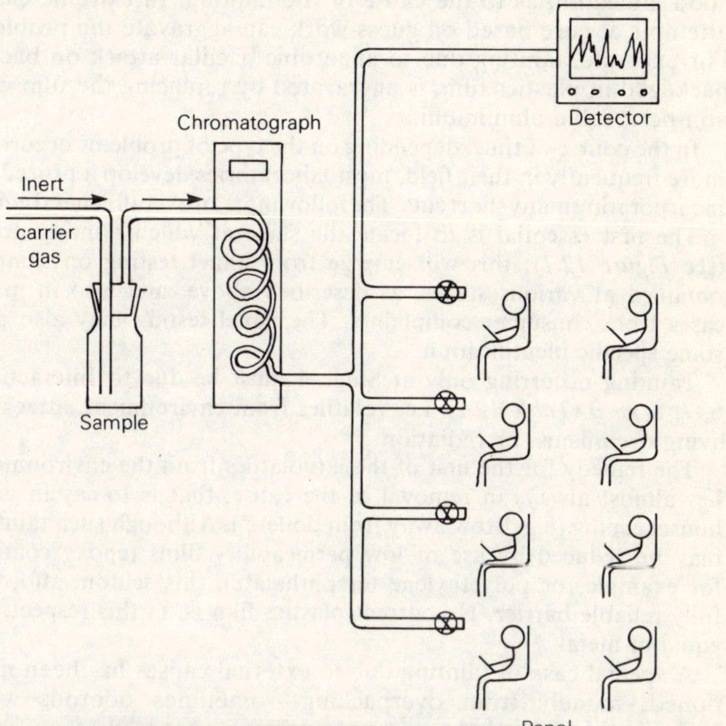

Figure 12.3. Combined panel/analysis test

The procedure has the great advantage of direct correlation. It presents problems, however, where the compound causing odour does not give a single identifiable trace.* In particular, many compounds give a series of peaks on a chromatograph, or several fractions on a mass spectrometer. Also some odours are themselves mixtures of chemical species. In both cases it may be possible to identify the compound by computer analysis. The procedure is a useful tool and has been successfully used where all other methods have failed. It is especially valuable where the offending compound is novel, or at least not well known or suspected.

*Also, it is essential to confirm in advance, absence of toxic hazard.

12.7.6 REMEDIES

It is almost impossible to deal with a tainting problem associated with a plastics film unless tests (as described above) have given a good indication as to the cause of the tainting. In extreme cases, attempts at cure based on guess-work can aggravate the problem. For example, tainting due to anaerobic bacillar attack on bacon, packaged in plastics film, is aggravated by replacing the film with non-permeable aluminium.

In the course of time, depending on the type of problems occurring more frequently in their field, most laboratories develop a procedure incorporating many short cuts. The following is an overall general guide.

The first essential is to locate the stage at which tainting arises (see *Figure 12.1*); this will emerge from panel testing on samples obtained at various stages, as described above, and also in many cases from consumer complaints. The panel testing may also give some specific identification.

Tainting occurring only at Stage 4 must be due to Interactions 6, 7, 8, or 9 (*Table 12.1*), i.e. volatiles from environment, attack by living organisms, or radiation.

The remedy for the first of these (volatiles from the environment) lies almost always in removal of the cause, that is to say in good housekeeping (e.g. 'stow away from boilers'). Although such tainting may be reduced by use of low permeability films (epoxy coating, for example, or polyethylene terephthalate), this seldom affords a fully reliable barrier. No current plastics film is, in this respect, the equal of metal.

A special case of tainting due to external causes has been mentioned, namely from overpacking—sometimes odorous wood (notably chlorinated phenol preservative), or paper products, more commonly printing inks. Again the remedy is removal of the cause. Inks must be chosen with tainting in mind; see section 12.4.8.

Attack by living organisms of external origin can be dealt with by ensuring adequate closure of packages. As already mentioned, the effects on toxicity and nutritive value are usually so great that they must be remedied and, in so doing, the consequential tainting is also remedied.

If tainting steps in at Stage 3, the relevant interactions are 2, 4, and 5, i.e. loss of volatiles from the food, or tainting by odoriferous components in the plastics film itself (but not present in the raw material).

With regard to the former (volatiles from food), if the material concerned is present in substantial amounts, e.g. water, it may be possible to improve the situation by choosing a film with a lower permeability. Where flavours are concerned, however, very low permeability films, e.g. coated cellulose or metallic laminates are virtually essential unless very short storage periods are acceptable.

Where tainting is attributed to diffusion of odoriferous material from film to food (and this does not occur at Stage 2), one should look first to processing conditions or manufacturing procedures. The most important causes are overheating during extrusion of the film, or contamination, e.g. by lubricating oils. There are, of course, many other possible odoriferous materials that may taint the film during processing or storage; in all cases the remedy is to remove the cause after identification.

If tainting occurs initially at Stage 2, similar remarks apply: the only difference being that the taint is strong enough to be detected even in the absence of food. The film itself is likely to be unsaleable.

If tainting can be detected at Stage 1 (raw material), then the cause is clearly established as being in the raw material. A description of possible troublesome components has been given already; once again, the remedy is their reduction or elimination. In some cases this is readily possible, e.g. changing an anti-oxidant, or ensuring low monomer content. Sometimes a catalyst change is called for, which may be feasible. Sometimes the cause is an essential catalyst, or essential processing chemical. In such—relatively rare—cases, the only remedy is to choose a different raw material altogether.

Masking and counteraction

In food itself, and even more so in cosmetics, toiletries, and hygiene products, it is not always possible to remove the cause of an odour. The problem may sometimes be remedied by masking or counteraction.

The difference between the two procedures is significant. Counteraction is the neutralisation of one odour by another: as a result of a biological mechanism not understood, the human nose detects little or no smell from mixtures (in the correct ratio) of, e.g. rubber or certain waxes and Balsam of Tolu; rubber and kerosene; pyridine and methyl salicylate; coffee and iodoform; etc.

Masking comprises the drowning of a weak odour by a strong one. Normally, of course, the strong odour is relatively pleasant and conceals a weaker unpleasant odour. There are commercially available strong, relatively pleasant, odours, e.g. 'Alamask', or oil of lemon.

Either or both methods could be used to remedy tainting from plastics films. For counteraction it would be necessary to identify the cause of taint with some precision, and the likelihood of then finding a counteractant would not be good. Masking does not call for knowledge of the cause of taint; it is a crude 'blunderbuss' method. Provided a masking agent can be found which imparts an acceptable odour to the packaged product the method is feasible. In actual practice, the method is very rarely used because of the consumer's very critical reaction to unfamiliar odours.

12.7.7 CONCLUSIONS

It is important not to deduce, even from carefully controlled experiments, generalisations wider than those encompassed by the trials. Tainting is often specific to individual foods; and a film which will taint one food may be totally satisfactory for use with another and vice versa. This reverts to a major point discussed at the beginning of this chapter, namely, that tainting is very much a subjective sensation. It must be related to the accepted standards for the food concerned. Prime examples of this are canned salmon whose taint (compared with fresh salmon) is so widely accepted that a deviation from it would be regarded as tainting in its own right; and orange squash packaging in polystyrene containers, where traces of styrene monomer are considered to enhance the flavour.

The chain of supply has been shown in *Figure 12.1*. This involves three parties at least; raw material manufacturer, converter, and food packager. Each has confidential data necessary for the evaluation of tainting (raw material formulations, film manufacturer's operating conditions and additives, composition of foodstuffs including preservatives and handling conditions). Any serious study of a tainting problem therefore has to involve all parties, preferably under a tripartite secrecy agreement.

12.8 FOOD TEXTURE

A conglomeration of sensory impressions, mainly visual but also including touch and even hearing, join in giving the overall hedonistic impression of food texture. For practically all solids or semi-solid foods, this is vitally important for consumer appeal and marketing, independent of health safety or nutritional value.

The differences between the same basic foods processed in different ways are obvious—sausages and hamburgers, *vs.* steak; but we are here referring to changes in a specific already finalised food. Examples include softening of cornflakes, agglomeration of instant coffee, loss of crispness of vegetables (an extreme example of which, admittedly not directly relevant to packaging film, is deep-frozen cucumber, which is inedible).

The subject is now a branch of psychophysics (or food technology), christened psychorheology. Instruments have been developed to measure strength, or weakness (required for breakfast cereals) of solid foods, fluid properties (e.g. viscosity, gooiness), etc.

The effects of film packaging may arise from external factors such as vapour transmission (e.g. increase in, or loss of, water vapour) or radiation transmission; or internal. The latter includes retention of vapour which would otherwise be lost, and mechanical compression, e.g. due to shrink wrapping.

It is clearly not possible to give general guidelines. Whilst some predictions could be made for single instances, e.g. loss of rigidity of breakfast cereals due to water vapour, it is usually necessary to test the specific packaging film with the specific food and assess the effects.

Despite the development of test equipment mentioned, the ultimate criterion has a major subjective element.

FURTHER READING

ADCOCK, L. H. and PEACOCK, B. W., 'Odour and Printing', *Packaging Technology*, 69 (May 1971)

AHMED, M., *Colouring of Plastics, Theory and Practice*, Van Nostrand Reinhold, New York (1979)

A.S.T.M.—E462(30), 'Sensory and Taste Transfer from Packaging Film'

A.S.T.M.—STP 434, 'Manual on Sensory Testing Methods'

BIRCH et al. (Eds.), *Sensory Properties of Food*, Applied Science Publishers, London (1977)

B.S.I.—PD 6459: 1971 'Guidance on Avoiding Odour from Packaging Materials'

B.S.S.—3755: 1964 'Methods of Test for the Assessment of Odour from Packaging Materials used for Foodstuffs' (Under revision)

JOHNSON, E. C., 'Combine GLC with Taste Panels for Flavour Formulations and Assessments', *Food Manufacture* (January 1970)

KANE, M. R. and MALLER, O., *The Chemical Senses and Nutrition*, Academic Press, New York (1977) (Nutrition Foundation Monograph)

MEYER, J. A., 'Identification of Odours in Plastics', Paper presented at SPE Meeting on Plastics in Packaging, 13–15 November 1978, Chicago

MONCRIEFF, R. W., *Odours*, Heinemann, London (1970)

NIEBERGALL, H., HUMEID, A. and BLOCHL, W., 'Aroma Permeability of Packaging Films and its Determination' Part I, Lebensmittel-Wissenscaft und Technologie, 11(1), 1–4 (1978)

13

Choice Criteria

Because there is no single film that will do every job it becomes necessary to make a choice. The best way to look at material choice is to consider the various properties that may be required for a particular end-use and then to look at the various materials available. Even with the wide range of materials available, however, it is often not possible to find one with all the properties required for a particular end-use. The answer lies in the use of composite materials formed by lamination or coating (see Chapter 19) and it is usually possible to obtain a composite film with almost any combination of desired properties, especially where aluminium foil or paper are considered as possible components. There are now thousands of composite films commercially available, in addition to the basic films, so that the choice facing the potential user can be a bewildering one.

The point at which any investigation must start is the question, 'What are the requirements for the envisaged end-use?' Once these have been determined, the choice of film (or laminate components) can be narrowed down to a relatively small number. The requirements for films to be used in building, agriculture and horticulture, etc., are usually straightforward compared with those for packaging materials so that the subject of packaging material choice will be dealt with here. The principles will, however, hold true in other end-uses.

Some of the main properties that have to be considered when looking at the various packaging end-uses are outlined below, followed by consideration of the main materials that can supply a particular property.

13.1 PRODUCT REQUIREMENTS

The product requirements can best be assessed by compiling a check-list of questions about the product. The following series of questions are given as a guide, together with some comments on the points raised.

(1) Is a transparent or opaque package necessary? Opacity may be considered necessary because of the sensitivity of the contents to visible or ultra-violet light or because the contents are not particularly prepossessing. If baked beans were to be packed in a pouch, for instance, it is unlikely that a transparent pack would be required! If the main cause of any degradation is ultra-violet light then it is possible to use a transparent coloured film or a transparent film in which ultra-violet absorbers have been incorporated. Where visible light has to be excluded, an opaque ply must be added. Degradation by ultra-violet or visible light may simply cause discolouration of the product but it can also cause chemical deterioration, with potential hazards, when the product is a foodstuff or a pharmaceutical product.

On the other hand, some contents do add to the sales appeal of the package and transparency may thus be considered necessary. Transparency is also useful, sometimes, to show the type or quantity of the contents. In these instances it is usually possible to produce a transparent laminate that fulfils any other requirements such as strength or barrier properties.

(2) Are the contents sensitive to oxygen pick-up? Many food products, including biscuits, milk powder, potato crisps and other snack items, contain fats which are sensitive to pick-up of oxygen, with consequent rancidity, giving off-flavours and off-odours. It is obviously essential in these instances to incorporate an effective oxygen barrier in the laminate.

In some cases it is so important to keep the food out of contact with oxygen that vacuum packaging or gas flushing are used. Vacuum packaging increases the pressure differential between the inside of the pouch and the external atmosphere and an even more efficient gas barrier is thus required. The stresses on the pouch and on the seals are also greater and these, too, will affect the choice of material. The use of gas flushing usually entails replacement of the normal atmosphere in the pouch by carbon dioxide or nitrogen and this means that barriers to these gases are also required. Fortunately, good oxygen barriers are usually also good barriers to carbon dioxide and nitrogen, thus simplifying the problem. The question of seal integrity is particularly important so that a material having good heat seal properties is required.

(3) Are the contents hygroscopic or are they sensitive to drying out? In both these instances a good moisture vapour barrier is essential and seal integrity is again an important factor. Biscuits and snack items are examples of products requiring protection from moisture pick-up while cigarettes and tobacco are liable to dry out and become stale.

(4) Do the contents of the package give off gases? An affirmative answer may mean the use of a vented package as even films with high transmission rates are not usually able to transmit large quantities of gas because of the relatively small surface areas involved. The simplest way to vent a package is to punch a few small holes in the laminate or film, but this is a two-way process and if the entry of gases or moisture vapour is to be prevented then some form of non-return valve may be necessary.

(5) Is mechanical protection required? This is important when fragile items are to be packaged. The pouch or sachet then needs to be stiff and rigid in order to give some sort of mechanical protection.

(6) Is odour or flavour loss (or pick-up) likely to be a problem? There are many products that pose problems of this sort. Perfumes, or cosmetic products containing perfumes, need an odour barrier to prevent perfume loss, not just to prevent a weakening of the perfume but also because the residual perfume may be considerably changed in nature. Perfumes are complicated mixtures of a large number of different compounds, with a wide range of volatilities. These compounds are painstakingly blended to give the desired result and the preferential loss of one or two components can completely ruin the overall effect. Highly flavoured foods require an effective flavour barrier for similar reasons. With goods having a rather more bland flavour the problem is likely to be pick-up of odours or flavours from outside the package but the requirements of the film or laminate in terms of barrier properties are the same.

(7) Does the product require to be refrigerated during its distribution and storage? If refrigeration, particularly deep freeze, is required then components with good resistance to low temperature must be used.

(8) Does the product have to be processed in the package? This can entail sterilisation immediately after packaging or heating by the customer, as in boil-in-the-bag foods. A heat-resistant material is hence required.

(9) Is there likely to be any interaction between the product and the package components? There are many possible ways in which a packaging material can affect the product. Products having a high grease content can affect certain films such as polyethylene so that if such a film is required for its contribution to other properties it must

be kept out of actual contact with the contents of the finished pack. Similar remarks apply to the use of untreated papers. Another possible chemical interaction is that between certain acids and alkalis and aluminium foil.

(10) What constitutes unsaleability for the particular product? This is sometimes very difficult to answer but it has an important bearing on the package selection. The trouble is that unsaleability may be due to one or more factors including changes in flavour, moisture content, texture and appearance. Since package cost is usually of extreme importance it is essential to pinpoint the most important factor in unsaleability and protect against that factor rather than spend too much money on other component plies designed to protect against things of far less importance. Is it permissible, for instance, for a slight loss of flavour to occur provided that the contents remain crisp? Conversely, would a slight loss of crispness be considered satisfactory provided that the flavour remained perfect? A different type of barrier material would be necessary for each of the two cases posed above.

(11) What length of shelf life is required? This is a most important question and should be answered realistically. Of course it is dangerous to take too low an estimate and so risk large-scale deterioration of the contents. It is equally wrong to take too high an estimate and risk pricing the product out of the market by paying for protection that is not needed.

13.2 MARKETING FACTORS

More information that will be of help in the choice of material can be gathered by compiling a check-list of questions concerned with marketing factors.

(1) What climatic conditions are likely to be encountered by the package? This, in conjunction with the nature of the contents, will have a large bearing on the barrier properties required.

(2) What system of distribution will be used? This may have a bearing on the barrier properties required and will certainly dictate the mechanical properties of the package.

(3) What type of retail outlet will the package encounter and are there any special display requirements? The latter may dictate the rigidity of the package.

(4) What package sizes are envisaged? The size of a package is

relevant not only because of rigidity requirements but also because the surface area to volume ratios affect the permeation of gases and water vapour.

(5) Are easy opening features to be incorporated? This may affect the overall protection given by the package and may also affect the choice of film in terms of performance of the opening device. A tear tape fitment, for example, does not work very well if the film has a high resistance to tear propagation.

(6) What are the printing requirements? If scuff-proof printing is required, for example, this can be achieved by reverse printing a transparent film, such as cellulose acetate, and laminating this to a sub-strate giving whatever basic properties are required. Special effects can be achieved using aluminium foil. The high reflectivity of the shiny side of the foil is very effective in conjunction with plain or coloured transparent lacquers, while the soft satiny look of the matt side can be equally effective in enhancing the quality of the printing.

13.3 MACHINE REQUIREMENTS

The behaviour of the web on various packaging machines is also important and must be taken into account when choosing a film or laminate. Where twist wrapping is required, for instance, then paper or regenerated cellulose are still the best materials. Most plastics films retain enough 'elastic memory' to untwist again to a great extent. Aluminium foil also retains a 'twist' very well but is best laminated to paper or coated with polyethylene in order to increase its flexibility and tear strength. A similar situation arises when the web is required as a wrapper. Aluminium foil has extremely good 'dead fold' properties—i.e. it will retain the fold and does not need to be secured. Paper and regenerated cellulose are less good in this respect but are still satisfactory for most uses. Plastics films normally lack dead fold properties and folds must be secured by heat sealing.

For vertical form-fill-seal machines, such as the Transwrap, the wrapping material should have good slip characteristics otherwise breakage may occur due to frictional drag between the web material and the filling tube or forming shoulder. Where polyethylene is in contact with the filling tube it is necessary to incorporate a slip additive in the film. Thin, clinging films, such as polyvinylidene chloride, are unsuitable for this type of equipment because of frictional drag.

Again in the vertical type of machine the packaging material should have a high melt strength because the product to be packed is dropped down the filling tube on to the cross seal while the seal is still hot. Low density polyethylene is an ideal material for this purpose. However, if unsupported low density polyethylene film is used then the normal heat sealing bars must be replaced by impulse heat sealers (see Chapter 15).

The web in vertical form-fill-seal machines should normally be thin and flexible enough to withstand turning through very acute angles, during the formation of the tube, without creasing or breaking. Thus, regenerated cellulose film and low density polyethylene and laminates such as cellulose film/cellulose film and cellulose film/polyethylene are suitable, as is polyethylene-coated paper. The use of aluminium foil-based laminates is usually excluded because of the rigidity of the aluminium and its tendency to crease and crack.

In horizontal form-fill-seal machines, conditions are not so critical from the point of view of slip characteristics because a filling tube is not used. A high melt strength is still necessary although the question of the product dropping on to hot seals does not arise. This is because dwell times during the sealing operation are usually short and there are normally no cooling stations. The remarks made concerning materials for vertical form-fill-seal machines apply in much the same way to the horizontal machines except the angles are usually less acute and the thinner types of aluminium foil laminates can be used. Low density polyethylene is not usually suitable as an unsupported film although some machines used for packaging textiles or bakery goods have been modified to take this material.

13.4 PROPERTIES AVAILABLE

Once the properties required by the packaging machine and by the product have been determined it remains to match these against the properties of individual plies. *Table 13.1* gives a summary of the various materials that can make a suitable contribution to the particular properties that may be required.

13.5 ECONOMICS

The question of material price obviously has a bearing on material

Table 13.1

Property required	*Suitable materials*
Opacity	Aluminium foil, paper, pigmented plastics films (carbon black is the most efficient pigment)
Transparency	Regenerated cellulose, most plastics films (blow-extruded h.d. polyethylene – translucent)
Water vapour barrier	Aluminium foil, polyvinylidene chloride, regenerated cellulose (moisture-proof grades) high density polyethylene, polypropylene, low density polethylene
Gas barrier	Aluminium foil, polyvinylidene chloride, polyester, rigid PVC, nylon, regenerated cellulose
High temperature performance	Aluminium foil, nylon, polyester, high density polyethylene, polypropylene, paper (dry heat)
Low temperature performance	Aluminium foil, paper, low density polyethylene, high density polyethylene, ionomer, EVA, PVC, polyester, nylon, polypropylene (oriented), polyvinylidene chloride
Grease resistance	Aluminium foil, regenerated cellulose, nylon, EVA, high density polyethylene, polypropylene, polyester, cellulose acetate, PVC, polyvinylidene chloride, ionomer, greaseproof papers
Heat sealability	Low density polyethylene, ionomer, EVA, nylon, PVC (by high frequency sealers), high density polyethylene, polypropylene, polyvinylidene chloride
Printability	Paper, aluminium foil, regenerated cellulose, polycarbonate, polyester, cellulose acetate, nylon (ionomer, EVA, polyethylene, polypropylene – if pre-treated), (PVC, polystyrene – special inks)

choice but it is not the only factor in the total economic picture. It must be remembered that total packaging cost is made up, *inter alia*, of the following:

(a) the package cost;
(b) storage and handling costs of the finished empty package;
(c) filling costs (including material wastage during setting up and running);
(d) storage costs of the filled package;
(e) transport costs for delivery of the filled package;

(f) insurance costs during transport;
(g) losses due to spoilage or breakage;
(h) effect of the package on sales.

The actual package cost (item 'a') may often play only a minor part in the overall cost build-up and it would be wrong, therefore, to attach too much weight to the relative costs of the packaging materials being considered. To take just one example; the use of a material giving greater protection to the product could well save much more than any extra cost involved in its purchase.

Another factor to be considered is the specific gravity of the plastic. The price paid for the granules or powder is based on weight but the area of film (of unit thickness) will be greater for plastics of lower specific gravity. A slightly more expensive material with a lower specific gravity could, therefore, be cheaper in use.

Part 3

Conversion of Films

14 Printing on Plastics Films

Films used for packaging are often required to be printed either as a means of brand identification or to impart instructions for opening the pack or using the contents, or to enhance the sales appeal of the package. Excellent results are now obtainable on most plastics films but it is very rarely possible to use the same inks and printing methods as are used for paper. Films such as the polyethylenes, polypropylene, EVA, etc., have to be pre-treated before printing in order to obtain satisfactory adhesion between the ink and the plastic. This is because their inert, non-polar surfaces do not permit of any chemical or mechanical bonding between them and the ink.

14.1 PRE-TREATMENT

From the above it can be seen that the object of pre-treatment is to produce a surface on the plastics film to which the ink can key. There are various processes but they are usually aimed at oxidising, the surface in some way. Pre-treatment methods include solvent, chemical, flame and electrical treatment but the last method is by far the most common one in commercial use.

14.1.1 SOLVENT TREATMENT

This method makes use of the swelling and partial solvent effect of certain hot organic liquids such as toluene or the chlorinated hydrocarbons, and is followed by hot air drying. It is more suitable for thicker sections such as injection or blow moulded articles but also has the general disadvantage that efficient vapour extraction plant is necessary.

14.1.2 CHEMICAL TREATMENTS

These methods are based on the use of strong oxidising agents, such as chromic acid, which attack the polymer surface to form carbon–oxygen bonds. The method is obviously a messy and dangerous one and is little used except for small numbers of awkwardly shaped mouldings which might be difficult to treat evenly by flame or electrical methods.

14.1.3 FLAME TREATMENT

This was one of the earliest methods for pre-treating film but is now more commonly used for bottles. In essence it consists of exposing the surface of the polymer to an oxidising flame for a short period of time, in the range 0.2–3.0 s. For films, the time would obviously be in the lower end of the range. Once again, carbon–oxygen bonds are formed, rendering the surface wettable and able to form an adhesive bond between the ink and the plastic.

14.1.4 ELECTRICAL TREATMENT

This method has gradually displaced flame treatment as the preferred method for films because their thin gauge makes it difficult to control the extent of flame treatment. Treatment is normally in-line with extrusion of the film. This is an advantage when anti-static and other additives are present because the film treatment can be carried out before the additives have had a chance to bloom to the surface and so upset the evenness of the process.

The film is passed between two electrodes, one of which is a metal blade connected to a high-voltage (10–40 kV), high frequency (1–4 kHz) generator. The other electrode is an earthed roller and is separated from the high voltage electrode by a narrow gap of about 1½–3 mm (0·06–0·12 in). The earthed roller is usually made from steel covered with a dielectric such as polyester film. The metal blade electrode should be slightly narrower in width (about 5–10 mm—0·2–0·4 in) than the film to be treated in order to prevent direct discharges to the roller. The electrical discharge is accompanied by the formation of ozone which oxidises the surface of the film, rendering it polar and receptive to inks. A certain mechanical roughening of the film may also occur due to the formation of micro-pits and this also helps to key the ink. The level of treatment is governed by the generator output and by the throughput of the wind-up unit. Both under and over-treatment are undesirable, the latter causing a powdering of the surface or even the formation of pinholes, brittleness and difficulties in sealing. The time between pre-treatment and printing should be kept to a minimum. This is because the effect of the treatment diminishes with time. Another reason is that the treated surface is sensitive to handling and dust pick-up.

14.1.5 TESTS FOR EFFICIENCY OF PRE-TREATMENT

One simple test for detecting whether or not the surface has been pre-treated is to run water over it. An untreated surface will immediately repel the water, whereas a treated film will retain the water film for up to several minutes. In between these two extremes a partially treated film will tend to show adjacent areas of good and bad adhesion. The test is satisfactory, therefore, only for determining whether the film has been pre-treated but does not show up over-treatment.

The peel adhesion test is an improvement on the above and is carried out by applying to the surface of the film a specified pressure sensitive tape (3M No. 851), using a roller. The peel strength is then measured using a tensometer. The higher the level of treatment, the higher the peel strength.

A variation on this test also utilises pressure sensitive tape (3M Scotch Tape No. 880) but this time it is applied to a sample of printed film. The tape is firmly pressed onto the surface then rapidly pulled off and the amount of ink removed is noted. Ink removal of not more than 1% is generally regarded as satisfactory. This test is widely used and acts as a useful control test but it does not always detect over-treatment.

14.2 METHODS OF PRINTING

There are four main types of printing process:

(1) Ink is forced through a partially masked screen as in screen printing.

(2) The ink is transferred to the film from a surface carrying the required design in relief, as in flexographic or letterpress printing.

(3) The ink is transferred to the film from an engraved surface, i.e. one on which the design has been recessed. An example of this is photogravure printing.

(4) The ink is transferred from a plain surface as in lithography. This works on the following principle: lithographic inks are of a greasy nature and the surface of the printing plate is normally grease repellent except in the area where the design has been etched to make it receptive to the ink. The process has rarely been used with polymer films so will not be mentioned further.

14.2.1 SCREEN PRINTING

This process is basically a stencilling process using a fine mesh screen to avoid the disadvantages of having unprinted areas where the design is connected to the surrounding material as in ordinary stencilling. The screens were originally made from silk but are now usually made from nylon or terylene.

Mesh sizes are of the order of 200–300 perforations per inch. Screens are prepared by photographic methods and masked so that they are porous only in the areas where the decoration is required. The screen is supported by a frame which keeps the screen taut and retains the ink supply. A flexible rubber squeegee is drawn across the screen and this forces the inks through the porous area of the screen onto the plastics film. The screen is then moved out of contact with the film which is then dried.

Screen printing allows the application of a much thicker layer of ink than other printing methods and it is thus suitable for cases where solid, glossy colours are required. Other advantages of screen printing include ease of training of operators, low cost of equipment and screen, low changeover times and economy on short run work.

The main disadvantage is that for multi-colour work it is necessary to use a series of screen stations with provision for drying between each printing station. However, if the colours do not overlap it is possible to use a split screen and print two or more colours at one pass.

14.2.2 LETTERPRESS

Letterpress is a printing process in which a raised type face applies the ink to the film surface. The ink has a pasty consistency which makes the process a difficult one to apply to plastics films except stiff films such as rigid PVC and regenerated cellulose which are not so susceptible to damage by the high pressures which have to be applied.

14.2.3 FLEXOGRAPHIC PRINTING

This is a high speed method and is the one most widely used for the decoration of plastics film. A thin, fast-drying solvent based ink is applied to the film surface by means of a flexible rubber plate (or stereo) with raised characters on it. This is mounted on a plate cylinder by means of an adhesive. Ink is transferred to the rubber plate from the ink fountain via a rubber inking roller and an anilox roller (*Figure 14.1*). The anilox roller is an engraved stainless steel roller which holds ink in the recesses of the design and acts as a metering device to the rubber plate. The process offers a combination of high speed (60–300 m/min—200–1000 ft/min) and comparatively low printing plate costs but stereo preparation does not lend itself to the reproduction of very fine half tones.

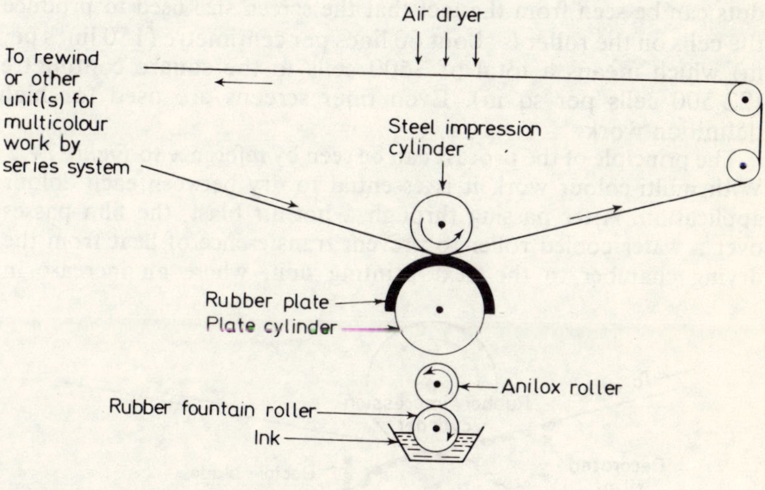

Figure 14.1. Flexographic printing process

The inks used in flexographic printing are mobile and the solvents volatile so that multi-coloured printing at one pass is possible using several printing heads on a single impression cylinder. Oven drying is essential to achieve high printing rates and up to six colours can be printed.

To summarise, therefore, flexographic printing has the advantages of high speed, relatively inexpensive plates and short changeover times but has the disadvantage of being unable to reproduce fine detail.

14.2.4 PHOTOGRAVURE PRINTING

This process consists in inking an engraved printing roller which then transfers the ink directly onto the film. The engraved design is made up of a series of tiny cells of varying depth so that differing amounts of ink are picked up by different parts of the roller. Excess ink is removed by a doctor blade. For all practical purposes photogravure printing is a continuous-tone process, for it is only in the light and some medium tones that the cell walls are visible after printing and then only under magnification. The size of the individual dots can be seen from the fact that the screen size used to produce the cells on the roller is about 60 lines per centimetre (150 lines per in) which means a total of 3600 cells to the square centimetre (22 500 cells per sq in). Even finer screens are used for high definition work.

The principle of the process can be seen by reference to *Figure 14.2*. With multi-colour work it is essential to dry between each colour application. After passing through a hot air blast, the film passes over a water cooled roller to prevent transference of heat from the drying chamber to the next printing unit, where an increase in

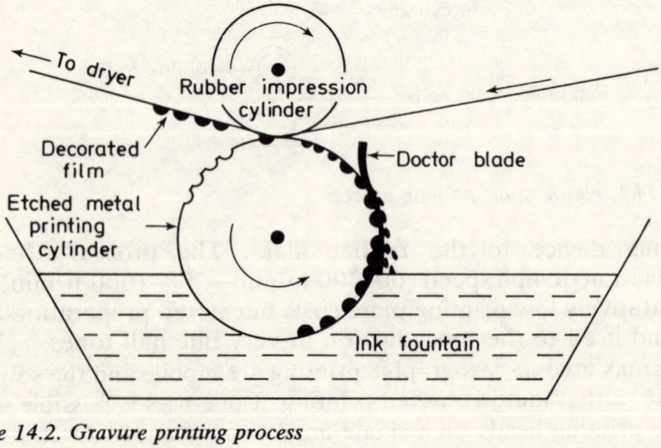

Figure 14.2. Gravure printing process

temperature would cause the volatile ink to partially evaporate and dry in the cells of the engraved cylinder before contacting the web. The cooling roller also hardens the ink from the previous station which was rendered tacky by the application of heat sufficient to melt the resin content of the ink.

The main disadvantages of photogravure printing are the high initial costs for the etched metal rolls and the printing speeds, which are somewhat slower than those obtainable with flexographic processes. Normal photogravure speeds are of the order of 18–120 m/min (60–390 ft/min). The main advantage of photogravure printing is that it can produce such high quality, multi-colour, fine detail printing.

14.2.5 HOT STAMPING

Hot stamping consists in the fusion and release of a special coating from a heat resistant carrier tape and its transfer onto the film. The stamping foil consists of a carrier tape (usually polyester film, glassine paper or cellulose film), a release layer which is a wax that melts at a predetermined temperature, a metal or pigment coating and an adhesive layer which acts as the bond between the imprint and the film surface. The foil is pressed onto the film by means of a heated male die which causes the release layer to melt and fuses the pigment to the film surface. Stamping foils can be made from a wide variety of different waxes, pigments, metals and adhesives and it is essential to inform the foil supplier of the plastics to be decorated so that he can supply the correct foil.

There are three variables which have to be considered in hot foil

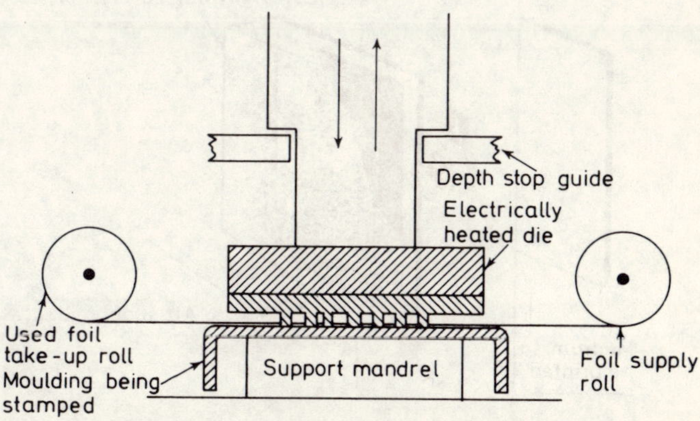

Figure 14.3. Hot stamping process

stamping, namely, temperature of the die, the contact pressure and the dwell time. These variables are all inter-related so that actual conditions for each polymer cannot be given. However, the normal temperature range is 100–225°C. The principle of hot foil stamping is shown in *Figure 14.3*. Fully automatic machines are available with speeds of from 500 to 5000 impressions per hour.

Very good definition is obtainable but the method is not suitable for large areas of print. Other advantages are that the print is dry and can be handled immediately after printing and no pre-treating is required even with polypropylene and polyethylene. In addition, the print has good opacity and the finish can range from glossy to matt, with a good range of colours. In this respect it should be noted that hot foil stamping can achieve decorative effects unobtainable by other techniques (e.g. metallic finishes such as gold and aluminium). Apart from the disadvantage of restrictions on area of coverage, the major drawback is the relatively high cost.

14.2.6 ELECTROSTATIC PRINTING

This is a comparatively recent technique and is only really important

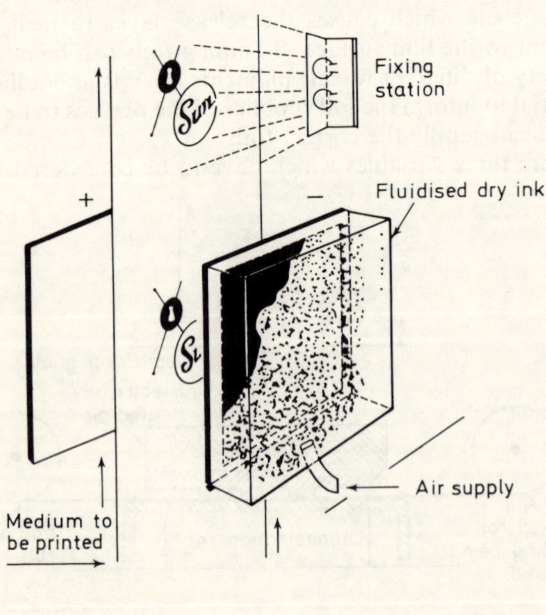

Figure 14.4. Electrostatic printing

as a method which allows the printing of awkwardly shaped objects. The technique utilises an electrically charged stencil consisting of a fine-mesh, electrically conducting metal screen on which the non-image areas are masked. The other component of the system is a conductive backing plate of the same relative size and shape as the screen and placed roughly parallel to it. A finely divided powdered ink is applied to the outside of the screen where it takes on the charge of the screen and is attracted, through the openings, towards the oppositely charged back plate. The ink sticks to the plate in a pattern that faithfully reproduces the design in the printing screen. In the same way, any article that is interposed between the plate and screen will also be printed. The image can be fixed by heat, solvent or vapour. The intercepting medium which is to be printed can consist of any material provided it does not interfere with the electrical field. In this respect, articles of high capacitance, such as polystyrene or polyethylene film, can give trouble unless the static charge on the surface is first dissipated. *Figure 14.4* shows the general principles of the process.

14.3 PRINTING INKS

The printing of plastics films poses many problems which are not present when printing paper. One of these, particularly relevant to the polyolefin films, is their inert surface and we have already discussed the pre-treatment methods which overcome this defect. There remain other problems, however, which are present in the case of all plastics films. The chief of these is the fact that their surfaces are smooth and impervious, unlike those of most papers which, however smooth in appearance, are discontinuous in nature. This means that paper can absorb printing inks and, in consequence, the inks can penetrate into the paper. This has two effects. First, the ink dries quickly and so there is less danger of set-off of ink between adjacent surfaces. Second, there is an ink residue left beneath the surface of the paper after drying as well as on it.

Because of the smooth and impervious nature of plastics films, printing inks have to be formulated so that they can attack the surface. This gives rise to another problem, namely, excessive attack on the films leading to swelling, loss of mechanical properties or tackiness. Solvent selection by the printing ink formulator is thus extremely important.

Before looking at the types of inks suitable for the various printing processes it will be useful to consider the different ways in which printing inks dry as this has an important effect on their suitability for a particular use. The main mechanisms by which an ink may dry are:

(1) Penetration.

(2) Penetration and oxidation.

(3) Oxidation.

(4) Evaporation.

(5) Chemical reaction (2-part inks).

(1) Inks that dry by penetration

Inks of this type are absorbed by the substrate as a liquid. They consist, usually, of a pigment (often carbon black) dispersed in a mineral oil and sometimes containing a little gilsonite (a type of bitumen) or rosin. Toners may also be added to increase or modify the colour. A common example of an ink of this sort is that used for news print. This type of ink cannot, of course, be used for plastics films for the reasons already mentioned.

(2) Inks that dry by penetration and oxidation

Drying by penetration alone is unsuitable, not only for plastics films, but also for the better quality papers where drying would be so slow that set-off would almost certainly occur. Inks are available in which the vehicle consists of a thin mineral oil blended with a drying varnish (based on, e.g. linseed oil). Such inks are stable while in the container, or on the machine, but when they contact the paper the thin mineral oil is absorbed leaving a stiff paste consisting of pigment and drying varnish. This paste is not dry but it is too stiff to cause set-off. Afterwards, oxidation of the linseed (or other vegetable) oil proceeds slowly, giving a hard film of printing ink. Dryers such as cobalt naphthenate, which accelerate the oxidation reaction, are also added.

(3) Inks that dry by oxidation

Where the printing surface has little or no absorbency, even inks of type (2) are unsuitable. Inks based on drying oils and pigments can be used but their adhesion is not usually good without the aid of absorption. A high proportion of a resin is added, therefore, and driers are again used to speed up the oxidation.

(4) Inks that dry by evaporation

Inks that dry by evaporation are known as heat set inks. They consist simply of a pigment dispersed in a resin solution. Solvents are chosen which evaporate rapidly, leaving behind the resin which acts as a binder for the pigment. The last traces of solvent are often difficult to remove because of the sponge-like action of the resin.

(5) Inks that dry by chemical reaction

Modern designs often possess the requirement for overall decoration of a sachet so that the printed area is in or near to the heat seal. Heat resistant inks are required in such instances and are based either on high softening point thermoplastics, such as the polyamides, or on thermosets such as polyurethane. Thermosets are more efficient but pose problems because the resin is formed *in situ* by chemical reaction of two liquids which are mixed immediately prior to use on the machine. The pot-life can vary up to a matter of days but even so, there usually remains ink on the machine which will set hard unless completely removed by washing down. It is often difficult, too, to estimate accurately the amount of ink necessary for a run and a substantial reservoir of ink may have to be wasted.

Silk screen inks may be of the oxidative drying type but the newer rapid drying inks are solvent based and dry by evaporation in which case they probably consist of a resin solution in a high boiling ketone.

Both gravure and flexographic inks are solvent based and are, therefore, the evaporative drying type. The solvents normally used for flexographic inks are alcohols or alcohol/ketone mixtures. Hydrocarbon solvents cannot be used because they would affect the rubber stereos. A typical flexographic ink, therefore, might consist of titanium oxide as the pigment, together with a coloured toner to give the required shade, plus a low melting point polymer in an alcohol or alcohol/ketone solution.

Gravure inks also contain a pigment plus a toner and a low melting point polymeric binder. The solvents used were formerly toluene and xylene but alcohol-based inks are also available and have the advantage of being less odorous.

Other general requirements for printing inks for use with plastics films are that they should be unaffected by the contents of the package and that they should be flexible. Flexibility is dictated by the fact that plastics films themselves are flexible and the ink would otherwise become detached if the film were to stretch or shrink while the ink did not.

14.4 ULTRA-VIOLET DRYING

As was mentioned earlier one of the problems in printing non-absorbent surfaces is drying of the ink and much work has been carried out to improve ink drying. One of the methods developed in the last few years is UV drying. The method depends on the use of special inks which are cured by UV radiation. The inks are 100% solids and the UV radiation does not 'dry', in the conventional sense of driving off solvents but a chemical reaction takes place giving almost instant drying (hardening). The major outlet for UV installations is for the printing and varnishing of carton board but it is also used for printing PVC which, being heat sensitive, benefits from this type of treatment.

14.5 INFRA-RED DRYING

The use of short to medium wavebands of infra-red radiation is one of the latest methods used for drying inks. Shortwave infra-red energy level transmission. Medium wavebands (2 μm to 4 μm) have penetration, less absorption by passage through air and a higher energy level transmission. Medium wavebands (2 μm to 4μm) have the advantage that they match the absorption spectra of many ink components and so can directly excite (and hence heat) the inks. However, there is some absorption and scattering by air.

Normal ink formulations can be used, the effect of infra-red drying being to heat the sub-strate as it passes under the dryer so that the stack of sheets is raised to a temperature of around 40° to 45°C. Ink penetration and oxidation are accelerated and drying is improved. Infra-red drying has been used for non-absorbent surfaces such as polyethylene/board laminates but the presence of aluminium causes problems because the shiny surface reflects some of the energy and so reduces the heating effect.

Infra-red drying is cheaper than UV curing and it utilises conventional inks. It does not, however, have the instant drying potential of UV curing.

14.6 VACUUM METALLISATION

Although not strictly a printing process, vacuum metallisation is a means of decoration and so is best discussed in this chapter. The fundamental principle of this process is the heating of aluminium to a high enough temperature to vaporise it. In practice this necessitates temperatures in the range 1500–1800°C. One way of achieving this is to hang short pieces of aluminium wire onto a tungsten filament which is then heated electrically. This is only suitable for short heating cycles and so is used for metallising plastics mouldings on a batch system. For longer cycles such as are needed for the metallisation of film, the aluminium wire is fed onto a block of metal, usually tantalum. The tantalum is heated by holding it in a carbon crucible across which is arced a high voltage current. Vaporisation of the aluminium causes minute particles to be ejected from the surface in all directions. The operation must be carried out in a vacuum in order that the metallic particles can reach the surface to be metallised.

The metallisation of continuous webs can be carried out in two different ways. One process involves unreeling and re-reeling of the film within the vacuum chamber, while the other consists in passing the film through vacuum-sealed slits with unreeling and re-reeling being carried out from outside the vacuum chamber.

Certain films may have to be de-gassed before vacuum metallising because they contain moisture or other volatile constituents, such as plasticisers, which could otherwise cause difficulties during the process. If the normal vacuumisation does not completely remove them before metallisation commences then their continued outgassing interferes with adhesion of the aluminium giving dull and incompletely anchored coatings. Even if the volatile constituents are completely removed they can make vacuumisation difficult and cause contamination of the pumping system. Films which need de-gassing include regenerated cellulose and cellulose acetate. Since this de-gassing may also remove other, perhaps important,

Table 14.1

15 μm OPP	*Permeability* Moisture vapour g/m²/24 h at 25°C and 75% R.H.	Oxygen cm³/cm²/s/cm Hg
Unmetallised	2.2	210×10^{-10}
Metallised	0.6	93×10^{-10}

constituents of the film a preferable method of solving the problem may be to seal the surface of the film with a lacquer. It is important, of course, that the lacquer itself should not contain anything likely to interfere with the vacuum. If coloured metallic effects are required, the film can be self-coloured or a coloured lacquer can be applied after metallising.

One of the most important things to note about vacuum metallisation is that any faults in the underlying film are not hidden by the very thin film of metal deposited. Indeed, faults may even be enhanced. The laquer film can help to hide sub-strate faults but can, of course, contribute faults of its own as, for example, 'orange peel', or 'fish eyes'. Correct tensioning during unwind or re-reeling is also important, especially with very thin plastics films because creasing will cause uneven deposition of the aluminium.

Although vacuum metallising is often carried out for purely decorative reasons, many applications are known that depend on other properties. Vacuum metallised films are used in electrical applications, for example, where a thin conducting layer on a thin insulator is required. The metallising process can also increase the barrier properties of a film as can be seen from Table 14.1, which gives results for 15 μm oriented polypropylene film. Claims for the reduction of moisture vapour transmission of polystyrene film by a factor of five times have also been made. Practical packaging tests using potato crisps have also confirmed the improvements in barrier properties conferred by vacuum metallisation. Vacuum metallisation improved the preservation of organoleptic properties after 33 days at high humidity (95% RH) and temperatures of 22°, 30° and 38°C. The pick-up of moisture vapour was also lower.

In lamination, unsupported metallised films may well be most competitive as alternatives to aluminium foil, which has to be paper-supported. Another plus point is that metallised film may be printed before metallising and the whole then sandwiched in a laminate. On the other hand, if it is desired to print directly on to the metallised surface then this presents no particular problems of ink adhesion or print quality other than those normally associated with aluminium foil. When sandwich printing is the method used, a depth and sparkle is added by viewing the print through the film. Either opaque or translucent inks may be used to give a metallic coloured effect.

A unique effect can be achieved by the vacuum metallisation of expanded polystyrene film. Instead of the bright metallic effects produced by other plastics films, a frosted effect is obtained and this is combined with an attractive soft and silky feel. The material may also be cold embossed, thus adding many other possibilities to its decorative appeal.

The more rigid films, such as oriented polystyrene and PVC can be vacuum formed to give a wide variety of finished products. An important factor when considering the use of vacuum forming is that if the metallised coating is on the top surface it will reflect some of the heat and so increase the heating cycle. Conversely, if the metallised coating is on the underside of the film it will reflect the heat back through the film, thus shortening the cycle.

A wide variety of plastics films are suitable for metallising, including polyester, polystyrene (both normal and expanded), polypropylene, unplasticised PVC and cellulose films. The first plastics film to be successfully vacuum metallised in this country was polyester and it was used for making Christmas trees and garlands. The market for Christmas trees has now largely changed over to unplasticised PVC because of the latter's better non-flammable performance. The other main requirements of the Christmas tree and garland applications are toughness and brilliance, with no tendency towards limpness. Metallised polyester can also be laminated to PVC and used in the manufacture of handbags, shoes and the decorative trim in cars, etc. One interesting use of metallised polyester film is in the manufacture of ultra-lightweight mirrors. The metallised film is stretched over a framework attached to a backing board in such a manner that the film is held clear of the backing board. In addition to being very light in weight, the extremely low heat content of the film means that condensation is greatly reduced because the film rapidly attains thermal equilibrium with the atmosphere. These lightweight mirrors have been used in aircraft washrooms.

Lacquered metallised polyester film is also slit into very narrow strips and used as decorative textile threads. Such threads are lustrous but strong, non-tarnishing and can be boiled or dry-cleaned. Hot stamping foils also utilise metallised polyester film. Among the best-known applications in this field is that of transferring the gold embossing on to UK commemorative postage stamps.

Electrical uses for metallised polyester film are also increasing. Its use in capacitors, for example, allows a great reduction in size over capacitors made from paper and aluminium foil. High fidelity electrostatic loudspeakers are also in service based on metallised polyester. The requirements for this application are high tensile strength, absence of plasticiser, high electric strength, chemical inertness, freedom from pinholes and dimensional stability over long periods of time and a wide range of climatic conditions. Metallised polyester film is also used for the so-called 'chromium' self-adhesive tape which often decorates motor scooters. Unlike normal chromium plating, these tapes retain their brilliant appearance virtually indefinitely.

Metallised PVC and polystyrene are other widely used films, particularly so in packaging applications such as decorative chocolate box inserts and biscuit trays. Metallised orientated polypropylene is another entrant in the packaging field and is the cheapest in terms of yield. It has wide possibilities in the field of gift wraps. Metallised films in general have obvious openings in the display field and many uses have already been found. The rigid or semi-rigid films give a bright mirror effect while the thinner films are used as drapes or for background. Metallised PVC and polystyrene are often used for vacuum formed letters while the frosted appearance of expanded polystyrene, mentioned earlier, adds yet another dimension to the art of display.

Space research has also made many calls on the unique properties of metallised films. Space suits, for instance, utilise them to reflect back external heat radiation and they are also used for the satellite balloons from which radar signals can be bounced back. Glamorous as these uses are, however, it is the bread and butter applications such as those in the packaging field that form the backbone of development. Chocolate box ribbons, labels for luxury items such as cosmetics and chocolate box lids are all examples of the way in which the packaging field has seized on the decorative effects of metallised films.

15
Sealing of Films

15.1 MECHANICAL METHODS

Although heat sealing of plastics films is the most common method used it is sometimes impractical for one reason or another. Where plastic bags are used in the home, for instance, it is the absence of heat sealing equipment which dictates other methods while in other cases it may be that the film does not heat seal readily, in spite of being a thermoplastic. Cellulose acetate is an example of this type of film.

One of the simplest methods of closing a plastics bag is with an elastic band or a wire tie and the latter is, in fact, used for fastening polyethylene bags for bread. Another method is to use staples for fastening. These are used particularly where a printed paper 'header' is attached to the bag and when a completely moisture vapour proof seal is not essential.

Heavy duty sacks can be stitched but such a seal is not usually sift-proof and the stitches can fail under stress. Cross-laminated film (see Chapter 19) is better in this respect because the lines of least resistance in each of the component films run at an angle to each other.

15.2 HEAT SEALING

Heat sealing, in one form or another, is the most common method of joining two pieces of film or for closing sacks, bags and sachets.

After the development of coated grades of regenerated cellulose, where the coatings are fusible as well as moisture proof, methods of heat sealing these films were developed. These usually consisted of heated metal jaws—often patterned, by embossing, to give the seals extra strength. Heated wheels were also used to give band seals. When polyethylene film first appeared, similar methods were tried but were unsuccessful because polyethylene sticks to hot metal. In laboratory work and some commercial small runs, it was found possible to use heated wheels and metal jaws by inserting a sheet of plain (uncoated) regenerated cellulose film between the metal and the polyethylene but this was not really a convenient method. A more satisfactory solution was found to be the coating of metal jaw sealers with a silicone rubber or with PTFE.

There is another difference between polyethylene film and regenerated cellulose film which adds to sealing difficulties, however. The latter has an infusible substrate and only the coatings melt when heat sealing is carried out. Polyethylene film is completely fusible and so there is a danger of loss of strength because the sealing pressure causes flowing and thinning of the seal. In addition, the seal is still molten when the jaws are opened and the seal is left unsupported so that there is a danger of the seal being ruptured.

One early way round the problem was used to make bags from layflat polyethylene tubing. The end of the tubing was held by two pieces of cold metal so that about a centimetre of tubing protruded. A bunsen flame was played on the protruding film which quickly fused and gave a heavy bead seal to the tubing. The metal jaws had the function here of conducting the heat away from the rest of the film. The method was, of course, slow as well as a potential fire hazard. Another method of overcoming these difficulties is the use of an 'impulse sealer'. In an impulse sealer the jaws are of light construction and have attached to them a resistance wire or ribbon covered with PTFE, which is heated electrically. Because of their small heat capacity the jaws heat up quickly and cool quickly when the current is switched off. Alternatively, models are available with water cooled carrier bars. However, these have the drawback that under humid conditions, condensation may occur on the bar and this can affect the contact between the bar and the heating unit or lead to deposition of water on to the film.

The sequence in impulse heat sealing is as follows. The bag is placed between the jaws, the jaws are heated for a certain time (the current is usually switched on by the action of the jaws on a microswitch), the current is switched off (by a timer) but the bag is retained by the jaws until cool. The jaws are then opened and the bag is removed. The cooling cycle is usually pre-set with a sequence timer and a red light is on during the heating and cooling cycle. Since the bag is held by the jaws while the seal cools there is no danger of seal rupture during withdrawal. There is a limit to the width of seal which can be made by impulse sealing because the wider the heating strip the longer it takes to cool and eventually the operation becomes uneconomic. However, even narrow impulse seals can be quite strong and are satisfactory for most uses.

Impulse heat sealers are intrinsically slower than heated jaw sealers because of the cooling sequence in part of the cycle. It is possible to incorporate various cooling methods in tandem with heated jaw sealers so giving heating and cooling without holding up the passage of the film during cooling. More specialised developments of the heated bar principle are rotary bar sealers and heated rotary band sealers.

When heat sealing laminated materials, it is generally preferable to use heated jaw sealers since laminates do not suffer from the disadvantages of single, fully fusible films. It is possible, therefore, to make use of the high speeds of the continuous heating method of sealing and embossed jaws can be used to give extra strength.

In general, a good heat seal, using heated jaw or impulse sealers depends on the temperature at the interface, contact or dwell time, pressure between the jaws and the nature of the film.

In the particular case of low density polyethylene film used on form, fill and seal sachet making equipment, the speed of cooling and the strength of the molten polymer are important factors. In the early days of such equipment, the filling speed was the rate determining factor but with increasing filling speeds, the limit is now set by the speed of sealing.

For materials sealing at the same temperature, the maximum speed of sealing is likely to be determined by the speed of crystallisation and this, in turn, is affected by density. The higher density grades crystallise at a higher temperature and so set up more rapidly. Of course, they also require higher seal temperatures but in practice sealing jaw temperatures are higher than the minimum required for a good seal and it is unlikely that a significant increase would be necessary for the high density grades.

Heat sealers have also been developed which work by applying a blast of hot air to the film surfaces. They are particularly useful for heavy duty sacks used for fertilisers or other products liable to cause contamination by powder of the two mating surfaces. Any dust particles are completely encapsulated by molten polyethylene and do not affect the heat seal strength.

Another method of heat sealing which is suitable for a range of thermoplastics films is hot wire sealing. The heat is applied by an electrically heated wire which serves to seal and sever the film in one operation in such a way that the seal is made on the trailing edge of one package and the leading edge of the next. It is used in bag and sack manufacture and in shrink-wrapping equipment. The seal so obtained is, of course, a narrow one and there is no bottom 'lip'. Such a seal may be quite strong in the sense of withstanding pressure from the contents but is susceptible to damage by large scale drop impacts.

Finally, a method of using infra-red radiation has been developed in Russia, with particular reference to the welding of PVC and low density polyethylene. Heating has been carried out using silicon carbide/clay rods with metallised ends for good electrical contact or a heating spiral of chromium steel wire in a glass tube. The efficiency of this method depends on the temperature achieved at the junction of the two pieces of film and this, in turn, depends on the amount of radiation absorbed by the material, part being reflected and part passing through. Maximum absorption for most plastics occurs at a wavelength of 3 μm and this corresponds to a surface temperature of 700°C. When thin films are welded, the rate of temperature rise and the final temperature are both dependent on the backing material on which the welding area lies. An appreciable amount of the incident radiation passes through the film and heats the backing pad which in turn heats the plastics film thus giving a better weld. Best results are obtained with compressed lamp black or black paper. Low speeds render this process uneconomic for commercial purposes but it is useful for hard-to-seal films such as PTFE.

15.2.1 SEALING OF ORIENTED FILM

The sealing of oriented films presents particular difficulties. When heated, the film naturally tries to return to its original, unstretched state in the vicinity of the heat seal and this leads to cockling of the seal. Highly oriented films are also liable to crystallise readily and if the cooling rate is slow then large spherulites are formed, giving brittle seals. Rapid cooling gives small spherulites and good strength at the seal. Secure clamping of the film and cooling of the areas adjacent to the heat seal can ameliorate these troubles but does not always cure them.

Another method which was designed especially for oriented polypropylene film is known as 'multi-point' sealing. As its name implies the heating jaws consist of a large number of very small points and the heating effect is, therefore, extremely localised thus preventing the tendency to shrinking of the whole heat seal area. Its disadvantage is that it is not a complete heat seal and air and moisture vapour can eventually diffuse through the seal.

Another more satisfactory method is to coat the oriented film with a polymer having a lower softening point and use the coating to effect the heat seal. Since the oriented film is not heated to its softening point the tendency to shrink is avoided. This method has the additional advantage that the oriented film gives support during the heat sealing operation provided that the heat seal temperature of the coating is far enough below the softening point of the oriented film substrate.

15.3 HIGH FREQUENCY HEATING

Films such as PVC and nylon 6.6 are difficult to heat seal by direct means because they tend to degrade at temperatures close to their softening point. Stabilisers have been developed which help in this respect but other methods of sealing are still preferable. For such materials one of the alternative methods of sealing is high frequency (or dielectric) heating.

The method works through the action of a high frequency current on the charged molecules of the plastic films. It is only applicable in those cases where the material is polar, i.e. it is capable of forming a dipole moment. Polyethylene, therefore, cannot be sealed by high frequency methods because it does not contain any inherent dipole moment. PVC, however, is polar and high frequency welding is widely used for the sealing of it. In essence, what happens is that the polar molecules try to align themselves with the oscillating high frequency field. Their rapid oscillation builds up internal friction which manifests itself as heat. This is sufficient to allow welding, especially under pressure. Because the heating occurs uniformly through the film there is not the same risk of over-heating occurring as when direct heat is used. In the latter case the surface temperature may well be too high before heat has penetrated to the centre of the film. In general, frequencies of around 50–80 MHz are used with outputs of up to about 25 kW. For very thin films it is essential to use the higher end of the frequency range and to lower the voltage. This precaution is taken to avoid 'flash' which occurs when the high frequency voltage is high enough to break down the insulation resistance of the film. In other words, since a given current produces a given amount of heat, it is safer to produce this current by using a high frequency and a low voltage in order to operate below the flash point.

One advantage of high frequency welding is the ease in which a seal can be made through a liquid. An example of this is in the manufacture of shampoo sachets where a completely filled cushion type sachet is required. A continuous tube is filled with the liquid and the individual sachets are formed by sealing across the filled tube.

An advantage of more general interest is the fact that shaped welds can be produced very simply by using appropriately shaped electrodes. The manufacture of sachets shaped to provide a convenient cut-off portion for dispensing is one example of the use of shaped welds while another is the encapsulation of expensive and delicate parts for automobile and aero-engines. The parts are placed in a vacuum formed sheet of transparent, rigid PVC and a corrosion inhibitor added. A similar moulded sheet is placed on top and the assembly is welded by an electrode shaped to the component thus giving a transparent, rigid, hermetically sealed pack providing perfect protection for its contents.

A further advantage of high frequency welding is that heating occurs throughout the film. With heat sealing the seal interface only receives heat by conduction from the heated surface which must *ipso facto* be at a higher temperature, since thermal conduction is so poor.

Relevant variables in the high frequency welding process are the heating time, pressure and the power supply. Heating time may be larger than theoretical for thin films because of their greater heat losses (due to large surface area and small heat content). In such cases, buffers may be used to retain as much heat as possible. Pressures should not be so high as to cause thinning but must maintain good thermal contact between the film surfaces. A cooling period, while the films are still under pressure improves the strength of the final seal. The importance of the power supply has already been mentioned as when avoiding flash in the sealing of the film.

15.4 ULTRASONIC SEALING

This method of sealing depends on the vibratory energy of an ultrasonic head (known as the horn) being transmitted through two pieces of plastic film in contact. At the interface between them, the mechanical energy is converted into heat and produces an almost instantaneous weld. Briefly, the principle of the method is as follows. The vibratory energy from the horn radiates outward from the horn tip and is then transmitted through the plastic as a wave motion. There may be some reflection of energy at the interface of the two pieces of film but this is usually small because the joint is under pressure due to the forces being applied at the horn. Most of the energy, therefore, passes through the lower piece of film. Here, the energy is reflected by the anvil on which the film is resting and tends to set the mating film surfaces vibrating against each other. Because of the high frequencies involved (about 18–20 kHz) the two surfaces are literally hammered together and the mechanical energy of motion is transformed into heat which melts the mating film surfaces.

The equipment consists of an ultrasonic generator feeding a 20 kHz signal into a transducer which transforms electrical energy into mechanical vibrations. These are amplified by means of the focusing tool or horn which, as already described, is in contact with the two pieces of film.

Ultrasonic welding is normally restricted to jointing thermoplastics of the same basic family. Members of the styrene family of plastics, such as basic and high impact polystyrene, ABS and styrene acrylonitrile give excellent results with ultrasonic welding and, in general, rigid plastics with a low melting point respond best to this method. It is a useful method for oriented films, including polyester film, because the high tensile strength of the oriented film is destroyed by normal heat sealing, whereas ultrasonic sealing produces little overall heating. Another advantage is that it can be used to seal packages containing greasy products because it can seal through surfaces contaminated by oils and greases.

In spite of the fact that normally only similar materials can be sealed by ultrasonic welding there are certain exceptions. One of these is the welding of plastics and paper or cloth. In these instances it is possible that bonding occurs because the high local pressure of the horn tip forces the fused plastic into the fibrous structure of the paper or cloth. In the case of plastics, ABS has been welded to acrylics and to polystyrene, the secret of success here being chemical compatibility and similarity of melting points.

Another example of the usefulness of ultrasonic welding is the sealing of multi-wall paper sacks with an inner polyethylene liner. The application is feasible because the heat is generated only at the welding interface and so the paper is not damaged. In ordinary heat sealing enough heat has to pass through the paper to melt the polyethylene and this may result in charring of the outer layer, especially when heavy gauges of paper are used. The main disadvantages of ultrasonic welding are, first, that sealing speeds are usually lower than those of heat sealing (although this may be overcome in time) and, second, possible health hazards due to the vibration. Some particularly sensitive people may be affected by ultrasonic frequencies and the upper level of audible frequencies which are also sometimes produced. Loss of hearing, headaches and cumulative tissue damage are possible through direct physical contact with ultrasonic devices so that operators should be provided with suitable ear protection and the equipment shielded with sound-absorbent material.

15.5 ADHESIVES

There is a tendency among plastics technologists to look upon adhesives as old-fashioned compared with ordinary heat sealing, let alone ultrasonics, infra-red and high frequency welding. In addition, there is the complication that plastics such as polyethylene and polypropylene have inert surfaces and are difficult to stick anyway. Nevertheless, adhesive sealing is still a very useful method of joining plastics, and can often be very much quicker than other methods. Inert plastics can be pre-treated in the same way as they are before printing, and a wide range of adhesives is now available for use with plastics.

Adhesive seals, of course, provide yet another answer to the problem of unsupported oriented films and are often used for polyester film. They are also useful where large area seals are required such as in block-bottom sacks, or for attaching valves to valved sacks. Finally, they are used for sealing plastics film and sheet to other materials such as fibreboard (as in blister packaging). Adhesives can be applied to plastics by similar methods to those used for other materials, including direct roll coating, direct roll kiss coating, reverse roll coating and air brush coating.

15.6 CHOICE OF METHOD

The choice of sealing method depends on many factors including the seal characteristics required, production speed required, the nature of the film to be sealed and the cost/effectiveness of the energy source or sealant. The effectiveness of the seal can be measured in a variety of ways, some of which are qualitative only but can be useful for production testing.

In the case of a simple sealed sachet, the exertion of hand pressure will show up any leakage by an escape of air. Another simple test is to insert a dyestuff on the inside of the seal and observe any leakage. A visual check under polarised light is also useful. Laboratory tests for assessing heat seal quality are described in Chapter 10 on 'Physical and Chemical Properties'.

16

Wrapping equipment

Automatic wrapping machines for packaging materials were introduced around the end of the nineteenth century and the beginning of the twentieth, and were designed to use paper. When metal foils were developed as packaging materials they were readily accepted by the packaging industry because in addition to their excellent end-use performance properties they were stiff enough to be fed to automatic wrapping equipment in the same way as paper. In fact for many applications the foil was laminated to paper.

Regenerated cellulose film was a later arrival on the scene and posed new problems to the machinery designer. It was thinner than most wrapping papers, for instance, although it possessed reasonable rigidity. The biggest difference, however, lay in the need for the provision of temperature-controlled heaters to seal the film. In addition, end-folds had to be held in place during the sealing operation otherwise they tended to spring apart before a firm bond had been achieved. With adhesive sealing it was usually possible to use an adhesive with enough initial 'tack' to hold the folds in place until the adhesive had dried.

16.1 WRAPPING WITH THERMOPLASTICS FILMS

Thermoplastics films have added two main problems of their own to the list of machinery designers' difficulties. Many films are too limp to be handled by the type of equipment so successfully developed for paper and cellulose film, while the problem of heat sealing is complicated by the fact that the entire film thickness melts on heating, whereas heat-sealable grades of cellulose film have a fusible coating only. The main substrate is infusible and gives support to the molten material on the surface. The ways in which these problems are being solved are best explained by considering the basic functions of a wrapping machine and the main categories of machine available.

The basic functions of a wrapping machine include the following:

(1) Feeding the wrapping material.

(2) Forming the pack.

(3) Closing the pack.

The machine may also include labelling attachments, label imprinters, a splicing unit, photo-electric cell registration, and coding devices but these do not affect the main functions of the machine.

16.1.1 FEEDING THE WRAPPING MATERIAL

This section of the machine contains the reel of wrapping material together with some device for cutting the individual wrapper to size and positioning the resultant sheet centrally, ready to receive the product. It will also contain some device for maintaining correct tension in the web of wrapping material. Too little tension will cause the web to wander from the prescribed path and this can cause mis-cutting or bad positioning of the sheet. The problem with thermoplastics films is that too much tension may cause appreciable stretching of the film causing, again, incorrect sheet cutting, or mis-feeding. The magnitude of the problem varies with the nature of the film and with its thickness, as well as with machine design. On some machines, tension controllers similar to those used for cellulose film have been found adequate but on others it has been found necessary to fit a pre-feed or 'pay-off' device which provides a loop, or reservoir, of film. This is then used to feed the machine, so avoiding having to pull the film directly off the reel. Even with all precautions, the tension control of very thin, extensible films, still presents difficulties and these are aggravated by the presence of static electricity. It is usually necessary, therefore, to fit some form of static eliminator in such cases.

The feed section can itself be classified under one of three main categories:

(1) Feed sections where the web of wrapping material is pushed into position for cutting into a sheet.

(2) Feed sections where some form of gripping mechanism pulls the web into position for cutting.

(3) Machines where the motion is continuous and the product is enclosed by the film before the web is cut, giving the completed pack.

The first type of feed mechanism will obviously cause difficulties with thin, flexible materials. Buckling of the film often occurs and can lead to stoppages on the machine. The main cause of the trouble is hold-up of the leading edge of the film as it is being pushed over the bed plate or support rails when it approaches the cut-off position. The problem can be partly solved by reducing the area of the surfaces where the web has to slide, prior to cut-off, and by adding some form of positively driven carrier belt to take the web to the cut-off position.

The second type of feed mechanism normally gives much greater control over the web than does the first type. Gripper fingers close over the leading edge of the web and then pull it for a pre-determined distance. A guillotine knife cuts a sheet of film which is either transported by the gripper fingers direct to the wrapping station, or is positioned for transfer to a second set of grippers. The latter system is particularly useful if the design of the rest of the machine necessitates the sheet moving at right angles to its original path in order to reach the wrapping station.

The speed of a pull-feed machine is normally less than that of a push-feed type but only if the wrapping material is rigid. Limp wrapping materials, as we have seen, can cause misfeeds and jamming of the machine. The actual overall speed then achieved on a push-feed machine is much less than its rated speed.

The third type of feeding mechanism is best considered as an integral part of the whole machine since its continuous action makes it difficult to separate the various functions.

16.1.2 FORMING THE PACK

The product is pushed up into the sheet of wrapping material by an elevating table and the sheet is formed as a sleeve round the product. The ends are then folded and tucked over. If the wrapping material is so flexible that the overlapping ends of the sleeve sag, then the end folds will not be neat and the whole package will look untidy.

Slip is a problem in the forming area because of the different interfaces involved and their varying requirements. Too high a slip between the product and the film may cause the sleeve of material to slide along the product but a high slip *is* required between the film and those parts of the mechanism which fold and tuck the film.

16.1.3 CLOSING THE PACK

Closure of the pack can be performed either by adhesives, labels, or by heat sealing. Although adhesives and adhesive labels are now available which are suitable for use with plastics films, the most common method of closing is heat sealing.

Heat sealing on automatic overwrapping equipment is based on two main principles, namely, resistance heating (heated metal jaws, plates or bands), and impulse heating, both of which were dealt with in the chapter on 'Sealing of Films'. Resistance heating is the most common, the simplest type of heater unit being a fixed metal plate (heated by a cartridge-type electrical heating element) over which the folded wrap is moved. Alternatives to the fixed heater plate include reciprocating and rotary heater units, the latter being more applicable to the continuous type of wrapping machine to be considered later. The advantage of reciprocating heaters over fixed heater plates is that they merely dab the folds of the package and so are less likely to disturb the folds by dragging. Another way of avoiding drag is to utilise heated plates or belts which move with the pack.

Other problems with heat sealing include heat transfer on fast machines where the dwell time is likely to be low (except in the cases mentioned above where moving platens or belts are fitted).

16.2 CONTINUOUS WRAPPING MACHINES

Such types of wrapping machines are particularly useful for the packaging of biscuits and sweet confectionery. A continuous tube of the wrapping material is formed round the product, the longitudinal seal being made by carrying the material past heater blocks, followed by cold rollers. The filled tube of material is kept moving by friction belt drives. The ends of each separate pack are sealed and cut simultaneously, usually by rotary crimping sealers.

The material feed problems mentioned when discussing the other types of wrapping machines are not normally encountered with continuous machines because the web of material is controlled at all stages of the wrapping operation and cut-off only takes place when the product has been fully enclosed.

16.3 POUCH MAKING EQUIPMENT

The manufacture of separate bags and sacks is dealt with in Chapter 17. In this chapter we deal with the subject of pouches, where these are formed, filled and sealed on the same machine.

The difference between these machines and the overwrapping ones just considered is that on overwrapping machines the product acts as the former, whereas on pouch forming machines the wrapping material is formed round a metal tube and the product is delivered independently, through the forming tube.

In general, the most important film properties are tensile strength, flexibility and slip. Both tensile and flexural strength have to be high because the film is pulled around the forming shoulder at very

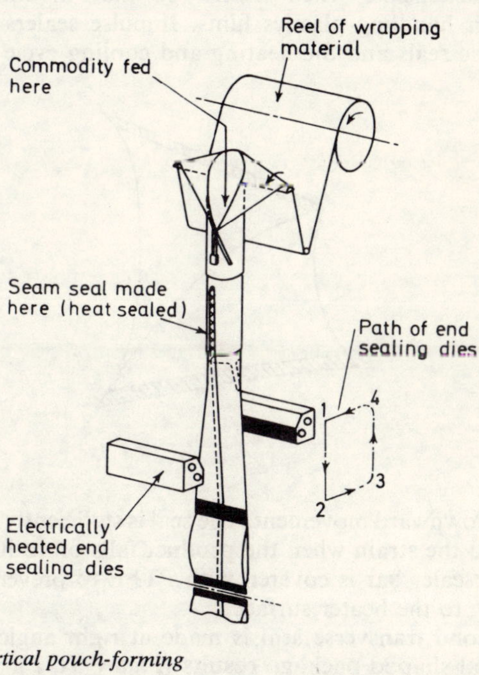

Figure 16.1. Vertical pouch-forming

acute angles. The entry from the feed section to the forming section should be made as gradual as possible, commensurate with the need to maintain sufficient control of the web to avoid wandering. Good slip is essential to ensure smooth passage of the film over the forming shoulder. The action of a typical vertical pouch forming machine can be seen from *Figure 16.1*.

The web of film is formed by passing it around a vertically mounted forming tube and then sealed by a vertically mounted sealer bar. Heated crimp jaws are horizontally mounted, and close on the formed web of film, so giving a transverse seal which forms the base seal of the pack. A measured weight or volume of the product is then allowed to fall through the forming tube into the partially formed pack. The crimp heat sealing jaws, while still closed, descend drawing the pack and the web downwards by a pre-determined amount. They then open and return to their original position, at which point they close again to form the top seal of the pack and the bottom seal of the one following. A cut-off knife then separates the filled pouch from the web and the whole cycle is repeated.

This type of machine was originally developed for materials which remained inextensible when heated so that modifications were necessary for handling plastics films. Impulse sealers are used for the transverse seals and the heating and cooling cycle is completed

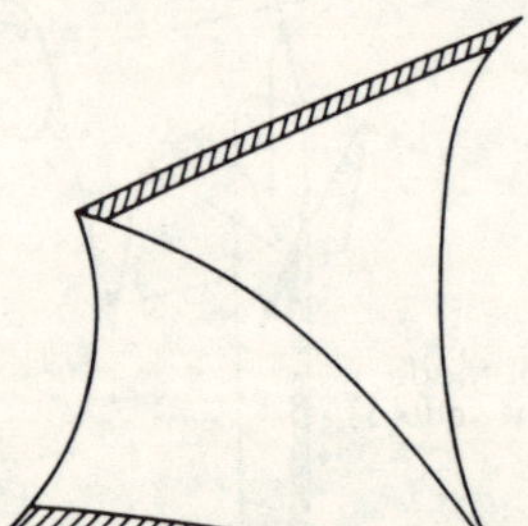

Figure 16.2

during the downward movement. The seal is sufficiently set, therefore, to withstand the strain when the product falls on to it. In addition, the vertical sealer bar is covered with PTFE to prevent the plastics web sticking to the heater surface.

If the second transverse seal is made at right angles to the first, a tetrahedral shaped package results (*Figure 16.2*).

Using a laminate of paper and low density polyethylene, such packs have found many applications in the packaging of liquids, including milk and fruit juices.

16.4 SACHET MAKING MACHINES

A range of machines is available which make sachets by passing the film over a triangular-shaped forming shoulder to give a folded film. In the horizontal version, vertical seals are made, giving a series of packets along the web of film. These are then cut into separate

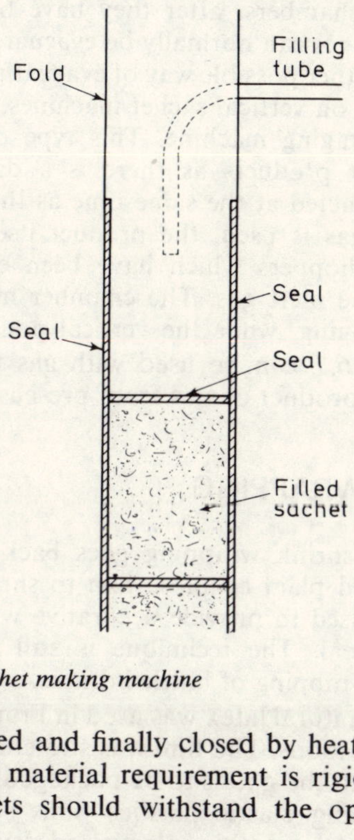

Figure 16.3. Vertical sachet making machine

sachets, opened, filled and finally closed by heat sealing along the top edge. The chief material requirement is rigidity, in order that the separated sachets should withstand the opening, filling and sealing operations.

In the vertical version, the folded film passes on both sides of a small diameter filling tube and is then sealed along three edges (i.e. one transverse seal and two side seals) and filled (*Figure 16.3*). The final closure is made at the same time as the bottom seal of the following sachet.

16.5 VACUUM AND GAS PACKAGING

Many products, particularly foodstuffs, need to be kept completely free from contact with oxygen during storage. The provision of a gas-proof wrapping material is not sufficient in such instances since there is usually enough residual air in the package to react with the contents. Two solutions are possible, namely, evacuation of the package, or replacement of the residual air by an inert gas such as nitrogen.

In the first method the packages are usually evacuated and then sealed in vacuum chambers, after they have been fabricated and filled. Several packages can normally be evacuated and sealed in the same chamber. Another possible way of evacuating certain packages, such as those made on vertical sachet machines, is to use a vacuum nozzle on the packaging machine. This type of vacuumisation is limited to granular products as there is a danger of powdered products being extracted at the same time as the gas is withdrawn.

When an inert gas is used, the product itself must usually be stored in bins or hoppers which have been evacuated and then flushed out with the inert gas. The chamber method may also be used for gas packaging, while the vertical pouch forming process shown in *Figure 16.1* can be used with gas flushing techniques provided that the product comes from pre-gassed hoppers.

16.6 SHRINK WRAPPING

The technique of shrink wrapping goes back to 1930 when the ability of dampened plain cellulose film to shrink to a tight wrap on drying was utilised to provide decorative wrappings and to act as a pilferproof seal. The technique is still widely used today, especially for the wrapping of bottled drinks.

Later, in 1936, natural latex was used in France as a shrink wrap to pack perishable foods. The film was stretched and then allowed to shrink back on to the goods to be packaged. This technique has been extended using elastic plastics films such as low density polyethylene, EVA and thin plasticised PVC. The process is referred to as stretch wrapping to distinguish it from processes based on the use of heat shrinkable films. The latter technique is the one normally referred to as shrink wrapping and is dealt with in sections 16.6 to 16.10. Stretch wrapping is discussed in section 16.11.

Shrink wrapping began around 1948 with the wrapping of poultry for deep freeze storage. The film used was polyvinylidene chloride. A film bag was slipped over the bird, a light vacuum was drawn and the mouth of the bag sealed, usually with a wire tie. The bag and contents were then immersed in a hot water bath when the bag shrunk tightly on to the contents. Apart from giving an extremely neat pack, shrink wrapping had the advantage that freezer burn (caused by intense dehydration of the surface of poultry) was prevented by the impermeable film, coupled with the close contour wrap. The latter also eliminated pockets in which ice crystals could form and obscure the contents.

The principle on which shrink wrapping is based is sometimes referred to as 'plastic memory'. In other words a film which has been stretched during manufacture (at a temperature above its softening point) and then cooled to 'freeze-in' the consequent orientation of the molecules, will tend to return to its unstretched dimensions when re-heated. The techniques of producing oriented films have already been described in Chapter 8. With most films other than polyvinylidene chloride, the temperature at which a suitable degree of shrinkage is obtained is above the boiling point of water so that hot-air tunnels have had to be developed for their use.

16.6.1 SCOPE OF PROCESS

The scope of shrink wrapping has expanded well beyond the original idea of contour wrapping frozen poultry although the shrink wrapping of items of food is still an important outlet. Because awkward shapes make little or no difference to the feasibility of shrink wrapping, the technique is particularly useful for the packaging of a premium offer with the product to which it is temporarily related. The equipment necessary for such a job is capable of handling subsequent offers within a fairly wide range of size and shape.

Figure 16.4. Shrink-wrapped bottles

Shrink wrapping is also advantageous in rounding off the contours of awkwardly shaped or complex sets of items so that they do not become dust traps when on display. However, one of the most rapidly developing fields open to shrink wrapping is the production of transit packs of cans, cartons or bottles, as an alternative to fibreboard cases. Cartons, of course, can be built up into a neat stack which is easily shrink wrapped, but cans and bottles are normally first collated on one or two fibreboard or plastics trays, before shrink wrapping (*Figure 16.4*).

16.6.2 TYPES OF SHRINK WRAP

There are two main types of shrink wrapping, namely, the sleeve wrap and the all-round or perimeter-sealed wrap. In the sleeve wrap, a tube of film, greater in length than the product to be packaged, is made by sealing longitudinally (see *Figure 16.5*). When this is passed through the shrink tunnel, the extremities of the sleeve shrink around the ends of the product (*Figure 16.6*). Less film is used than in a complete over-wrap, and the holes at each end can be used for opening the pack. Sleeve wraps are particularly suitable

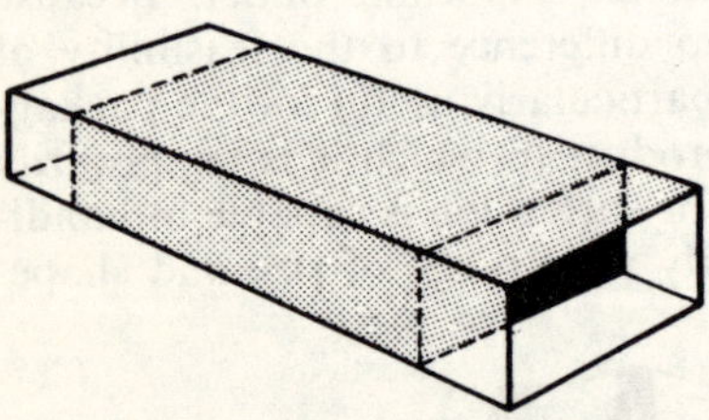

Figure 16.5

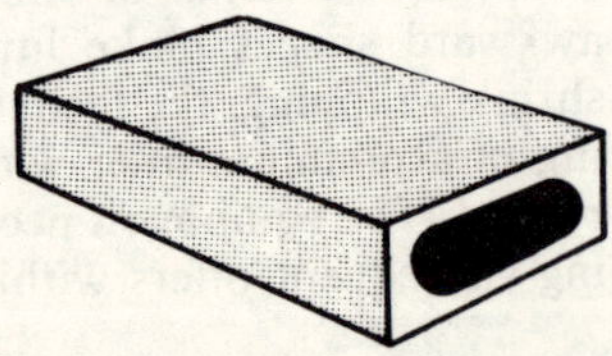

Figure 16.6

for securing premium offers or for making carry-home packs of, say, three or six unit packages, but they are also used for transit packs or one or two dozen cans or bottles. Another use for the holes at either end is as hand-holes for carrying. The all-round, or completely sealed, pack is a more efficient barrier against dust or moisture pick-up but it may be necessary to make a few micro-perforations in the film to allow the escape of entrapped air during the shrinking process.

With cans, a sleeve wrap may be preferable to a complete over-wrap because the cans could still hold residual moisture from the cooling water used after processing. In a complete wrap, condensation will eventually occur but with a sleeve wrap, the resultant ventilation should prevent this, provided that storage conditions are reasonably dry.

A sleeve wrap for holding, say, a premium offer and the product will require the use of film with a high degree of shrinkage in one direction only. For transit packs, where either a sleeve wrap or a perimeter wrap may be used, the type of film necessary will be dictated more by the proportions of the pack.

When a sleeve transit wrap is used, the shrinkage may be unbalanced to a small degree, but not to the extent indicated above for collating two fairly small items. Where the height of a pack is out of proportion to its length and breadth, the difference between machine direction shrinkage and transverse direction shrinkage should be at a minimum. For products requiring an all-round seal, the shrinkage should again be more or less balanced.

16.7 SHRINK WRAPPING EQUIPMENT

The equipment necessary for shrink wrapping transit packs of cans or bottles will usually consist of collating equipment, tray erector, tray loader, film wrapper and sealer, and a shrink tunnel, together with the necessary conveyors.

16.7.1 TRAY ERECTION

Plastic trays do not require erection because they are thermoformed in one piece. For fibreboard trays, either manual or mechanical erection may be used. However, for high speed lines manual erection can be expensive in terms of man-power, and mechanical erection will normally be used.

16.7.2 FILM WRAPPING AND SEALING

A range of equipment is available, from hand operated to fully automatic. Basically, however, there are two main types. One is used specifically for all-round sealing and utilises centre-folded film (i.e. film folded in half, in the machine direction) and an 'L'-shaped hot wire sealer. The object to be wrapped is placed within the folded film and the 'L' sealer brought down in front of it. This seals transversely in front of the pack (and also cuts the film), at the same time sealing the side of the pack (opposite to the fold). The pack then moves forward, another object is placed in the folded film and the operation is repeated. This seals the trailing edge of the first pack and the leading edge of the following one. Each subsequent

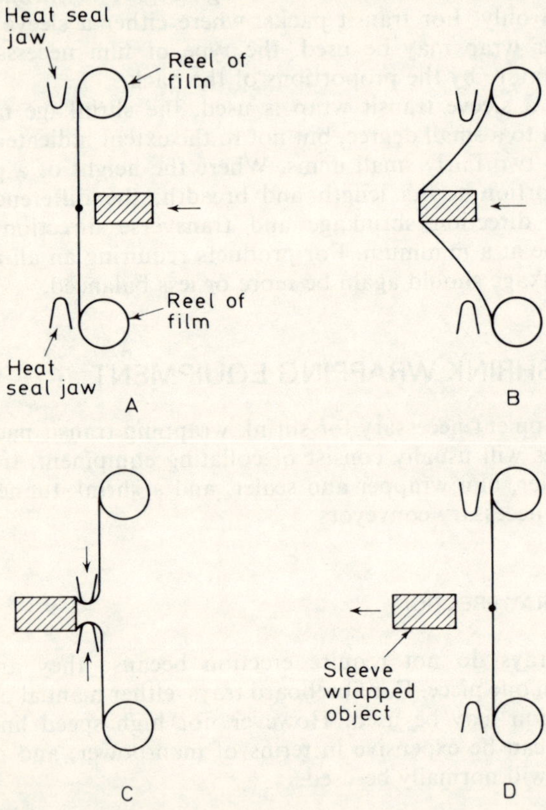

Figure 16.7. Sequence of operations for sleeve wrap

operation of the 'L' sealer, seals the trailing edge of one pack, and the leading edge and side of the following one.

For sleeve packs the usual method is to use two reels of flat film, located above and below the sealing table. The ends of these two reels are first sealed together and the object to be wrapped is pushed through the curtain of film so formed, beyond the heat-sealing and cutting jaws. These are then operated to seal the pack, cut it from the web of film and reform the curtain (*Figure 16.7*). If a complete overwrap is required, side seals can be added further along the line.

16.7.3 SHRINK TUNNELS

Shrink temperatures are often not far from the melting point of the film. Temperature control is an important factor, therefore. The normal means of heating a shrink tunnel is by hot air, and heat losses are minimised by fitting PTFE or asbestos curtains at each end of the tunnel.

16.8 PROPERTIES OF HEAT-SHRINKABLE FILMS

In general, the orientation of a thermoplastics film improves its impact and tensile strengths, its clarity and its flexibility. Gas and moisture vapour permeabilities are also reduced in some cases. On the debit side, elongation is reduced and the heat-sealing range is narrowed.

16.8.1 SHRINK TEMPERATURE

The range of temperature over which a film will shrink is important commercially. Films with low shrink temperatures require simpler equipment and also pose fewer problems with the packaging of heat-sensitive goods. In addition, a film with a wide softening range is preferable to one with a fairly sharp melting point because the latter renders temperature control of the oven more critical.

16.8.2 DEGREE OF SHRINKAGE

The maximum amount of shrink available in commercially obtainable films varies from around 20% to around 75%. The percentage shrink of a particular film also increases with the temperature of shrinkage. Films with a steep shrink/temperature curve are more difficult to handle because of the closer temperature control necessary. Polypropylene, for example, has a steep shrink/temperature curve and a temperature variation of 10°C (i.e. ±5° tolerance in the tunnel) could cause the degree of shrink to vary by up to 20%.

The degree of shrinkage necessary depends on the particular application. For tightening-up a loosely wrapped package only a very little shrinkage is necessary, whereas the contour wrapping of a complex-shaped article requires a high degree of shrinkage.

For printed films, the problem is not so much the total degree of shrinkage as the amount in different directions. Balanced orientation obviously gives the least distortion under normal conditions but there may be complications if the product is of irregular shape. In such cases the only course of action may be to choose a simple print design which is relatively unaffected by distortion.

16.8.3 SHRINK TENSION

This is the stress exerted by the film when it is restrained from shrinking at elevated temperatures. A certain amount of shrink tension is necessary in order to give a tight package but care must be taken when wrapping articles which could be crushed or distorted. High shrink tensions are really necessary only when the film is intended to become a structural part of the packing.

16.9 PALLET OVERWRAPPING

Shrink tunnels have been built which are large enough to accommodate complete pallet loads, and this has opened up a wide range of applications for shrink wrapping. The concept of shrink wrapping complete pallet loads was first developed for the packaging and transit of refractories. These are fragile and are often oddly shaped, thus making them extremely difficult to transport by normal methods. Using 150–200 μm thick, low density polyethylene film, a tough contour wrap can be obtained which is firmly anchored to the pallet. Other advantages of a shrink film overwrap for such pallet loads include protection against dirt and weather conditions, elimination of strapping damage, easy detection of pilferage, and a reduction in abrasion and crushing due to movement during transit.

The concept has since been expanded to include pallet loads of container glassware, bricks, and a wide range of products packed in sacks or fibreboard cases. Speeds of up to 300 packs per hour have been claimed. The shrink tunnels are usually heated by gas but electric ovens have also been used. Other heating developments include the use of a vertically reciprocating bank of heaters, and portable hot air heaters (rather similar to giant hair dryers).

16.10 GENERAL ADVANTAGES AND PROBLEMS

Many of the advantages of shrink wrapping have already been mentioned but a general summary may be useful. For the packaging of awkwardly shaped foodstuffs like hams, poultry, and so on, shrink wrapping provides a convenient and economical method for preventing freezer burn (in deep freeze storage) and for equipping the foodstuff with a second skin which protects it from contamination by dirt and bacteria. For transit packs of cans, bottles or cartons, shrink wrapping can often be cheaper than a conventional fibreboard case, as well as providing an attractive display pack. Two advantages of particular relevance to the retailer are the reduction in packaging material to be disposed of at the end of the day, and the fact that a shrink wrap takes up progressively less space on the warehouse shelf as the contents are removed, whereas a fibreboard case takes up the same amount of space whether full or empty.

However, there are still a number of problems which can arise under certain conditions. Some are a matter of user education but others are more technical. The latter can quite often be solved technically but not necessarily economically. With the packaging of cartons, for example, care must be taken where these are coated externally with wax or low density polyethylene. Bonding sometimes occurs between the shrink film and the carton coating, particularly if there are local hot spots in the tunnel. The trouble can usually be eased by reducing the dwell time in the tunnel and by tightening up the heat control.

Inconsistent results can sometimes be traced to temperature variations within the tunnel. This may not necessarily be due to the temperature control but may be caused by running the packs through with too little space between them. Sometimes, too, different temperature settings are necessary for the same film if it is to be used for different products. Cartons and cans, for example, will need different temperature settings because of their different heat conductivities.

One of the problems of transit shrink wraps is that of mixed loads, where shrink wrapped goods are carried in the same vehicle as, say, wooden boxes or metal containers. This problem can only really be solved by separate stacking.

The question of excessive shrink tension has already been mentioned. This can lead not only to crushing and distortion of flimsy cartons but may lead to trouble with items such as polyethylene bottles. If these are filled with a detergent, stress cracking may occur if the bottles are under stress. Even if the shrink tension of the film is not excessive, stress may be induced in the bottles when they are stacked. Where such bottles are carried in a fibreboard water case, the case itself usually takes part of the weight involved in stacking.

16.11 STRETCH WRAPPING

The principle of stretch wrapping, as mentioned earlier, is that the film is stretched around the article to be wrapped and is then heat sealed. The residual tension in the film gives a tight contour wrap (the simplest analogy is that of an elastic band).

The main films used in stretch wrapping are low density polyethylene, PVC and EVA, the choice depending on factors such as appearance (clarity, sparkle, etc.), protection required (gas barrier, water vapour barrier, etc.) and the susceptibility to damage by compression of the articles being wrapped.

One of the factors which initially held back the commercial development of stretch wrapping was the lack of suitable equipment and in the early days fragile items, such as cartons containing tea bags, were damaged by the excessive film tensions used. Suitable equipment is now available, however, and the advantages of stretch wrapping can be exploited. These advantages, *vis-à-vis* shrink wrapping, are dealt with below.

(a) Energy saving – This is perhaps the most obvious advantage, arising from the fact that no shrink tunnel is necessary in most cases. Apart from the savings in capital cost of a shrink tunnel, the savings in running costs can also be considerable. The direct savings in energy are also important.

When a sleeve wrap, extending beyond the length of the pack, is used it may be required to bring the ends in by shrinking. In such instances, a short end-closing tunnel may be used or the packs may be passed between two hot air guns. In either case, the energy used is less than in a normal shrink wrapping set-up.

(b) Savings on yield – It is often possible to use a thinner gauge for stretching wrapping. Typically, where a 50 μm film might be used for shrink wrapping, it can often be replaced by a 38 μm film in stretch wrapping.

(c) Savings in space – In general, stretch wrapping takes up less space than does shrink wrapping and this can lead to important cost savings. A typical figure would be 100 units of area for stretch wrapping compared with 160 units for shrink wrapping. With the high (and rising) costs of factory space this can also lead to appreciable cost savings.

(d) Installation and commissioning costs – These are usually less for stretch wrapping.

(e) Use of standard film – The use of stretch wrapping allows the use of standard film widths for a range of different product collations. This enables cost savings to be made in the ordering of film.

(f) Use of double wrapping – One technical advantage of stretch wrapping is in double wrapping, particularly in the overwrapping of pallet loads of pre-shrink wrapped packs. If the pallet loads are subsequently shrink wrapped there is the possibility of welding or laminating occurring between the pallet wrap and the film covering the individual packs. The use of dissimilar films can sometimes solve this problem but the use of stretch wrapping as the outer wrap solves the problem completely because no heat is used to obtain the final tension.

16.11.1 TENSIONING OF THE FILM

For collation stretch wrapping, much of the equipment used is based on that developed for shrink wrapping. One of the main differences is that tensioning of the film is more important than in shrink wrapping so that adjustable torque units have to be fitted. In addition, packs have to be held in position to avoid movement during the stretching, clamping and sealing operations.

One way of stretch wrapping pallet loads is by spiral winding so that a standard width of film can be used, irrespective of the pallet load dimensions. Basically, pallet stretch wrap equipment consists of a turnable on which is placed the pallet load to be wrapped. To one side of this there is a vertical pillar, up and down which moves a carriage holding a reel of film. After placing the pallet load on the turntable the film is secured to the bottom of the pallet, the turntable is set revolving and a suitable number of turns of film is given to provide suitable anchorage. The film carriage is then taken up and down again until the requisite number of layers has been applied.

As mentioned earlier, one of the factors influencing the choice of film is the fragility of the objects to be wrapped. By this is meant the liability to deformation of the product or packs and this, in turn, governs the amount of tension which can be tolerated during wrapping. Of the three films already mentioned, low density polyethylene gives the highest initial stress for a given percentage deformation (when used in stretch wraps) while PVC gives the least. However, all give figures well above the initial stress of low density polyethylene shrink wraps. This is not the whole story, however, for the percentage stress retained after 24 hours also varies with the different films and processes so that after 24 hours, PVC stretch wraps have slightly less retained stress than low density polyethylene shrink wraps and even EVA and low density polyethylene stretch wraps have a reasonably low stress.

Some typical figures for initial and retained stresses are given in Table 16.1.

Table 16.1

	Stretch wrapped			*Shrink wrapped*
	LDPE	EVA	PVC	LDPE
Initial Stress (kg/m^2) at 20% deformation.	100	70	65	28
Stress retained after 24 hours (%)	55	60	30	95
Stress retained after 24 hours (actual) (kg/m^2)	55	42	20	26

17

Bag and Sack Manufacture

The methods used for making bags from plastics film depend on:

(1) The nature of the film.

(2) Whether flat or tubular film is being used.

17.1 NATURE OF THE FILM

The most important influence that the nature of the film has on the choice of bag making method is the type of seal to be employed. The various types of sealing available for plastics films were discussed in Chapter 15 but the ones most likely to be used for film bags are adhesives, heat sealing and high frequency welding. In general, the division has been the use of adhesives for cellulose films (especially for film faced paper bags), heat sealing for polyethylene and polypropylene bags and high frequency welding for PVC. Adhesives are also likely to be used for films such as polyester where a big loss of strength occurs at the heat seals due to crystallisation. These rather rigid compartments are breaking down somewhat because of developments in adhesives for plastics such as polyethylene which enable the faster and more flexible machines based on adhesive sealing to be used for a wide range of films. Thus, heavy duty sacks with adhesive seals have been made from low density polyethylene as have small bags from high density polyethylene film.

One of the earliest methods used for making polyethylene bags was to blow extrude lay-flat or gussetted tubing, cut it into lengths then place about 10–12 of these tubular portions between two iron bars, with a few millimetres of each tube portruding. A girl then passed along the iron bars and melted the protruding ends with a bunsen flame, thus giving bags with a beaded seal at one end. This was a slow and dangerous method but it did give a strong seal and was especially good for gussetted bags because the end seal consists

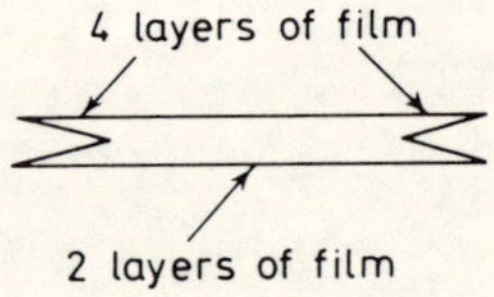

Figure 17.1

partly of four layers of film and partly of two. With normal heater bar methods of sealing, the change from two to four sections is a source of weakness (*Figure 17.1*) because it is difficult to supply adequate pressure to the two layers without excessively squeezing the four layers.

PVC film is normally fabricated by high frequency welding techniques. These are particularly valuable where a shaped bag is required and bags of quite intricate profiles can be made quickly and easily by using the appropriately shaped electrodes.

17.1.1 BAGS MADE FROM TUBULAR FILM

The simplest bag is one made from lay-flat tubing with a simple end seal. Most low density polyethylene bags are made in this way because of the immense popularity of the blow extrusion process for low density polyethylene. The end seal is usually made either by a hot wire or by using an impulse sealer (see Chapter 15). If printed bags are required, the whole process is normally carried out in-line

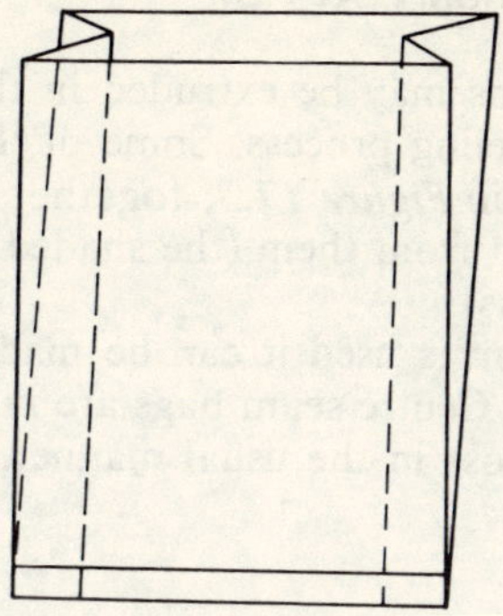

Figure 17.2

i.e. the tube is extruded, pre-treated with a corona discharge, printed, sealed and separated. Similar techniques are used for the manufacture of simple open-ended heavy duty sacks. Where a block bottom bag is required, gussetted tubing is used instead of lay-flat (*Figure 17.2*).

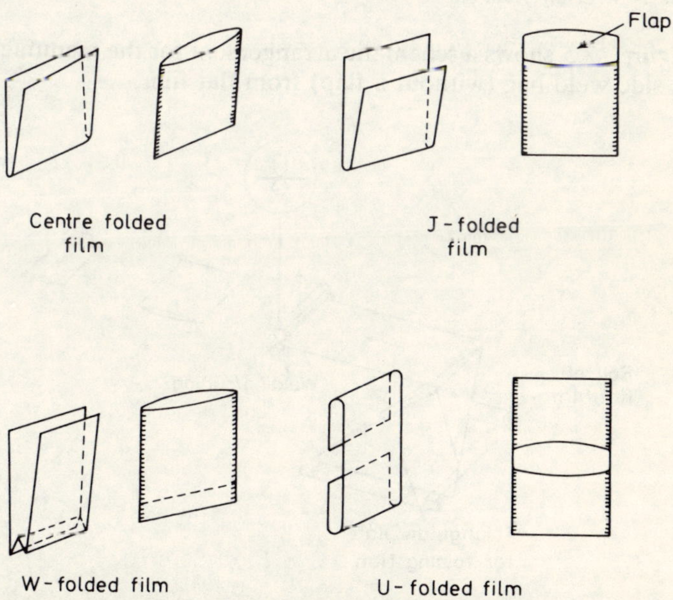

Figure 17.3. Side weld bags and envelopes from folded flat film

17.1.2 BAGS MADE FROM FLAT FILM

Although plastic films may be extruded in the flat, they are often folded during the reeling process. Some of the ways in which film is folded are shown in *Figure 17.3,* together with the types of bags which can be formed from them (the shaded areas in the diagrams denote heat seal areas).

If flat unfolded film is used it can be made into side seam bags or centre seam bags. Centre seam bags are made by forming a tube and then sealing across in the usual manner. A centre seam bag is shown in *Figure 17.4*.

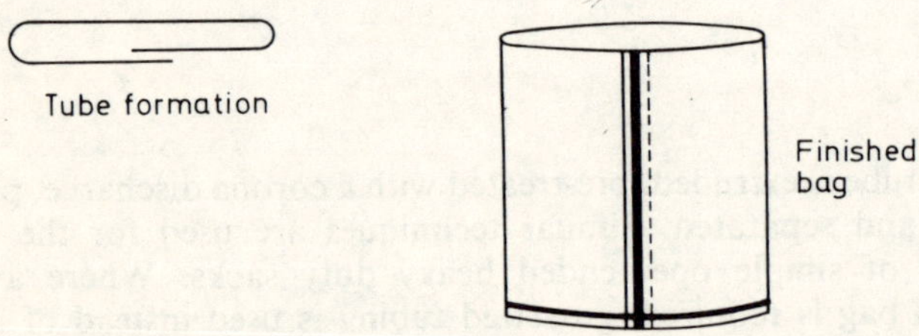

Figure 17.4. Centre seam bag

Figure 17.5 shows a schematic arrangement for the manufacture of a side weld bag (without a flap) from flat film.

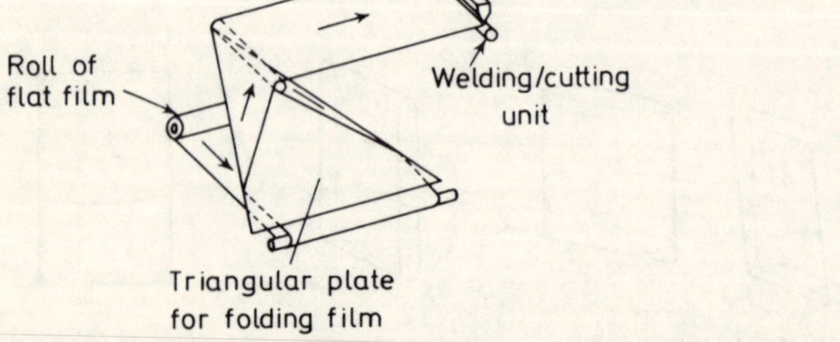

Figure 17.5. Manufacture of side weld bags

The roll of flat film is unwound and run round the roller, passing under the triangle. This gives a simple fold to the film before it travels to the heat sealing/cutting arrangement. This consists of a heated knife (thermostatically controlled) which cuts down on to the rubber roll covered with polytetrafluorethylene. The effect is to seal the trailing edge of one bag and the leading edge of the one following, while at the same time the two bags are separated.

The same type of equipment can be used for making bags from centre folded or J-folded film in which case, of course, the triangle

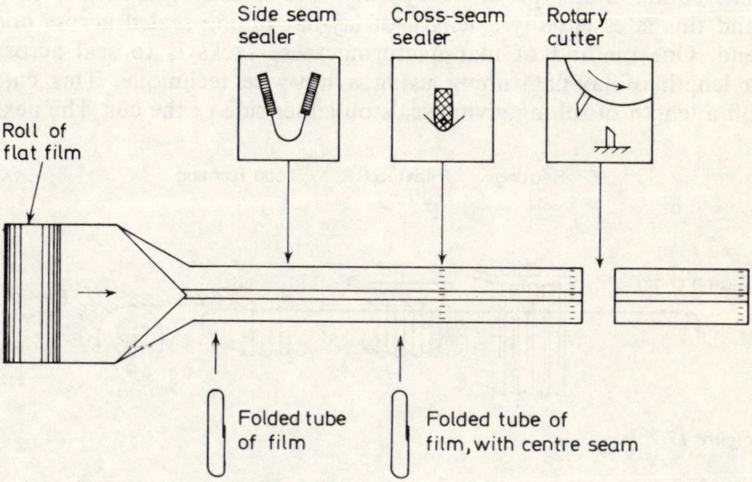

Figure 17.6. Manufacture of centre seam bags

would be omitted. A schematic drawing showing the principle of equipment for making centre seam bags is shown in *Figure 17.6.*

After being unwound from the roll, the flat film is folded to form a flattened tube. The overlapping edges pass over a heated metal arc thus forming the centre seam. The transverse seal is made by an electrically heated bar covered with a glass fibre fabric impregnated with polytetrafluorethylene. The bags are separated by a rotary cutter which cuts against a stationary blade. The type of sealer/cutter used in making side weld bags would not be suitable here because the thin bead seal so formed would be weak at the point where the number of plies changed from two to four.

Formers such as those mentioned above should not be highly polished when films with low rigidity and slip (for example, low density polyethylene) are used, otherwise drag or 'seizure' will occur. With these films, formers should be given a matt surface by sand blasting or by using an acid etc.

Cellulose films are more rigid (have less 'cling') and have sufficient slip to pass over formers more easily.

17.2 HEAVY DUTY SACK MANUFACTURE

The commonest type of heavy duty sack is the open mouth sack and this is essentially a length of lay-flat tubing sealed across one end. One method of manufacturing these sacks is to seal across a length of lay-flat tubing using a hot-wire technique. This cuts off a length of tubing giving seals on either side of the cut. The next

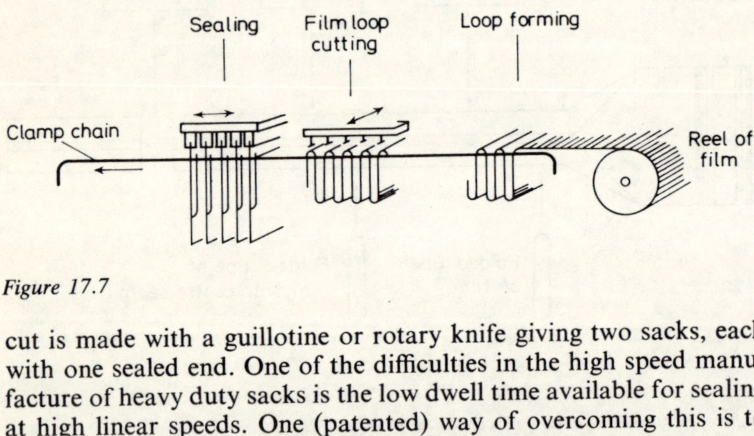

Figure 17.7

cut is made with a guillotine or rotary knife giving two sacks, each with one sealed end. One of the difficulties in the high speed manufacture of heavy duty sacks is the low dwell time available for sealing at high linear speeds. One (patented) way of overcoming this is to feed the film into a loop forming system in which the high speed of the film is brought to the relatively low speed of the chain in which the film loops are vertically clamped (*Figure 17.7*). Each loop of film represents exactly the length of bag which is to be made and is easily adjusted. While they are in the loop formation, the bags are cut (4–6 at a time) in one movement which is synchronised with the chain. Heat sealing is then carried out by sets of sealing bars coated with PTFE. Even at high speeds, dwell times of up to 1.5 s are possible, so allowing accurate temperature control without overheating and possible weakening of the seal. After sealing, the bags remain suspended from the carrier beams until the seals have cooled.

When open mouth sacks are filled, they tend to give pillow

shaped sacks which are difficult to stack. This can be overcome to some extent by providing a few micro-perforations at the end of the sack so that the internal air can escape and this point is dealt with in more detail in Chapter 20 on 'Packaging'. The provision of micro-perforations is only a partial answer to the problem, however.

Another disadvantage in many applications is that filling is geared to that of valved paper sacks and such filling equipment is not suitable for open mouth plastic sacks. Many ingenious designs have been developed for valved plastic sacks but most of them have been expensive because of the difficulty in fitting the valves on automatic sack making equipment. The latest answer to this problem is based on similar constructions to those used in the manufacture of valved paper sacks and also incorporates block bottoms to the sacks, thus giving a filled sack which can be easily palletised. The main innovation has been to substitute adhesion sealing for heat sealing thus allowing rapid, automatic construction of the sacks.

The basic steps in the manufacture of block bottom valve sacks are as follows:

(1) Lay-flat tubing is printed on both sides (if required) and the tubing cut into suitable lengths and stacked in batches.

(2) The stacks of sack blanks are placed in a rotary feeding machine which separates, aligns and feeds individual blanks to the pasting station. Here the adhesive is applied along both cut edges on both sides of the tube. The adhesive is a two component one consisting of resin and hardener and it is made up just prior to use.

(3) The solvents in the applied adhesive are driven off by an application of hot air.

(4) Score marks are made to give triangular pockets which assist later folding operations, and the bottoms of the tubes are opened.

(5) The open bottoms are arranged in the appropriate configuration (see *Figure 17.8*) ready for valve insertion at one end.

(6) The valve, which consists of a thin low density polyethylene lay-flat tube (lay-flat width equal to width of block bottom) is unwound from a reel, pre-treated for adhesion, cut to lengths and applied under pressure to one of the pre-pasted triangular pockets as shown in *Figure 17.9*.

(7) The bottom flaps are then folded over and overlapped. The

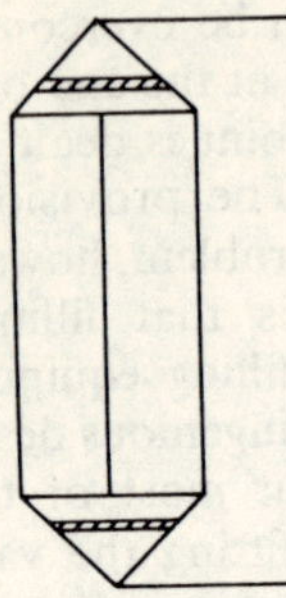

Figure 17.8

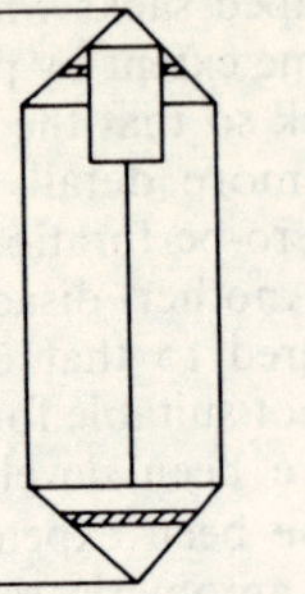

Figure 17.9

pre-pasted areas in the triangular pockets come face to face in this operation and so stick rapidly and securely (*Figure 17.10*).

(8) Pre-pasted and cut rectangles of film are placed on the sack bottoms and pressure applied (*Figure 17.11*).

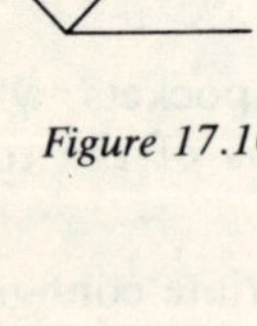

Figure 17.10

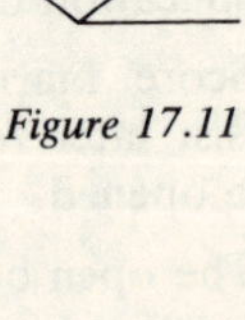

Figure 17.11

FURTHER READING

PAINE, F. A. (Ed), The Packaging Media (Part 3, Chapter 4), Blackie, Glasgow (1977)

18

Thermoforming

Thermoforming, as the name implies, consists essentially of forming a heat-softened sheet or film, either into or around a mould. Most thermoforming techniques are applicable to sheet and film, and it is sometimes difficult to group the applications under the rather arbitrary headings of film (250 μm or under) and sheet (over 250 μm), given in Chapter 1. The principles are the same, however, so that these will be described, together with applications for either film or very thin sheet (i.e. close to the film/sheet borderline).

18.1 METHODS OF THERMOFORMING

There are basically three types of thermoforming, namely, vacuum forming, pressure forming and matched mould forming.

18.1.1 VACUUM FORMING

In its simplest form, vacuum forming equipment consists of a vacuum box with an air outlet and a clamping frame, a mould, a heating panel and a vacuum pump. The mould, which is partly hollow underneath and is perforated, is placed over the air outlet. The thermoplastics sheet is then placed over the open top of the vacuum box and securely clamped by means of the frame, giving an airtight compartment. The sheet is heated until rubbery, the heater is with-

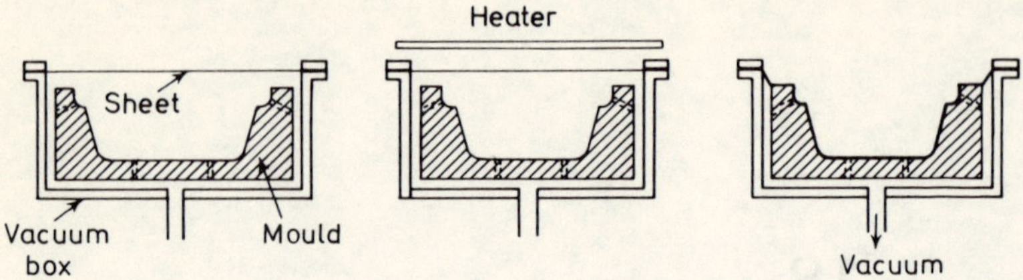

Figure 18.1. Vacuum forming sequence

drawn, and the air in the box is rapidly evacuated by the vacuum pump. Atmospheric pressure above the sheet forces it down into close contact with the mould where it is cooled sufficiently to retain its shape. The clamping frame is then released, the formed sheet is removed from the mould, and the excess material trimmed off. The sequence is shown in *Figure 18.1*.

There are a number of variations on the simple technique outlined above, including drape forming and plug assist. In drape forming the mould is mounted on a piston within the vacuum box. The piston

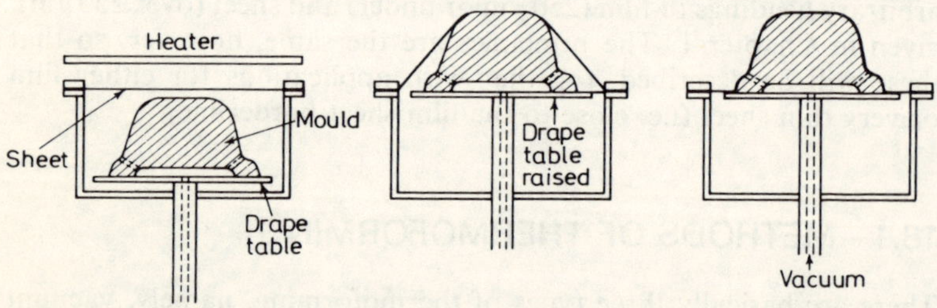

Figure 18.2. Drape forming sequence

rises and pushes the mould into the heat-softened sheet, immediately prior to the vacuum being applied (*Figure 18.2*). This gives a certain amount of pre-forming, and so lessens thinning of the sheet at the corners of the mould.

Plug assist is particularly valuable for deep forming and consists of a hydraulic ram carrying a rough pre-form mould. This is pushed down into the top of the sheet, immediately prior to the vacuum being applied (*Figure 18.3*).

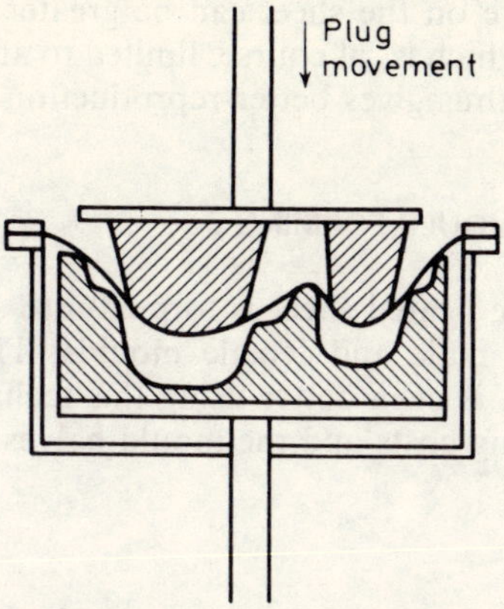

Figure 18.3. Plug assist preforming

Skin packaging

An interesting variation on vacuum forming utilises thin film rather than thick sheet. The articles to be packaged act as their own moulds, the process being known as skin packaging. The articles are placed on a porous board, the whole assembly being then placed in the skin packaging equipment. Film is drawn from a reel and clamped in a frame set above the loaded board. A cutter severs the film from the reel, and a heater moves above the film. After a set period the heater retracts and the frame descends, draping the hot film over the articles on the board. A vacuum is then applied beneath the board and the film is drawn around the articles to be packaged and into intimate contact with the board, forming a permanent bond.

Low density polyethylene film is widely used for this application while ionomer film is also very suitable. The film should be flexible enough to follow the contours of even the most irregularly shaped objects and have a high melt strength. The latter is essential if pin-holes and puncturing at sharp corners are to be avoided.

18.1.2 PRESSURE FORMING

This is similar to vacuum forming with the exception that a positive air pressure (relative to atmospheric) is applied to the top surface of the sheet. As in vacuum forming, this has the effect of forcing the heat-softened sheet into contact with the mould. The main advantage is that the pressure on the sheet can be greater than in the case of vacuum forming which is, of course, limited to atmospheric pressure. Pressure forming thus gives better reproduction of mould detail.

18.1.3 MATCHED MOULD FORMING

In this method the heated sheet is formed into shape by trapping it between matched male and female moulds. The mould detail, as one would expect, is even better using this technique but it is more expensive in tooling costs and the mould halves have to be made to tight tolerances.

Form/fill/seal

Thermoforming techniques are well adapted to form/fill/seal operations. In one type of machine, blisters are formed to a profile roughly similar to the article to be packaged, the article is fed into the blister and the blister sealed to a reel of fibreboard. The individual packages are then trimmed and separated. These machines are used for the packaging of high sales volume items such as reels of cotton, razor blades, etc.

The other main area of thermoform/fill/seal is foodstuffs, including items such as cream, jam and other preserves, sauces, cheeses, etc. The thermoplastics web is passed through a forming stage, the formed sheet is then indexed under a filling head and the filled compartments are lidded by heat sealing another web of material on top. The filled and lidded containers are then cut and separated. The reel of material used for lidding is normally pre-printed and may be a thermoplastic or a thermoplastics coated paper or aluminium foil.

A somewhat similar operation is the packaging of tablets or pills into 'push-through' blister packs. A multi-cavity mould is used and the tablets or pills are automatically distributed into the formed cavities in the sheet. The aluminium foil lidding web is then brought down over the filled plastics web and heat sealed to it. The tablets are eventually dispensed by pushing them from the plastics side, through the aluminium foil.

The plastics web must be rigid enough to protect the tablets during distribution and retailing but flexible enough to allow the tablet to be pushed through the aluminium foil. The aluminium foil should have a hard temper (i.e. be fairly brittle) so that it breaks easily when the tablet is pushed against it during dispensing.

18.2 MACHINE VARIABLES

The machine variables of greatest importance are heating, cooling, and mould design. Other items include trimming and, sometimes, printing.

18.2.1 HEATING

The usual form of heating employed is infra-red, normally the ceramic type having a heating wire element sealed into it. Many machines have heater panels in which a large number of separate elements are wired up in 'chessboard' fashion. This allows more heat to be supplied to certain parts of the sheet, as required.

The heaters are positioned fairly close to the sheet in order not to waste heat but too close an approach must be avoided otherwise a heater pattern may be impressed on to the sheet surface.

18.2.2 COOLING

All plastics are poor conductors of heat and so the moulded sheet should be removed from the mould as soon as it is dimensionally stable. The cooling of thick mouldings may often be assisted by air blasts or water-mist sprays but this is not normally necessary with thin sheet and film.

18.2.3 MOULDS

For small runs and prototype work, thermoforming moulds can be made from plaster or wood and are, hence, very cheap. For long run production work, however, epoxy resins or aluminium are preferred.

Deep and sharply angled corners should be avoided as they cause local thinning of the sheet. The sides of the mould must be tapered to facilitate removal of the moulding, particularly with male moulds. The forming of undercuts is possible using moulds and removable sections but this means a more expensive mould and a longer moulding cycle.

18.2.4 TRIMMING

Trimming may be carried out by a variety of equipment ranging from a simple knife to a complicated cutting press. Roller presses, for example, are widely used in conjunction with a cutting board.

18.2.5 PRINTING

Printing of the finished formed containers is usually possible but may be a slow process, and hence expensive. Printing of the sheet before forming can also be carried out, the design being distorted in such a way that the appearance after forming is as required. One way to determine the distortion necessary is to form a sheet on which a grid pattern has been printed while the sheet was flat. The amount of distortion on forming, at various points on the surface, can then be determined.

18.3 MATERIALS AND APPLICATIONS

Most thermoplastics can be thermoformed but the ease of forming varies considerably. In addition, the end-use must be taken into account when considering the choice of material. The polymers most commonly used for thin-wall thermoforming are considered, briefly, below.

18.3.1 PVC

PVC sheet has good clarity, rigidity and chemical resistance. Very good formings, with good mould detail, can be obtained from even the thinner sheet gauges. It is used for blister packs and for many food packaging uses, particularly chocolate and biscuit box inserts, and for margarine tubs. Because PVC is self-extinguishing it is particularly useful in applications such as aircraft food trays.

18.3.2 TOUGHENED POLYSTYRENE

This is the most commonly used thermoforming material in the United Kingdom. It is easily formed and gives fairly rigid containers with a good depth of draw. One of the largest outlets is for vending and other disposable drinking cups. It is also used for forming a wide range of tubs, trays and box inserts. In the USA a polystyrene cigarette pack has been produced that illustrates its good drawing properties and retention of mould detail. Wall thickness was around 375 μm and outputs of 20 000 packs an hour were claimed, using a 24 impression mould.

18.3.3 BIAXIALLY ORIENTED POLYSTYRENE

Toughened polystyrene is translucent and cannot be used where good clarity is required. Normally, the clear basic polystyrene is too brittle for use but if the clear sheet is biaxially oriented it is effectively toughened and can be thermoformed to give crystal-clear containers. It is difficult to thermoform because of the strain present in the sheet but it can be pressure formed on special equipment.

18.3.4 EXPANDED POLYSTYRENE

Expanded polystyrene sheet can also be thermoformed, matched moulds being preferred for deep or complicated draws. The main problem with this material is that the foam structure gives it an even lower heat conductivity than normal. It is difficult, therefore, to heat the sheet right through without overheating the surface. The sheet is, therefore, normally heated from both sides at once. Expanded polystyrene sheet is opaque and has an attractive surface sheen. Applications include trays for pre-packaging fruit, vegetables and meat, as well as for the transport of delicate electronic components.

18.3.5 ABS (ACRYLONITRILE/BUTADIENE/STYRENE)

The thermoforming properties of this material are similar to those of toughened polystyrene (which it resembles chemically) but it is more expensive. It is also tougher and can sometimes be used in thinner gauges than toughened polystyrene, so reducing the price differential.

18.3.6 LOW DENSITY POLYETHYLENE

This material can be thermoformed although it has a tendency to sag when heated, and the finished formings are also rather limp.

18.3.7 HIGH DENSITY POLYETHYLENE

Thermoforming is easier with this material than with low density polyethylene because of its increased rigidity and hardness. Double sided heating is usually desirable because the sheet expands appreciably on heating to give folds which may weld together, given time.

In addition to fast heating accurate control of temperature is essential. In the USA high density polyethylene has been used for the manufacture of margarine tubs.

18.3.8 POLYPROPYLENE

Polypropylene has greater stiffness and is more resistant to stress cracking than are the polyethylenes, and it is more resistant to high temperatures. The forming properties of polypropylene are similar to those of high density polyethylene and, in general, the same methods can be used for both materials.

Applications are usually in areas where the thermoformings need to be resistant to boiling water or to moderate sterilisation temperatures. At the other end of the scale, polypropylene is not so suitable as high density polyethylene because the impact strength and other physical properties decrease rapidly with decreasing temperature (below about 0°C).

18.3.9 CELLULOSE ACETATE

Cellulose acetate possesses excellent clarity and is readily thermoformed, giving rigid mouldings particularly suitable for blister packaging. It is rather expensive but is used where high class formings are required. Temperature control is critical because too high a temperature causes surface defects such as blistering, while too low a temperature leads to 'milkiness' in the finished forming. One disadvantage of cellulose acetate is its dimensional instability in conditions of changing humidity, due to moisture pick-up from the atmosphere. Cellulose acetate formings are easily fabricated either by adhesives or by solvent welding.

18.3.10 CELLULOSE ACETATE/BUTYRATE

This material is similar to cellulose acetate but has superior dimensional stability. Like cellulose acetate it possesses good rigidity and clarity and is easily fabricated by adhesives and by solvent welding.

18.3.11 POLYCARBONATE

Polycarbonate can be thermoformed but needs higher forming

temperatures than the other materials previously mentioned because of its higher softening point. It possesses great clarity and a very high impact strength. It is particularly useful for the manufacture of blisters for hard, sharp articles such as hardware.

18.4 COLD FORMING

A big disadvantage of thermoforming is the time and money wasted in double heating, i.e. heating of the granules followed by reheating of the sheet. Thermoforming of the newly extruded sheet can be carried out but this introduces problems in matching the speeds of the separate processes. Changeovers are also expensive because the extruder must be shut-down as well as the moulding process.

Work has been carried out, therefore, on the forming of cold or only slightly heated sheet using techniques adapted from sheet metal working. These techniques are more applicable to sheet rather than film but some success has been achieved with the more rigid materials such as ABS, polycarbonate and rigid PVC.

FURTHER READING

BEADLE, JOHN D. (Ed), *Plastics Forming*, Macmillan, London (1971)
MILES, D. C. and BRISTON, J. H., *Polymer Technology, 2nd edition*, Chemical Publishing Co. Inc., New York (1979)

19 Lamination

Strictly speaking lamination is the process of bringing two or more webs (films, papers or foils) together and bonding them with an adhesive or by heating. However, it is convenient to include coating and co-extrusion techniques in this chapter so that lamination is here considered to be any method of producing a layer of composite film.

19.1 COATING

The coating process is important not only as a means of modifying film properties but also as a means of applying the adhesive in adhesive lamination. The various coating methods will be dealt with under this heading, therefore, but will be referred to again when dealing with adhesive lamination.

Coating processes can be divided into two main groups. In the first, an excess of coating is applied to the web and the surplus is removed by means of a rigid knife, flexible blade, smoothing bar, or air-jet knife (air doctor). In the second group, a pre-determined amount is applied to the web by means of a calender, rollers, brushes, or by curtain coating. Extrusion coating is another method but this is of particular importance in the application of low density polyethylene to various substrates and will be dealt with separately.

19.1.1 EXCESS APPLICATION SYSTEMS

Flexible blade coaters are more commonly used for applying high solids clay-based coatings to paper and board than for use with polymer coatings. However, the growth of synthetic art papers where

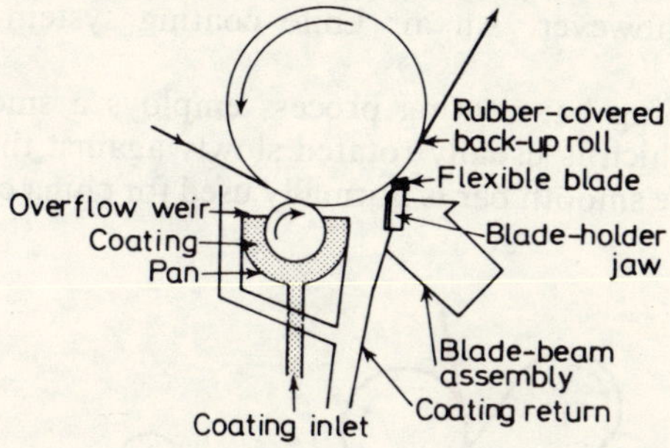

Figure 19.1

a plastics film is covered with the type of coating mentioned could well create an interest in flexible blade systems. A flexible blade/roll coater is shown in *Figure 19.1*.

Rigid knife systems usually operate with a roll or blanket backing (see *Figures 19.2* and *19.3*). They are normally used with viscous coating materials, a particular example being the coating of fabrics with PVC dispersions.

Where low to medium viscosity coatings are to be applied, the air doctor is preferred. The excess of coating material applied is comparatively small (about 1:1 to 1.6:1 as compared with up to 15:1

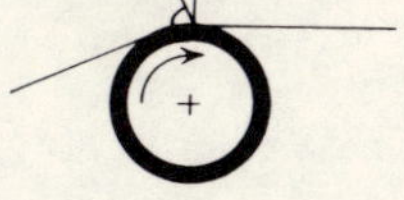

Figure 19.2. Knife over roller spreading

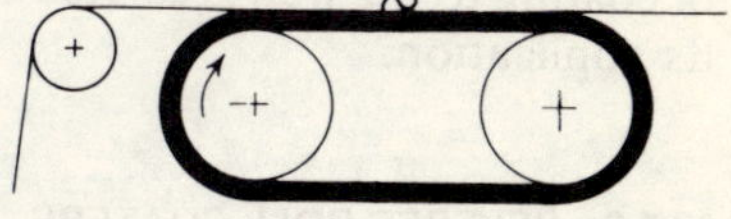

Figure 19.3. Knife over blanket spreading

for flexible blade coaters) and the main determining factor for coating weight is the air knife pressure. Other factors include air stream angle, web speed and the physical properties of the coating.

The air knife method is widely used for coating a variety of webs, including plastics film. It has been generally restricted, however, to the application of water based coatings, because of difficulties in handling large volumes of air mixed with flammable solvent vapours, and surface drying of the coating by the air knife. New equipment and techniques are expected to lead to improvements in this respect, however. An air knife coating system is shown in *Figure 19.4*.

The smoothing bar coating process employs a smooth or wire-wound bar, which is usually rotated slowly against the direction of web travel. The smooth bar is normally used for china clay coating of

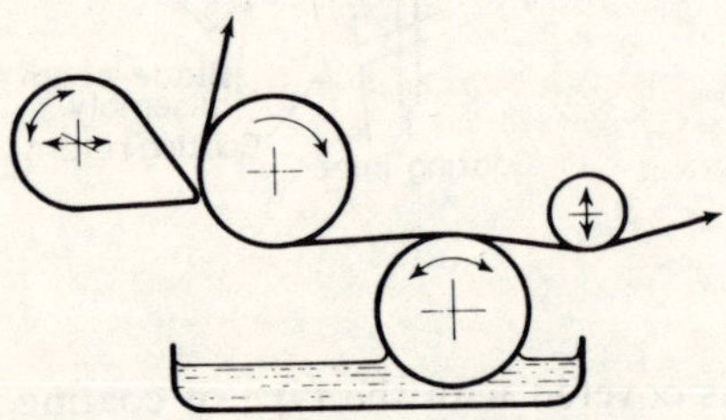

Figure 19.4. Air doctor coating

fibreboard while the wire-wound bar can be used for hot melt coating. It is difficult to achieve an even coating weight with this method and it is liable to produce occasional scratch marks on the coating.

19.1.2 PREDETERMINED SYSTEMS

In this type of system the application of a predetermined amount of coating to the web is carried out by pre-metering the coating before its application.

19.1.3 REVERSE ROLL COATERS

Reverse roll coaters are capable of metering and applying uniform films of coating to a wide variety of surfaces, using a wide range of coating solids and viscosities. The simplest form of roll coaters, such as the single roll and two roll reverse coaters, have only a limited metering action, however, and are not recommended when precise coating control is required. In a one roll reverse coater, the roll rotates in a bath containing the coating and the web passes over the top of the roll in the opposite direction to the rotation of the roll. The two roll coater merely adds a rubber backing roll, run at line speed (*Figure 19.5*).

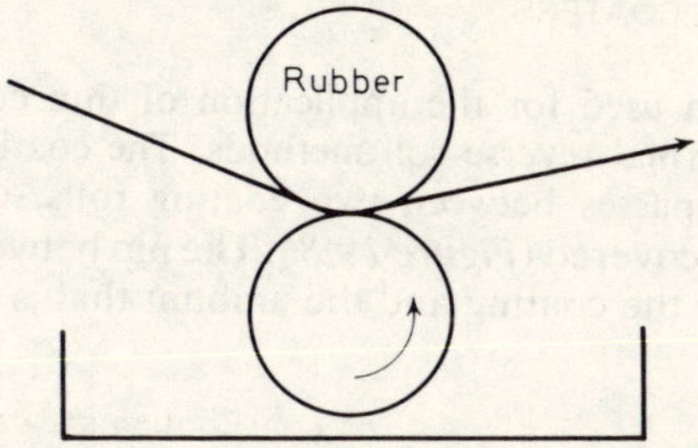

Figure 19.5. Two roll reverse-roll coater

In three or four roll coaters the extra rolls serve to meter the coating more precisely (*Figure 19.6*). Coating weight is primarily controlled by the metering roll gap and the rate of rotation of the

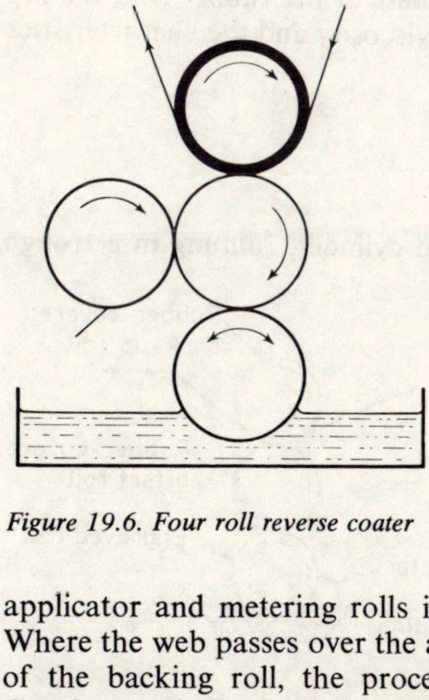

Figure 19.6. Four roll reverse coater

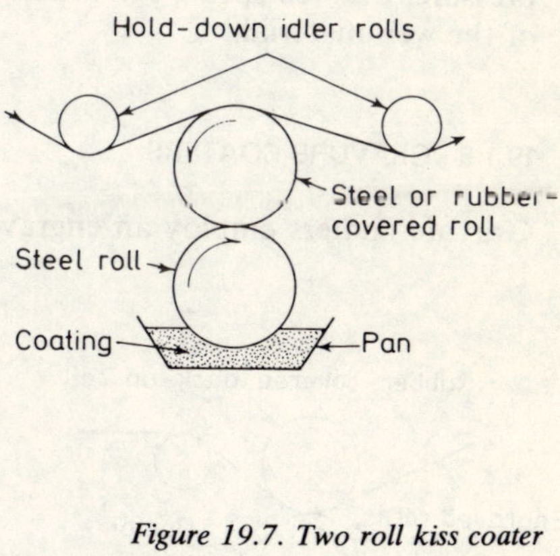

Figure 19.7. Two roll kiss coater

applicator and metering rolls in relation to the speed of the web. Where the web passes over the applicator roll without the assistance of the backing roll, the process is referred to as 'kiss' coating. Good control of web tension and complete contact with the applicator roll are essential for uniform coating. These are usually achieved by hold-down idler rolls on either side of the applicator roller (*Figure 19.7*).

19.1.4 NIP ROLL COATERS

These are often used for the application of thin coatings, difficult to apply by normal reverse roll methods. The coating is applied to the web as it passes between two coating rolls, one of which is usually rubber covered (*Figure 19.8*). The nip between the two rolls is flooded with the coating and the amount that is actually applied

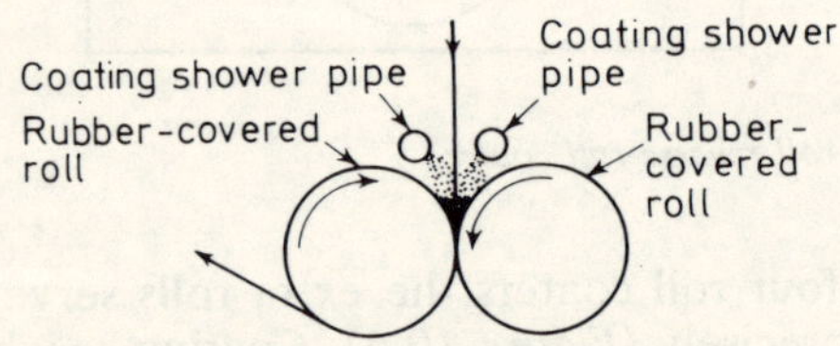

Figure 19.8. Two roll nip coater

to the web is controlled by the hardness of the rubber roll, the nip pressure, the web speed, the coating viscosity and the characteristics of the web material.

19.1.5 GRAVURE COATERS

Gravure coaters employ an engraved cylinder, running in a trough

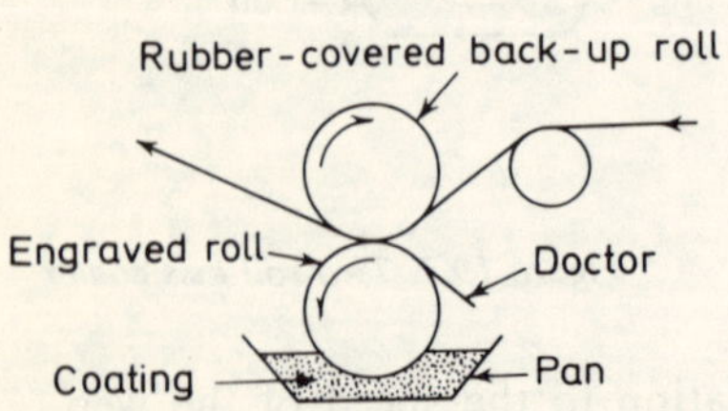

Figure 19.9. Direct gravure coater

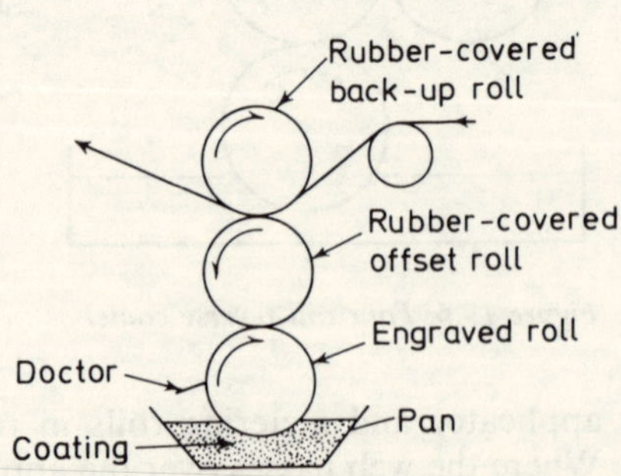

Figure 19.10. Offset gravure coater

of coating material, to meter the transfer of coating to the web. The engraved roll is doctored with a thin blade to ensure that an even amount of coating is left in the engraved cell (*Figure 19.9*). Direct gravure coaters consist of two rolls only. The engraved roll transfers the coating directly to the web at the nip formed by the engraved roll and a rubber backing roll. This has the disadvantage that unless there is good flow-out of the coating, the engraved pattern is transferred to the web. This is eliminated by using a three roll system as in the off-set gravure process, shown in *Figure 19.10*.

Coating weights are generally limited in the gravure process but precise control is possible. One disadvantage is that the engraved cylinder must be changed when a change in coating weight is desired.

19.1.6 CALENDER COATING

Calenders have already been described in Chapter 8 in connection with the manufacture of film (particularly from PVC). The calender can be used as a coater merely by passing the web to be coated into

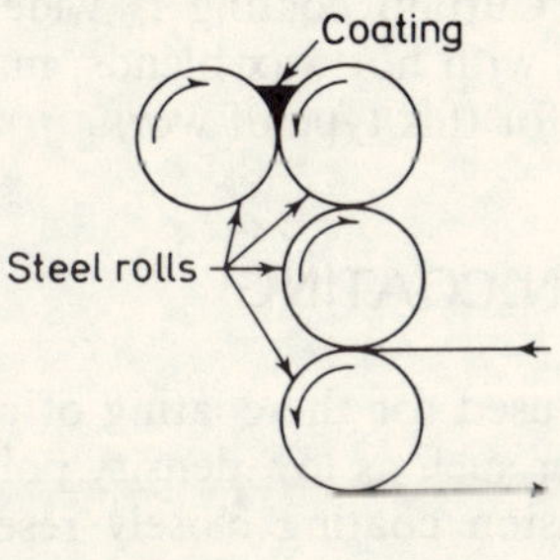

Figure 19.11. Four roll calender coater

the nip between the last two rolls. A four-roll calender coater is shown in *Figure 19.11*. This method of coating is generally restricted to heavy coating weights and is often used for applying high viscosity vinyls to paper or fabrics. The thickness of coating is controlled by the gap between the last two rolls.

19.1.7 CURTAIN COATING

A typical curtain is shown in *Figure 19.12*. A laminar film of the coating material is extruded through a slotted orifice or over a weir.

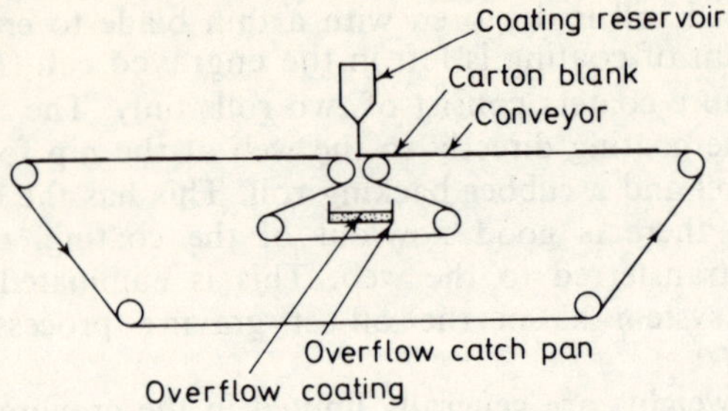

Figure 19.12. Curtain coater

Curtain coating is suitable for moderately thin, evenly flowing materials, such as lacquers, hot melts and some latices. It gives good coverage of irregular surfaces and is particularly suitable for carton blanks. Coating thickness is governed by the orifice dimensions or the flow rate over the weir, the web speed and the coating formulation. The thinness of the falling curtain is limited by the need for it to be continuous and this also limits the speed of the coating line. Curtain coating is widely used for hot melt coatings, particularly with hot wax blends, many of which are now specially formulated for this type of work.

19.2 EXTRUSION COATING

Extrusion coating is used for the coating of a wide variety of substrates with a polymer such as low density polyethylene. The equipment used for extrusion coating closely resembles that described in Chapter 8 for the chill roll casting of film. Basically, the only

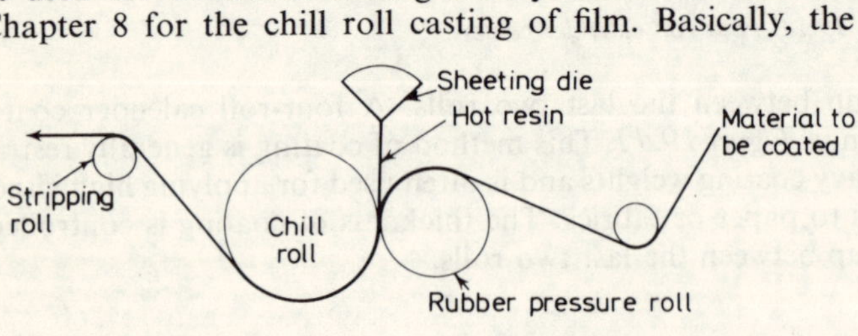

Figure 19.13. Hot extrusion coater

additions necessary are a carrier unit for the web to be coated and a pressure roll, situated on the first cooling roll (see *Figure 19.13*).

The molten polymer is then extruded as a continuous film onto the web at the nip formed by the cooling roll and the rubber-covered pressure roll. Control of web width is achieved by fitting adjustable deckles to the slit die.

One of the most important features of an extrusion coating unit is the cooling roll as it controls the uniformity of coating properties and the surface finish of the final material. Uniformity of coating properties depends on the maintenance of a uniform surface temperature over the whole area of the roll. Roll finishes can vary from a highly polished chromium plate finish to a rough texture designed to give a matt finish to the coating.

Adhesion of the molten polymer to the substrate is obviously important and this is promoted by mounting a pressure roll behind the substrate at the point where the melt first contacts the cooling roll. This roll is usually constructed of a heat resistant rubber. Optimum conditions for effective adhesion are obtained by adjusting the pressure of the roll on the substrate and the position of the pressure roll. Insufficient pressure will lead to poor bond strength of the coating while too great a pressure will give rise to surface defects such as the formation of a small ridge of molten polymer. Another factor affecting the adhesion of the molten polymer to the substrate is the distance between the extruder die and the nip between the two rolls. This area is known as the air gap or draw distance and it is in this region that draw down and neck-in occur. Neck-in is the reduction in width which occurs in the molten film as it leaves the die. Surface oxidation of the melt, which promotes adhesion, also occurs in this region. The separation should, therefore, be sufficient to permit oxidation to occur but must not be large enough to cause an appreciable reduction in melt temperature since this would reduce penetration of the polymer into the substrate. The air gap must also be kept as small as possible to prevent excessive neck-in.

Draw down and neck-in are both important variables in the extrusion coating process and are both related to polymer elasticity. Draw down is the reduction in the thickness of the extruded polymer between the extruder die and the chill roll. Ideally, the required characteristics of an extrusion coating polymer are low neck-in and high draw down. Low neck-in means a higher utilisation of substrate width while high draw down leads to economies by virtue of attaining as thin a film as possible without pinholing or tearing. The thickness of the applied film is governed by the width of the die opening and the relative speeds of the film and the substrate. In practice, a compromise must be accepted since neck-in and draw down are both reduced by an increase in polymer elasticity.

High temperatures are required for extrusion coating in order to reduce melt viscosity. Low density polyethylene is inclined to give some odour under these conditions and this is undesirable for food packaging use. Development work on adhesive primers has enabled temperatures to be reduced with a consequent reduction in the problem of odour. Adhesive lamination is another method used. The absence of taint, here, depends on the solvent used and the efficiency of the drying process.

Other developments in extrusion coating include the use of a perforated roll instead of the normal chill roll for porous substrates such as paper. A vacuum can then be applied behind the paper thus achieving greater adhesion of the coating to the paper at higher speeds. There is also an increase in the quality of the coating. Developments in equipment design seem to be favouring smaller extruder diameters with high power input and longer screw-metering sections. Better methods of controlling temperature are also being developed including infra-red radiation pyrometers.

The extrusion coating process has been used mainly for the coating, with low density polyethylene, of paper, aluminium foil and regenerated cellulose film. In the case of regenerated cellulose film a single side coated film is used and the polyethylene is applied to the uncoated side. It should be noted here, too, that extrusion coating can be adapted to the production of a triple laminate. In this case the extruded polymer acts as the bonding agent between two substrates.

Extrusion coating utilises expensive equipment and is a high output operation. For economic running, therefore, it must be used continuously with as little down time and as few polymer changes as possible.

19.3 ADHESIVE LAMINATION

Adhesive lamination is adaptable to short as well as long runs and is well-suited to the one-step production of laminates having more than two components. Laminating processes can be divided into two major categories, namely, wet bonding and dry bonding.

19.3.1 WET BONDING

Wet bonding uses solvent or aqueous-based adhesives and can only be used when one or more of the webs is permeable to the water or other solvent used in the adhesive formulation. In general, wet bonding is not usually successful with plastic films even when

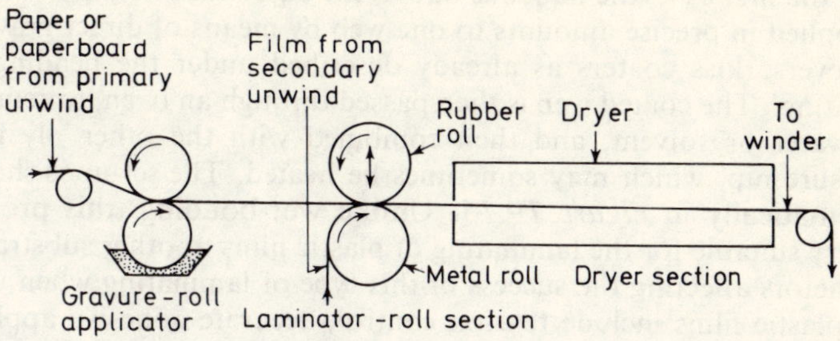

Figure 19.14. Wet bonding of paper to film

laminating them to paper. However, wet bonding using organic solvent-based adhesives, has been carried out in some instances, and even aqueous-based adhesive lamination can be carried out for films such as cellulose acetate when bonded to paper. The cellulose acetate is fairly permeable to water vapour and so aids drying out of the water after laminating. In any case, the finished laminate must be run through a drying oven to speed up the drying process, whereas when laminating fibre board to fibre board the water is usually absorbed into the two webs and additional drying is unnecessary. A typical set-up for wet bonding of paper to film is shown in *Figure 19.14*.

19.3.2 DRY BONDING

The dry bonding process incorporates either, (1) the use of an aqueous or solvent-based adhesive film that is dried prior to laminating or, (2) a hot melt adhesive, based on wax or on one of a range of polymers.

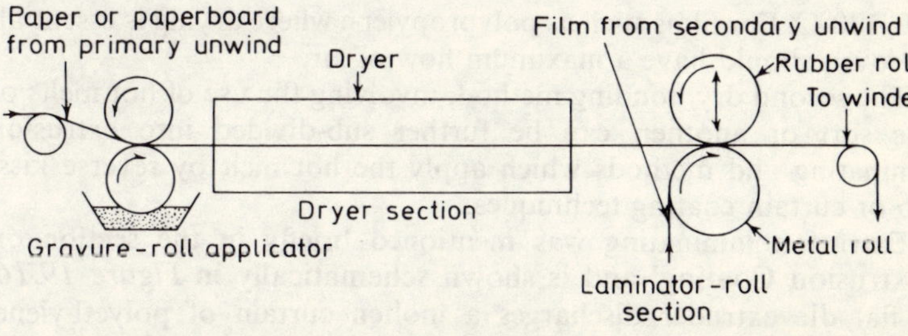

Figure 19.15. Solvent dry bonding

In the first case, the aqueous or solvent-based laminating adhesive is applied in precise amounts to one web by means of direct gravure or reverse kiss coaters as already described under the heading of 'Coating'. The coated web is then passed through an oven, to remove all water or solvent, and then combined with the other ply in a pressure nip, which may sometimes be heated. The set-up is shown schematically in *Figure 19.15.* Unlike wet bonding, this process is very suitable for the laminating of plastic films to other substrates.

Factors affecting the success of this type of laminating when used for plastic films include tension control, accurate adhesive application and accurate control of drying. Film tension should normally be kept to a minimum and will depend on the distance the film has to be pulled through the laminating equipment and on the sharpness of any change in direction as it passes over the various rolls. Effective control of tension is easiest in those machines where both the unwind station and the rollers in the drying tunnel are in direct line. Where a reverse roll coating system is used the reverse roll should not run too fast otherwise drag will occur, thus introducing increased film tension. Smooth and accurate application of the adhesive is extremely important and failure in this respect will probably lead to delamination. Both gravure coaters and reverse roll systems should give adequate control of adhesive application but adhesive viscosity is another factor and this should be kept as constant as possible.

Drying oven performance is particularly important when laminating plastics film where solvents cannot be absorbed into the film. Excess solvent remaining in the adhesive at the nip stage is a major cause of delamination. Films vary in respect of solvent absorption and cellulose acetate, for instance, is able to absorb a slight amount of solvent without serious effects during lamination, whereas oriented polypropylene is much more critical in this respect. The temperature stability of oriented polypropylene is lowered in the presence of certain adhesive solvents and shrinkage can occur at temperatures of around 100°C whereas in the absence of solvent it would withstand 130–140°C. For films such as polypropylene where drying is essential, the oven should have a maximum flow of air.

The second dry bonding method, involving the use of hot melts of one sort or another, can be further sub-divided into extrusion laminating and methods which apply the hot melt by reverse kiss, dip or curtain coating techniques.

Extrusion laminating was mentioned briefly in the section on 'Extrusion Coating' and is shown schematically in *Figure 19.16.* A flat die extruder discharges a molten curtain of polyethylene, polypropylene or some other thermoplastic into the nip between the two webs to be laminated. The heated adhesive is cooled by passing

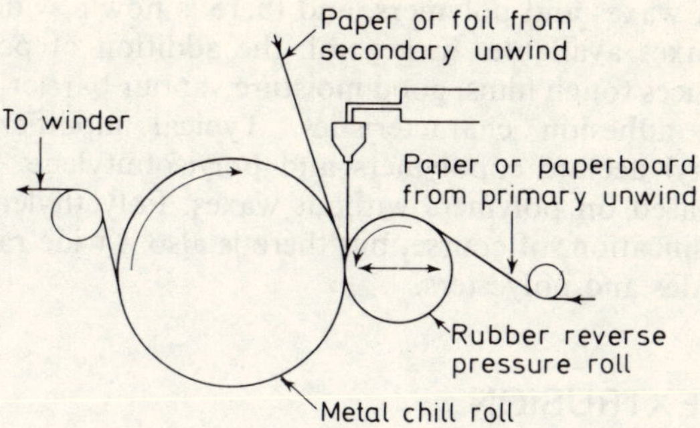

Figure 19.16. Extrusion laminating

the laminated sheet over a specially designed combining and cooling roll section. This cooling section replaces the drying section generally needed in wet adhesive bonding. Extrusion laminates are good barriers to water and water vapour and are usually tough and flexible. Typical laminates produced by this method are paper/polyethylene/aluminium foil or paper/polyethylene/regenerated cellulose film.

The equipment used for roll application of hot melts is shown in *Figure 19.17*. After the hot melt has been applied by a reverse-kiss coater to one of the webs, the two webs are brought together at the nip between a pneumatically loaded rubber roll and a fixed steel roll. The laminate again passes through a cooling section and is then wound up. The earliest hot melt adhesives were waxes of one sort or another and these are still widely used. Microcrystalline waxes

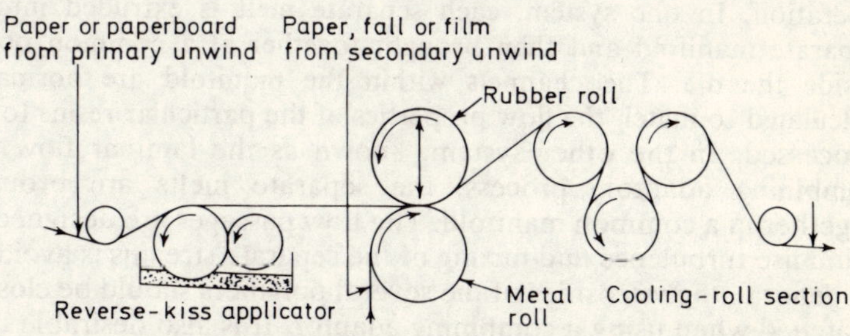

Figure 19.17. Hot melt laminating

are the most popular for this work but paraffin wax is still used. With more critical performance requirements came developments in blends of waxes and polymers, and there is now a wide range of modified waxes available. In general, the addition of polymers to waxes produces tough films, good moisture vapour barrier properties and better adhesion characteristics. Typical modifiers include ethylene/vinyl acetate copolymers and polyisobutylene. Some hot melts are based on polymers without waxes. Polyethylene belongs to this classification, of course, but there is also a wide range based on polyamides and polyesters.

19.4 COEXTRUSION

This method of producing composite films differs in one important detail from all those already described, inasmuch as it can be adapted to produce a composite film in either lay-flat tubing form or as a flat film. The process for making tubular composite films is often referred to as coaxial extrusion.

Basically the process consists in coupling two or more extruders to a single die head. It is essential that precise control is available over screw speed, melt pressures and temperatures, and power capacity of the extruder drives. Size of extruders used will depend on the ratio of the components in the laminate. Extruders of similar capacity are usually employed when a wide range of composite film thicknesses is required, but for composites of widely different polymer content, different sizes of extruders are usually combined, so minimising capital cost.

However, the success of the technique lies chiefly in the design of the die and the way in which the individually extruded melts are brought together, prior to being extruded as a multi-layer film. There are, essentially, two different methods of carrying out this operation. In one system, each separate melt is extruded into a separate manifold and then brought together at a common point inside the die. The channels within the manifold are normally calculated to match the flow properties of the particular resins to be processed. In the other system, known as the laminar flow (or combining adaptor) process, the separate melts are brought together in a common manifold. The flow passages are designed to minimise turbulence and mixing of the separate streams is avoided. In general, the viscosities of the several polymers should be closely matched when using a combining adaptor. It is also desirable that the viscosities match at about the same temperature to eliminate problems associated with maintaining different melt temperatures

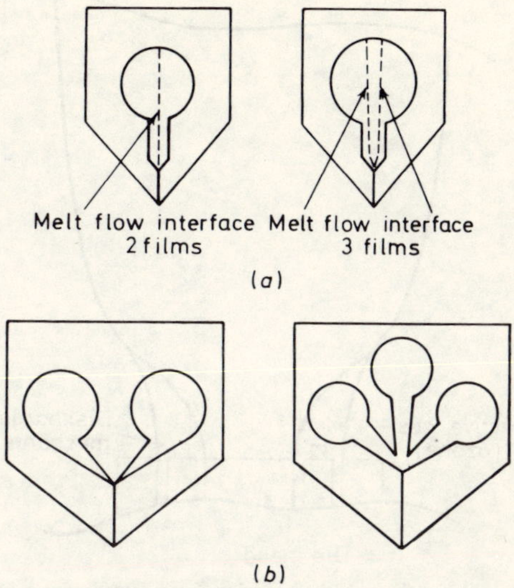

Figure 19.18. Coextrusion using (a) single manifold dies (b) multiple manifold dies

in the die.

The combining adaptor allows great flexibility in operation. It can extrude two or more layers relatively easily and it can extrude a polymer in a much thinner layer than would be possible if it were extruded as a single film. However, it is not possible to maintain a large temperature difference between layers (which is possible in the multiple manifold process). There are also many combinations of materials that do not flow together and produce a uniform product. A diagrammatic representation of each method is shown in *Figure 19.18*.

In the case of coextrusion (of tubular film) there is also a third process. This involves the coextrusion of two films from separate annuli in the same die head. The separate films are brought together below the 'freeze line' of both bubbles, i.e. while still molten (*Figure 19.19*). Provision is made in the die for the space between the films to be saturated with a reactive gas which 'treats' the adhering surfaces and improves the bond strength. The most commonly used reactive gas, for this process, is ozone, formed by an electrical discharge mechanism. This method is particularly valuable where unlike materials, such as nylon and polyethylene, have to be combined.

Coextrusion, generally, is an important addition to the various

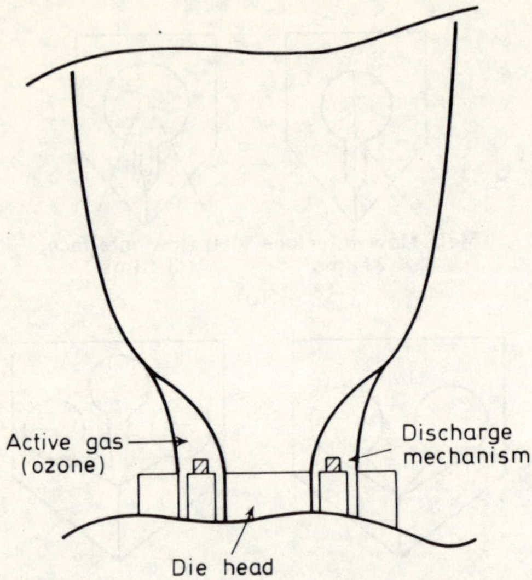

Figure 19.19. Coaxial extrusion with provision for treating with active gas to improve adhesion

techniques used for producing composite films. The main advantages of coextruded films over conventionally produced laminates are lower cost, a lower tendency towards delamination and a greater flexibility in obtaining a wide range of properties. The possibilities of cost savings may not be immediately obvious. After all, the equipment needed is bound to be more expensive than that necessary to produce the separate films. Extrusion speeds, too, will probably be lower although developments in equipment design may well alter this. The first possibility of cost savings lies in the fact that in multi-layer extrusion, the finished laminate is produced in one operation. In normal laminating procedures, of course, the respective single films have first to be extruded, followed by reeling-up and storage. Later on, these reels have to be un-reeled and run through the laminating process and this obviously adds to labour costs. Cost savings on material are also possible. The minimum thickness of any particular ply in a conventional laminate is often dictated by its handling characteristics on laminating equipment rather than by the barrier or other properties required. However, very thin plies indeed can be produced by coextrusion so that the optimum thickness can be chosen for the job. In this connection it is interesting to note that three layer films have been produced with a thickness of less than 25 μm overall.

Apart from cost savings in the production of the laminated film, there are also possible savings in bag making. Multi-layer lay-flat film can be produced by coaxial extrusion. It is then possible to make a bag of laminated construction merely by making one cross-seal. Coaxial extrusion, then, provides the only method of producing a bag of laminated construction without side or centre seams.

The lower tendency towards delamination is possibly due to the fact that the interfacial surfaces of coextruded films do not oxidise because they never come into contact with the air. Normally, such oxidation does occur on the surface of any polymeric film as it emerges hot from the extruder. Tiny crystals are then formed on the surface and these interfere with the subsequent bonding during lamination. The position is not always so simple, however. During early work on producing coaxially extruded laminates of such dissimilar films as polyethylene and nylon, difficulties in bonding were encountered. The problems were solved as mentioned earlier by introducing a gaseous oxidative medium in the annular space between the coaxially extruded plies. The paradox may perhaps be explained by the fact that in the case of the coextruded film, the bonding occurred between the two oxidised surfaces while the polymers were still molten so that no actual crystals were formed. Another way of overcoming difficulties in bonding two particularly incompatible films is to use 'bridge' materials which are compatible with both of the former materials. In some cases this may involve a graduated 'bridge' leading to finished laminates of five or seven materials. This sort of approach is only possible economically because of the very thin layers which are possible using the coextrusion process.

The advantage of flexibility in obtaining a wide range of properties depends on the possibilities (already mentioned) of producing very thin component plies. Apart from the obvious factor of cost, there is often a limit to the thickness of a laminate which is set by handle-ability (stiffness, etc.) or even clarity. In such cases the number of different properties which can be built-in by using different single plies is also limited.

There are also disadvantages to the use of coextrusion. One of these is the difficulty of utilisation of the scrap produced during the extrusion process. When extruding a single ply film, a fair proportion of scrap film can be fed into the extruder and used again since with most plastics there is no danger of chemical degradation. With coextrusion, however, the problem is that a mixture of polymers is produced with no intervening stage, and separation of the plies, even when possible, would not be economic. This disadvantage will, of course, be aggravated in the case of short runs. Coextrusion will, therefore, be best for long runs with steady running conditions. On the other hand, one can visualise certain composite films where the trim could be used to produce a blended core with more specialised polymers as the outer plies. There is, of course, a limit to this type of scrap utilisation.

Another disadvantage is the fact that laminates with sandwich print cannot be produced. This is a particular disadvantage in food packaging when a scuff-proof decoration may well be required but where printing cannot be on the inside of the ply in contact with the food.

Coextrusion has already found many applications, particularly in the USA. One of the first was the use of a three-layer film consisting of 16 μm low density polyethylene/15 μm polypropylene/16 μm low density polyethylene for bread wrapping. The polypropylene core was incorporated in order to eliminate 'burn-through' during heat sealing. Later, a cheaper film was introduced using only 2.5 μm polypropylene between 2 × 11 μm layers of low density polyethylene. The necessity for such laminates eventually disappeared with the changeover to prefabricated polyethylene bags utilising wire ties instead of heat seals.

In Holland, a laminate of white and black polyethylene was developed for milk sachets. The black polyethylene was considered necessary to prevent loss of vitamins by light while the white layer was used for the outside of the sachet in order to give a more acceptable appearance. Such a laminate could have been produced as a flat film by normal laminating techniques, but many users required bags made from lay-flat tubing in order to eliminate possible leakage at centre seals. The coaxial extrusion process, was, therefore, the only one which could be used.

More sophisticated laminates have been produced in the USA, one example being a two-layer laminate consisting of high density polyethylene and a blend of low density polyethylene with ethylene vinyl acetate (EVA). This particular laminate is used for shipping sacks. The high density polyethylene supplies stiffness to the sacks and helps them to retain their shape during filling while the EVA widens the sealing range of the high density polyethylene and supplies extra impact strength. An ionomer/low density polyethylene laminate has apparently replaced a cellulose film/cellulose film laminate for certain confectionery lines. The bags are allegedly wrinkle-free and tougher than the cellulose film one and grease resistance is still good.

One laminate with a particularly good combination of properties is that made from low density polyethylene and nylon. It is a good barrier to both water vapour and gases, the nylon confers particularly good grease resistance and odour impermeability and the laminate is useful over a wide temperature range. In the UK, it has been used for sachets to hold putty, quick frozen cream and smoked salmon. In Germany it has been used, or suggested, for such diverse applications as vacuum packaging of bacon, cheese, sausages and fish, the packaging of greases and ointments and the packaging of liquids such as fruit juices, essences and milk. An interesting variation is blow-up packaging where the package is pressurised with air or inert gases. This not only provides the right sort of atmosphere but also acts as a cushioning pack giving protection to the contents. The nylon/polyethylene laminate gives the gas barrier properties and strength required for this particular type of pack.

The newer plastics are also being used in coextrusion. As well as the ionomer based one already mentioned, there is one based on TPX. TPX has a particularly high melting point (240°C) and it has been put forward as the outer layer in laminates utilising low density polyethylene or other low melting point polymers as the heat seal medium. Because of the low heat conductivity of polymers it is necessary to use quite high temperatures if heat sealing is to be carried out in a short space of time. The use of a high melting point polymer, such as TPX, as the outer ply eliminates the danger of breakdown of the outer ply and sticking to the heat sealer jaws.

Foamed plastics can also be utilised in coextrusion techniques and some have already been developed. One example already produced in the USA, is a 3-layer film with a core of foamed low density polyethylene sandwiched between two layers of unfoamed low density polyethylene. It should also be noted that it is possible to extrude multi-layer coatings on to substrates such as paper or aluminium foil. In the thicker gauges, multi-layer films can be used for making thermoformed containers and again, some work has already been carried out.

More straightforward coextruded laminates (but ones that have good potential in the field of packaging) have also been developed. These are (a) polyester, and (b) biaxially oriented polypropylene, with heat sealable layers on the outside. The coextrusion based on oriented polypropylene, in particular, is being used in the packaging of moisture sensitive products such as potato crisps. The coextruded heat sealable surface consists of a copolymer of propylene with a small amount of ethylene. The copolymer has a lower melting point than that of the homopolymer and with careful temperature control a completely satisfactory heat seal can be obtained.

19.5 CROSS-LAMINATED FILM

One of the latest innovations in multi-structural films is a lamination of two high density polyethylene films, with each layer's uniaxial orientation (which runs at a 45° angle from the edge of the film) at right angles to the other's orientation.

The material possesses a very high tear strength by virtue of the cross-lamination and thinner gauges than normal can be used. The high tear resistance also allows the use of conventional stitching equipment with a crepe-tape binder for sealing of the bags.

A similar laminate is based principally on polypropylene but may contain other polymers. It is manufactured by cold lamination and biaxial orientation (using a patented cold roll technique) or from two or more cross-oriented layers of co-extruded film. The material can be heat sealed, sewn, eyeletted, taped and nailed and has good frictional properties. Uses include sacks, pallet covers and tarpaulins.

Part 4
Applications

20 Packaging

Packaging forms by far the largest outlet for plastics films and it is worth considering why. Packaging is not all that easy to define but one fairly concise definition is that, 'Packaging must protect what it sells, and sell what it protects'. Packaging has, therefore, to supply both protection and sales to the product being sold. Plastics films are able to carry out both these duties very effectively and if one particular film cannot be found to supply the combination of properties required, there are simple techniques available for combining two or more films to give a satisfactory result.

The primary function of packaging, of course, is to contain the product and here the versatility of plastics films with regards to sealing methods is extremely valuable. As discussed in Chapter 15, plastics films can be heat sealed, stapled or stitched, bonded with adhesives or, in some cases, sealed with ultrasonics or high frequency heating.

The versatility of plastics films also extends to types of packaging. Thus, they can be used for straight wrapping, shrink wrapping, sachets, pouches, bags and heavy duty sacks, skin packaging and blister packs. In addition, the more rigid films, such as PVC can be vacuum formed into trays, tubs and box inserts. In some cases, plastics films have introduced new packaging concepts, as in the case of shrink wrapping, and the range of form/fill/seal sachets and pouches.

The range of use for plastics films in packaging is so wide that it becomes difficult to avoid making a catalogue of the various applications. The subject will be broken down, therefore, by considering the different plastics separately.

20.1 LOW DENSITY POLYETHYLENE

This particular polymer accounts for about 75% of the total usage of thermoplastics films in packaging, and over 56% of the total of thermoplastics films plus regenerated cellulose. It is thus the most important of the packaging films.

20.1.1 HEAVY DUTY SACKS

The biggest single outlet for low density polyethylene film is in the field of heavy duty sacks. The first products to be packed in these sacks in the UK in the early 1960s, were compound fertilisers. The existing package was a multi-walled paper sack, the usual construction for a 50 kg sack being five paper plies with an inner ply of 25 μm low density polyethylene. The main advantage seen for the all plastic sack was that the filled sack could be stored in the open air thus freeing expensive storage space for the farmer. In addition, the farmer could then buy fertiliser out of season and so qualify for an early delivery rebate. Finally, the plastics sacks of fertilisers could be kept in the fields where they would later be required, thus saving labour in distributing sacks from a central storage point. The original polyethylene sacks used 250 μm thick film and were dearer than the existing paper sacks. Nevertheless, work went ahead and a market now exists for about 65 000 tonnes/annum (1979 figures) of low density polyethylene for sacks. The fertiliser sacks in particular are now cheaper than their paper counterparts and are only 200 μm in thickness.

There were many difficulties encountered during the development of this now very successful market and it is interesting to look at some of these. One of the first difficulties was that of handling. When the open mouth sack was filled and sealed it assumed a pillow-like shape which was difficult to stack. Stack instability was often thought to be due to a lower coefficient of friction for low density polyethylene, but in fact polyethylene on polyethylene has a higher coefficient than has paper on paper and it was soon recognised that the main trouble was the small area of contact with the pillow-shaped polyethylene sacks, against that given by the block-shaped paper sacks. One early solution that was proposed was to place the empty sack in a block shaped jig, fill, and seal all but a small proportion of the seal. A vacuum was drawn in the sack and the seal completed. This gave a brick-like package but the taut film was easily punctured and the sack was then more floppy than the normally filled one. A more acceptable solution was found to be the provision of a few micro-perforations along the top of the bag. These holes allowed the air inside the sack to escape and the contents of the bags were consolidated under the influence of impacts during handling and movements on conveyor belts, etc. The holes were small enough not to let in moisture except in heavy rainfall. This meant that bags could not be stored under completely unprotected conditions, but a light tarpaulin or sheet of polyethylene film over a stack is sufficient for normal protection. This is still better than paper sacks which lose strength even under conditions where liquid water is absent but where the relative humidity is high. There are also many situations where filling equipment demands valved sacks so that these had to be developed for polyethylene sacks as well as for paper. Today, sacks are available with fitted valves (consisting of thin polyethylene flaps covering a slit aperture) and block bottom construction so that neither filling nor stacking is a problem.

Advantages of the plastic sacks, apart from outdoor storage and lower cost, include higher resistance to tear propagation and visibility of contents (where necessary).

There are now many other uses for low density polyethylene sacks in addition to the packaging of fertilizers. Peat is packed in such sacks, so is coal. Another large market is the packaging of plastics granules themselves. The market was slow to develop because of difficulties in stacking, palletisation and transport but the advent of block bottom, valved sacks has solved many of these problems.

20.1.2 SHRINK WRAPPING

Shrink wrapping is another large outlet for low density polyethylene film and one that is still growing. Shrink wrapping is based on the fact that a plastics film which has been stretched during its manufacture will revert to its original dimensions when heated. The plastics film is loosely sealed round the object to be packed, therefore, and then heated, when a tight transparent wrap is obtained. The first application for these techniques was in 1948, using a film bag of vinylidine chloride/vinyl chloride copolymer for the packaging of oven-ready turkeys for deep freeze storage. The turkey was placed in the bag, a vacuum drawn and the bag closed with a wire tie.

The bag and contents were shrunk by plunging in a hot water bath. Shrink wrapping is now carried out using a heated tunnel (see Chapter 16). Low density polyethylene film has been used mainly in the field of transit packaging, where a shrink-wrapped package of, say, one or two dozen tins or jars, collated on one or two trays, is used instead of a fibreboard case. The mechanical protection given by the shrink wrap is not as great as that given by a well designed fibreboard case, but is usually adequate. There is also the psychological point that a shrink wrapped package can be seen to be fragile and is usually handled more carefully as a result. Other advantages of shrink wrapping, this time to the retailer who receives the goods, are:

(1) There is less packaging material to dispose of, after use.
(2) The package takes up progressively less space on the warehouse shelf as the goods are sold, whereas a fibreboard case takes up the same amount of room whether it contains one can or two dozen.

Low density polyethylene film is also used for the shrink wrapping of whole pallet loads. The tight, even wrap gives a more stable load than wrapping with paper plus strapping and, in addition, confers a large measure of weather-proofing on the load.

Other, miscellaneous applications for low density polyethylene shrink wrapping include rolls of wall paper, cucumbers and other awkwardly shaped objects. Low density polyethylene is used for all these applications, because it is cheap, tough, waterproof and has sufficient clarity for its purpose. In cases where a clearer or more sparkling film is required, then polypropylene or PVC are likely to be used.

Some of low density polyethylene's shrink wrapping markets are being taken over by stretch wrapping. Pallet wrapping is one market where there has been appreciable movement from shrink wrapping and some collation applications are now also using stretch wrapping techniques.

20.1.3 PRODUCE PRE-PACKAGING

Low density polyethylene is widely used in the pre-packaging of fresh horticultural produce. A tough film is required for heavy items, such as potatoes, while its use for green vegetables such as lettuces, brussel sprouts, etc., is due more to its moisture vapour barrier properties and its low cost. There are complications in the packaging of fresh produce, however. Respiration of the produce continues, after harvesting, so that the package must allow ingress of oxygen and egress of carbón dioxide. Although low density polyethylene is not a particularly good barrier to gases, its permeability is not great enough to allow the free transfer of oxygen and carbon dioxide required. In practice, a few small holes are punched in the film. These are sufficient to prevent a build-up of carbon dioxide in the bag but do not appreciably affect the total water vapour barrier properties and drying out of the produce does not occur. The ventilation holes often have another function which is to prevent condensation of moisture on the inside of the bag, with consequent loss of visibility of the contents.

20.1.4 TEXTILE PACKAGING

Low density polyethylene has long been used for the packaging of textiles and similar items. In the case of woollen items, particularly, the soft feel of the film is an advantage inasmuch as the texture of the product can be felt by the prospective purchaser without opening the bag. High clarity grades are required, and the fact that the impact strength is lower, does not matter in this type of application. In some cases there has been a changeover to cast polypropylene film because of its better clarity and sparkle and it is likely that the tubular quench grades of polypropylene film, being cheaper than cast polypropylene in most cases, will accelerate the trend towards polypropylene. For larger items, such as blankets or bolts of cloth, it is likely that low density polyethylene will remain the preferred material.

Allied to textile packaging is the use of low density polyethylene film in laundering and dry cleaning. With dry cleaning, very thin film is used to protect the cleaned garment from dust pick-up and handling, and special equipment is used to dispense the film and place it round the garment while it is on a hanger.

20.1.5 FROZEN FOODS PACKAGING

The toughness of low density polyethylene at low temperatures and the fact that it can be formed into bags on a form/fill/seal basis has led to its use in frozen food packaging. The first use was for the packaging of frozen peas and was dependent on the development of individually frozen peas. Frozen peas were originally packed in cartons, the carton being filled and then frozen between plates. The peas were then received by the housewife as a solid block. In the case of the form/fill/seal bags, the peas are frozen in a blast of air and remain loose enough to be filled into a polyethylene bag formed from flat film round the filling mandrel. This type of packaging is cheaper than the carton and is preferred by the housewife because it is easier to use a portion of the peas and return the rest to the deep freeze.

The use of polyethylene film has spread to frozen soft fruits, brussel sprouts and other items able to be handled by this type of filling operation.

20.1.6 GENERAL FOOD PACKAGING

The general inertness of low density polyethylene film has naturally led to a widespread use in food packaging. A large tonnage of film is used for made-up bags sold direct to the housewife for use in the home, usually for the storage of food in the refrigerator.

Another market is for the packaging of sliced bread. Penetration of this market by low density polyethylene film was extremely slow, partly because of the low cost of the normal material—waxed paper—and partly because of the difficulty of handling a limp, totally fusible film as opposed to the stiffer waxed paper in which only the wax melted on heat sealing, the paper remaining as a rigid substrate throughout the operation. The penetration of low density polyethylene film into this field did not really begin until equipment was developed for packaging sliced loaves into ready-made-up bags. These were closed with a wire tie, thus overcoming the problems of heat sealing. The bags could be attractively printed, without register problems, while another plus point from the housewife's point of view was that re-closure was easily accomplished.

Interestingly enough, there is now a move back to waxed paper in some instances because many consumers look upon waxed paper as the traditional material for sliced bread. The return to this wrapping material would appear to be part of a general wave of nostalgia.

Low density polyethylene is also being used for the wrapping of sandwiches for retail sale and for a wide range of groceries, including sugar.

20.1.7 MISCELLANEOUS NON-FOODS

Low density polyethylene bags of various shapes and sizes are used for an incredibly wide range of goods, ranging from paper goods to hardware, from toys to domestic appliances and from records (the inner bag) to boots and shoes and leather goods generally. In thicker gauges, the film is used for making drum liners for a number of liquid and solid chemicals and as protective wraps inside wooden crates for heavy engineering equipment.

20.2 HIGH DENSITY POLYETHYLENE

One of the fastest growing outlets for high density polyethylene film in packaging is dealt with in Chapter 23 under the heading of 'Plastics Papers'. This is the use of high density polyethylene ultra-thin film for the wrapping of meat, fish, meat pies, cut flowers, etc., and as retail bags for supermarkets and chain stores as a replacement for the traditional paper bag. The advantages include much better barrier properties (particularly to moisture and greases), strength and lightness in weight.

One of the properties in which the high density polyethylene is superior to low density polyethylene is the softening point, which is above the boiling point of water. This makes it suitable for boil-in-the-bag foods and high density polyethylene has been used particularly for fish fillets, such as kippers and haddocks. The general advantages of boil-in-the-bag foods is that they are easy to prepare, eliminate washing up of messy saucepans and prevent cooking smells. The latter is particularly valid in the case of kipper and haddock fillets.

20.3 POLYPROPYLENE

Cast polypropylene films (or the various blown and quenched tubular films) find their main outlets in textile packaging. These outlets are relatively small, however, and the large volume markets require the superior properties of oriented polypropylene film (OPP). These superior properties include better clarity, better barrier properties and a higher impact strength. The superior impact properties of OPP over cast PP are especially noticeable at low temperatures. One of the problems of using OPP as a packaging film is that of heat sealing which causes the film to shrink again to its original dimensions at the heat seal. There are basically two solutions to this problem. One is the multi-point system of heat sealing which uses heat seal bars consisting of a large number of raised points. The area of heat seal contact is thus extremely small at any point and cooling is rapid. The disadvantage is that the heat seal area is not completely gas or vapour proof. The second solution is to coat the polypropylene film with another polymer of lower melting point and utilise this coating as the heat seal medium. The polypropylene itself is not heated sufficiently to cause shrinkage. An advantage of this solution is that the coating can be chosen to contribute to the barrier properties of the film. Vinylidine chloride/vinyl chloride copolymers are often used for this purpose in the same way as they are used for the coating of regenerated cellulose film (in the MXXT grades). Another type of coating, becoming more popular, is a propylene/ethylene copolymer, as mentioned in Section 19.4 on Coextrusion (see Chapter 19).

Coated OPP grades are used for the wrapping of biscuits where extremely good barrier properties, to oxygen and moisture vapour, are required. They are also used for the packaging of potato crisps and other snack items which are similarly sensitive to oxygen and moisture vapour pick-up. The wrapping of cigarette packages is a large field of potential use but so far it has not proved possible to run OPP grades on the extremely fast machinery developed for regenerated cellulose films. However, some success has been achieved with one or two of the minor brands where slower wrapping machinery is used.

Oriented polypropylene films have been used for shrink wrapping where extra sales appeal is required. They are more expensive than low density polyethylene films so that it is only in fields where greater clarity and sparkle are considered essential, that the use of polypropylene can be justified. One general field is in the shrink wrapping of collations of cosmetic or toiletry items. Polypropylene film is also used for the over-wrapping of single cartons in order to give protection against pick-up of moisture vapour. This is not shrink wrapping although it is sometimes the case that the cartons are quickly passed through a heated oven in order to tighten the wrap and remove wrinkles. One of the advantages of polypropylene film in such applications (compared with regenerated cellulose film, which it replaces) is that the moisture vapour barrier properties are intrinsic to the polypropylene and not conferred only by the coating. With moisture proof grades of regenerated cellulose the barrier properties of the package are often reduced because of damage to the coating at edges and corners of the carton.

20.4 POLYVINYL CHLORIDE

Thin plasticised PVC film is widely used in supermarkets for the shrink wrapping or stretch wrapping of trays containing cuts of fresh meat. Requirements for such a film are quite stringent. It must have a high enough oxygen permeability to allow the formation of oxymyoglobin which gives the desired purple 'bloom' of fresh red meat. It must also be tough, able to withstand low temperatures, be shrinkable and have good clarity and gloss. The low oxygen permeability of rigid PVC is increased by plasticisation and the toughness of PVC allows thin gauges to be used, thus increasing the gas permeability still further. A similar application is the shrink wrapping of supermarket trays of fresh produce such as tomatoes and apples. The relatively high moisture vapour transmission rate of PVC is useful in preventing condensation on the inside of the film.

In thicker gauges, plasticised PVC film is used for the manufacture of sachets for individual shampoo packs. Transparency, toughness and resistance to detergents are all relevant properties in this application. In addition, it is possible to seal the sachets, through the liquid, by high frequency methods. PVC sachets have also been used for the packaging of anti-freeze and automotive lubricating oils.

In some countries, notably Italy, PVC has been used for the manufacture of heavy duty sacks for fertilisers. But, as already mentioned, this market has been taken by low density polyethylene in the UK. One difficulty with early bags made from PVC was loss of plasticiser, when the sacks were stored in the open air, and this led to embrittlement, particularly at low temperatures. Formulations which overcome this problem are more expensive. Difficulties were also experienced with hot filling of fertilisers in the United Kingdom, where PVC sacks collapsed.

Unplasticised homopolymer and copolymer grades of film are widely used in the field of thermoforming because of their rigidity, toughness and excellent reproduction of mould detail. Copolymer grades are frequently used for blister packs while both types have been used for insert trays, as used in boxes of confectionery and biscuits. The film is normally around 75 μm thick and can be supplied clear or pigmented. Where necessary, decorative effects can be achieved by vacuum metallisation.

Thermoformed PVC tubs have been used for the packaging of margarine but in Holland, PVC is being displaced by polypropylene on economic grounds because the scrap produced is easier to reprocess in the factory. On the other hand PVC has also taken over new markets such as individual portion packs of preserves which first utilised polystyrene. The PVC packs are thermoformed and then lidded with multi-colour printed aluminium foil for sales appeal.

20.5 POLYVINYLIDENE CHLORIDE

PVDC is used as a shrinkable film for the wrapping of poultry, hams and similar food items and for the in-store wrapping of cheese. The cheese is placed on a piece of film which is then stretched around the cheese and heat sealed with the aid of a hot plate. In addition to the in-store wrapping of cheese, PVDC is used to shrink wrap cheese to give it a longer shelf life. The pre-packaging is then carried out at a central pack-house. The larger quantities which are then handled at one site have allowed the development of semi- or fully-automatic machines.

The use of PVDC (with its low gas permeability) for hams, bacon boiling joints and a range of other cooked or cured meats is dictated by the need to maintain a vacuum in order to prevent bacterial growth and discolouration. Vacuumised bags of PVDC have also been used for the curing of cheeses. The PVDC prevents dehydration and rind formation and also produces a softer cheese. Rolls of PVDC film are used in catering establishments and in the home for the wrapping of food to preserve freshness. The film is used as a stretch wrap, being pulled round the food and then allowed to relax to give a tight wrap.

20.6 ETHYLENE/VINYL ACETATE COPOLYMERS (EVA)

One application for EVA films which makes use of its stretch properties is for bags for the packaging of frozen poultry. A tight contour wrap is achieved by stretching the bags over a former, placing the poultry inside the bag and then allowing the bag to relax round the bird. Unlike normal shrink wrapping, no subsequent heating is necessary. The other properties of EVA that are useful here are low temperature flexibility and retention of impact strength.

EVA films have also been used for the lining of bag-in-box containers. They are particularly suited for the packaging of liquids by virtue of their very good heat seal properties, their impact strength and their long flex life. In the USA a copolymer with a high vinyl acetate content has been chill roll·cast to give a flexible, high clarity film which apparently competes with flexible PVC as a meat wrap.

An interesting development makes use of the fact that EVA can tolerate high filler loadings. Films have been produced with a high percentage of conducting carbon and it is thought that these will be useful for the packaging of fine powders, such as gunpowder, where static electricity build-up could be an explosive hazard. In a similar connection a British Standard has been produced for high carbon black—loaded PVC flooring for use in hospital operating theatres to give a 'conductive' floor in areas of high explosive hazard.

20.7 IONOMERS

Packaging uses for ionomer films are somewhat restricted by the high price of the material compared with normal low density polyethylene. However, there are some uses where its special properties are required, even at a higher price. Skin packaging is one applicational area, particularly where a high resistance to puncturing is necessary. The extremely good adhesion of ionomer films to porous board surfaces is also useful for this application.

Good draw down characteristics make ionomer films suitable for a wide range of deep draw thermoforming. Ionomers are also particularly suitable for the extrusion coating of many other substrates because they can be drawn down to very thin coatings (down to 12 μm) without pinholing. Such thin coatings are still feasible from the end-use performance point of view for packaging greases or oils, or foods with a high fat content because of the much higher grease resistance of the material. There are a number of applications which make use of the high grease resistance. In the USA, for instance, ionomer coated material has been utilised in form/fill/seal pouches for a range of convenience foods with a high added fat content.

Ionomers are also becoming increasingly used in coextruded films because of their grease resistance in thin gauges.

Because of its very high resistance to puncturing, ionomer film is also used for the packaging of hardware. A 50 μm single film is suitable for even very sharp objects but it is rather expensive. Similar puncture resistance, but at lower cost, can be obtained by laminating 25 μm ionomer film with 25 μm low density polyethylene film.

20.8 POLYCARBONATE

Polycarbonate has been thermoformed into heat-and-serve food trays and, in Japan, the film has been used for boil-in-the-bag packs. Both of these applications make use of polycarbonate's heat resistance. In the USA, polycarbonate film, coated with low density polyethylene, is being used for skin packaging. A tough, transparent, high lustre pack is obtained and is particularly suitable for protecting sharp objects such as hardware.

Polycarbonate films are also available in thicknesses from 20 to 250 μm, especially for food packaging at elevated temperatures. The uses envisaged are retort pouches and microwave oven cookware.

20.9 NYLONS

Nylon films have been used for the vacuum packing of foodstuffs, in particularly cheese slices, because of their low gas permeabilities. Their high softening points have led to their use for boil-in-the-bag packs, another valuable property, here, being their low odour transmission. A similar use is for the packaging of surgical equipment for steam sterilisation.

Because of the grease resistance of nylon films they are used for the packaging of engineering components which have been protected with a film of grease. A nylon/polyethylene laminate is used for the packaging of putty and certain fish packs such as frozen prawns, the nylon being present to give oil and grease resistance and as an oxygen barrier.

20.10 POLYSTYRENE

20.10.1 BI-AXIALLY ORIENTED FILM

In thin gauges, below about 75 μm, oriented polystyrene film is used for window cartons. In addition to its transparency it has excellent dimensional stability. Such films are also used for the overwrapping of lettuces and other fresh produce where a 'breatheable' film is required. Because it has been stretched during manufacture, the film is also suitable for use in shrink wrapping.

In thicker gauges, oriented polystyrene has been used for blister packaging as well as for the thermoforming of clear vending cups and other types of rigid tubs. Such tubs have been used for desserts, preserves and automatic vending of complete meals and have been suggested for many other food packaging applications.

In the USA, a large market has been building up for the manufacture of transparent trays for the pre-packaging of fresh meat. The oriented polystyrene trays have made inroads into the meat tray market because the housewife has expressed a preference for see-through trays where both sides of the cut of meat can be seen.

20.10.2 TOUGHENED POLYSTYRENE

The incorporation of synthetic rubbers to increase the flexibility and impact strength of polystyrene affects the clarity, and even in gauges down to about 100 μm thick it is no more than translucent. It is thermoformed into a wide variety of tubs, trays, vending cups and other types of packaging. Tubs are used for dairy products such as cream, yoghurt, cheese, butter and ice-cream as well as for the ready diluted orange squashes sold in cinemas and railway buffets. Toughened polystyrene trays are also contenders for the pre-packaged fresh meat market, while it has become increasingly popular for egg pre-packs.

The growth of unit portion packaging also owes a lot to toughened polystyrene sheet because of its ease of forming and sealing, its inertness in contact with food and its low price.

In the field of non-food items it has been formed into cigarette packs and sterile packs for disposable surgical items.

20.10.3 EXPANDED POLYSTYRENE

Expanded polystyrene sheet is thermoformed into a wide range of packaging articles including apple box interleaves which have been used particularly for apples exported from South Africa. Less bruising of the apples occurs because the material is resilient and the impact loading is distributed over a larger area of the fruit.

In the UK expanded polystyrene sheet is also widely used for supermarket trays for fresh meat and as trays for fish and chips and other take-away meals. The material is grease resistant and is also an efficient thermal insulator.

Egg pre-packs are now being made in expanded polystyrene as well as in moulded pulp and in toughened polystyrene. Most of the applications for expanded polystyrene sheet or film are in the realm of thermoforming, such as the ones mentioned above. One use for thin film (up to 250 μm) is as a non-skid mat under some airline meal trays. The film is usually attractively printed and covers the whole tray surface so that the contents are prevented from slipping.

20.11 POLYESTERS

Shrinkable grades of polyester film have been used for the vacuum wrapping of cooked meat products where a low gas permeability (particularly low oxygen permeability) is essential. Polyester bags have also been used for long term storage of frozen poultry.

In the USA thermoformable laminates have been developed consisting of 12 μm polyester plus 50 μm low density polyethylene. The polyethylene gives heat sealability, better barrier properties and printability.

Packaging uses for the heat stable film include boil-in-bag packs (laminated to low density polyethylene) and the overwrapping of oily products. Laminates of polyester with polyethylene or PVDC are also used for vacuum or gas flushed packs of coffee.

In the UK the high cost of polyester film has inhibited its use in packaging although its very high strength allows it to be used in thicknesses down to 6 μm.

A novel use for polyester film is for the packaging of processed cheese. Molten cheese is pumped into a continuous polyester film tube which is flattened and heat sealed at intervals. The continuous ribbon, containing the hot cheese, is cooled with water and cut to give an individual pack for each slice. Machines are available to produce 100–550 slices per minute. The film has good oxygen and water vapour barrier properties and is generally acceptable from the food contact point of view.

20.11.1 ACRYLIC MULTI-POLYMER

This is still a relatively new material and in the UK its high price has impeded its acceptance as a packaging material. In the USA the uses for thermoformed containers have been confined to those markets where its resistance to oils and greases and its good impact strength are particularly relevant. Such applications include tubs for margarine and peanut butter. In the case of peanut butter the product is particularly sensitive to oxygen pick-up, leading to rancidity, so that acrylic multi-polymer's low oxygen permeability is also valuable.

20.11.2 BAREX

Barex is the trade name for a material made by Vistron Division of Standard Oil of Ohio. It is made by copolymerising a 75:25 mixture of acrylonitrile and methyl acrylate in the presence of a small amount of a butadiene/acrylonitrile elastomer. This material was originally developed as a bottle blowing material for carbonated drinks. It has good clarity, excellent gas barrier properties and high resistance to creep. In addition, it has good impact strength and is insoluble in most organic solvents.

It is also produced in the form of film and sheet and when laminated to other materials such as low density polyethylene it is suitable for themoforming containers for cheese and meat.

In the USA, acrylonitrile polymers and copolymers are under attack because of carcinogenicity (determined on test animals at certain intake levels) and acrylonitrile polymers have been banned from beverage packaging. It is unlikely that a similar ban will follow in the UK or the EEC generally unless new toxicity data emerges specifically related to humans.

20.12 VINYL CHLORIDE/PROPYLENE COPOLYMERS

These copolymers are, essentially, modified PVC's, the propylene content being up to 10% by weight. Applications for the film are expected to lie in the fields of food wrapping (particularly fresh meat) and for the overwrapping of cigarettes. It may also take the place of PVC in some laminates and in the field of blister packs. Somewhat thicker sheet can be thermoformed into tubs or trays, suitable for use with food, and having good impact strength.

20.13 CELLULOSE ACETATE

Cellulose acetate film has many uses in the fields of packaging and display. The fact that it permits the passage of water vapour and gases while remaining impervious to liquid water makes it particularly useful where a 'breathing' film is required as, for example, in the packaging of fresh produce. It is also used in the manufacture of window cartons because it can easily be stuck to the board, giving good seals even at high machine speeds.

Cellulose acetate film has long been used in the field of print lamination, for display cards, journal covers, etc., but some of these applications are now being lost to polypropylene which is more dimensionally stable. Polypropylene is also cheaper but some of this advantage is lost because special adhesives have to be used.

Cellulose acetate sheet is utilised in two different ways. Containers can be made by cutting, creasing and welding (either by solvents or by high frequency heating techniques) or by pressing and deep drawing. These containers find a wide use for the packaging of fancy assortments of sweets and chocolates or for the display of flowers, especially orchids and roses. Cellulose acetate is also easily thermoformed and is widely used for the manufacture of blister packs.

20.14 REGENERATED CELLULOSE

20.14.1 NON-MOISTUREPROOF FILMS

These films are used where protection against dust or grease is required but where mould growth may be a problem. Meat pies, for instance, require a highly moisture vapour permeable film in order to avoid condensation and softening of the pastry. Cakes with a high moisture content and little or no fruit or sugar, and cakes containing fondant, also require a permeable film. If heat sealability is required then grades such as QM (a coated, semi-permeable grade) would be used in preference to a plain uncoated grade (P). Non-moistureproof grades are also suitable for chocolate coated sweets and are generally the preferred type for wrapping fresh sausages. Uncoated films are also made into packages by the use of wire ties, plastic clips, adhesives or adhesive tapes, or stapled to paper 'headers'.

Plain film is used as the base for pressure sensitive tapes and also for the multi-packaging of bottles or jars. The film is dampened before application and forms a tight skin-like wrap after drying out. PF, which is a more flexible form of P, is used for the twist wrapping of sweets where moisture pick-up is not a problem.

20.14.2 MOISTUREPROOF FILMS

The basic type of nitrocellulose coated film is MS (moistureproof, sealable, transparent) and it has a wide range of uses. As an over-wrap it is used for cigarettes, confectionery, pharmaceuticals, bread and stationery. Flexible grades are also available which are used for twist-wrapping of hygroscopic sweets, or for use in dry atmospheres. Special grades of moistureproof films with an added softener have also been developed for the overwrapping of frozen food cartons.

For the packaging of fresh meat a single side-coated film is used. The uncoated side is placed in contact with the meat, thus moistening the film and making it more permeable to oxygen, preserving the red colour of the meat. The coating on the outside of the film reduces drying out of the meat with consequent loss of weight.

Single side-coated film is also used when extrusion coating with low density polyethylene is required, because a better key is obtained to the base film, than to the nitrocellulose coating. The polyethylene coated cellulose film is used for vacuum packs of bacon, cheese, coffee, etc. The cellulose film provides the necessary gas barrier while the polyethylene provides a strong heat seal and extra moistureproofness.

20.14.3 HIGHLY MOISTUREPROOF FILMS

PVDC coated cellulose films are characterised as MXXT with the suffixes S (for solvent coated) or A (for aqueous dispersion coated). The added moistureproofness of MXXT grades makes them especially suitable for very hygroscopic products such as biscuits and potato crisps. The copolymer coating also gives an increased resistance to abrasion and an extra sparkle. The abrasion resistance is useful for products such as confectionery containing nuts. Single side copolymer coated films are also available for use in lamination or for polyethylene extrusion coating.

20.15 LAMINATES

Most of the packaging applications so far outlined are based on single films but there is also an enormous range of applications where no single film satisfies the requirements and laminates are, therefore, used.

Many plastics films are laminated to paper or aluminium foil to provide heat sealability. Paper is an ideal material in many respects, having excellent printability, stiffness, foldability and general behaviour on packaging equipment. It is a poor barrier to water vapour, however, and the addition of, say, low density polyethylene or polypropylene upgrades it in this respect.

Regenerated cellulose film is also laminated to plastics films, particularly low density polyethylene. Such laminates are used for the vacuum packaging of a range of foods including coffee, cheese slices and bacon. Cellulose film supplies the gas barrier properties required, with low density polyethylene giving moisture vapour barrier properties and good heat seal strength.

The potential number of combinations available by using the various plastics films, together with paper and aluminium foil, is enormous. The choice can be narrowed by carefully considering the requirements of the product to be packaged. The possible components can then be identified and the most economical combination determined.

Potentially one of the largest uses for foil laminates is the retort pouch. This is a flexible pack designed to be filled with a food product and fully heat processed. The food then has a shelf-life of two years or more at ambient temperatures. The market, then, is that held at present by the metal can (and, to a lesser extent, the glass bottle). The main requirements are excellent barrier properties (to oxygen and water vapour) and maintenance of seal integrity at sterilisation temperatures. One lamination that has been used successfully consists of an outer layer of polyester film and an inner layer of modified high density polyethylene or a copyolymer of ethylene and propylene. Sandwiched between these two films is a layer of aluminium foil. The foil provides the barrier properties necessary while the inner film provides a heat seal with high integrity. The outer film has to provide abrasion resistance and printability.

Retort pouches give high product quality because the cross-section of the pouch is small in relation to its total capacity. The heat processing time necessary for product sterility is thus relatively short. This leads to improvements in product texture, colour and flavour retention. The advantage over frozen food packs is that frozen or chilled storage is not required.

21 Agriculture and Horticulture

In the UK, agriculture is generally taken to cover farming activities such as the rearing of livestock and the growing of major crops, e.g. wheat, sugar beet and potatoes. Horticulture, on the other hand, covers the growing of crops similar to cucumbers, tomatoes, lettuces, various fruits, and flowers. In some cases the division of crop growing into agriculture or horticulture is a matter of scale with agriculture being comparatively large scale with a small labour content whereas horticulture is smaller in scale, rather more specialised as regards crops grown, and has a higher labour content per unit area under cultivation.

For the purpose of examining the uses of plastics films, it is convenient to treat them together. The subject of packaging (of, for instance, produce and fertilisers) is dealt with in the chapter on packaging. Other uses for plastics films in agriculture and horticulture can conveniently be dealt with under the following headings: growing, disease and pest control, conservation of feedstuffs and crops, water conservation, rearing of livestock and buildings.

21.1 GROWING

One of the largest outlets for plastics films in agriculture and horticulture, worldwide, is for crop protection. In temperate climates, such as the UK, however, there are special factors which have to be taken into account when comparing plastics films with glass. Normal thicknesses of film are much more transparent to infra-red radiations than is glass. This means that there is much greater heat loss at night in a film structure than in a glass covered one, due to escape of the infra-red (heat) radiation from the soil. Film thus gives less protection against frost than glass and a heated film covered greenhouse would require a higher fuel consumption to maintain an equivalent temperature.

The greatest use of plastics films for crop protection, then, is in countries such as Italy, the Southern States of the USA, and Japan where infra-red loss is not so important. However, low density polyethylene film is a useful addition to existing glass houses as a cheap form of 'double glazing'. The polyethylene film is used as an inner lining and considerably reduces heat loss in cold weather. This is an important consideration in these days of energy shortages. The use of tinted or pigmented film can obviate whitewash shading in summer to prevent plant scorch. The surface of the film may have to be treated to prevent condensation with consequent drip damage to plants. The inner layer of film usually requires renewal every year but the reduction in heat loss (up to 60% in many cases) can yield fuel cost savings which more than compensate for the cost of film.

Where plastics films are particularly valuable is in providing temporary protection for those crops that would not normally bear the cost of full protection under glass. One example is the plastic equivalent of the glass cloche. If double layer protection is required a simple way to attain it is to use flattened lay-flat tubing. Simple constructions are normally based on arch-shaped tunnels, rather than the triangular glass ones and are supported by wire hoops. The edges of the film are held in place by covering with soil. Such simple form cloches are usually dismantled at the end of the growing season and the film burnt, together with any waste plant material. The wire hoops are collected and used again. Exposure to the elements of the thin film used causes embrittlement and weakening so that re-use is not normally practical. Protection of the film by the use of UV stabilisers and antioxidants could increase its life but the process of collection, re-rolling and storage for another year is not usually economic.

Large tunnels based on low density polyethylene film are also used. The semi-circular supports are sturdier and the film is secured to the framework. Because of the greater cost involved it is worthwhile in this instance to use stabilised film and so obtain a second year's usage.

A larger scale example is the air supported structure used by one grower for the protection of watercress beds. Four strips of heavy gauge polyethylene, each nearly 100 m (330 ft) long, were heat sealed together to form a single sheet about 91 × 11 m (300 × 36 ft) in size. The sheet was anchored to the low walls separating the various watercress beds and then inflated to form a dome nearly 4 m (13 ft) high. With this type of structure, watercress can be grown economically all the year round whereas a regular glass-house would be too expensive. Other air supported greenhouses have also been constructed in the form of a semi-cylinder with quarter spheres at each end. Polyethylene is the most popular film for this purpose in the UK but PVC, nylon reinforced PVC, polyethylene terephthalate and polyvinyl fluoride have been used in other countries. Similar techniques have been used in Israel to reduce loss of water.

Another way of utilising plastics films as temporary protection for crops is to cover a lightweight structure with film. The whole assembly is then lifted by several men and put into position over the crop. This method is particularly useful for flowers to give protection from heavy rainfall at flowering time.

Plastics films are also widely used for mulching. Black pigmented polyethylene film is used and has the advantage over traditional materials, such as straw, that it needs less labour to lay it and is much more efficient as a barrier. Crops for which it has been used most successfully include strawberries, lettuces and cabbages. The film is laid on top of the prepared soil and holes are made in the film through which the seedlings are planted in the soil. The edges of the film are anchored by burying them in furrows of earth, usually by means of a tractor with special attachments. The black polyethylene film conserves soil moisture, greatly reduces weed growth and maintains the soil temperature at a higher than usual level by absorbing solar heat and reducing evaporative cooling. In the case of strawberry growing the film mulch also keeps the fruit off the ground and so prevents mildew attack. Increased yields and earlier cropping (with consequent higher prices at the market) make the project an economical one. Black pigmented polyethylene does not degrade photochemically as rapidly as clear film and so can normally be used for a second season. An alternative approach has been adopted in Israel during the growing of banana shoots where the film's main purpose is the retention of water. In Israel's climate the use of black pigmentation to raise day-time soil temperatures is not necessary. Without pigementation the film would not last more than one season but still has to be gathered up and disposed of. By using film specially treated to make it degradable by sunlight it is claimed that after three months the film had almost completely disappeared and was absorbed by the soil. By this time the banana shoots were well established.

Plastics film mulches in general are particularly valuable in areas where irrigation and the spraying of weed-killers is difficult or uneconomic.

An alternative method of mulching utilises a double layer of film. A bottom layer of black polyethylene film is laid down and lightly covered with loose soil. A thinner layer of clear polyethylene is laid on top and anchored at the edges with more soil. Better results are obtained because when the sun is shining the clear polyethylene allows the solar rays to pass through and be absorbed by the black film, which then transmits the heat to the soil. At night, when the black film would normally lose its heat to the atmosphere, the heat is retained by an insulating layer of air between the two films.

The use of polyethylene film also allows plants to be grown in soil-less areas such as balconies, loggias, etc. Polyethylene bags, filled with a suitable growing medium, are sealed and then sold in retail outlets. They are particularly suitable for the growing of tomatoes because the bags can be placed in the most suitable place to catch the sun for ripening of the fruit.

Polyethylene film has also been used to line mushroom trays. At the end of the season the growing medium is easily removed from the trays and leads to their more efficient sterilisation.

The growing of grass has been aided by the use of clear film (usually polyethylene) to cover the seed when sown. This raises the daytime soil temperature, protects the seed from birds and retains the soil moisture.

Finally, an idea from the USA for conserving soil heat is to make up long (3·7 m—12 ft) polyethylene bags, part filling them with water, and then laying them between plant beds.

21.2 DISEASE AND PEST CONTROL

Plastic films are widely used in the important area of soil sterilisation. Sterilisation may be by steam or by gases such as methyl bromide. Using methyl bromide, the procedure is to lay the film (polyethylene, PVC, and nylon reinforced PVC, etc.) over the soil, anchoring the edges by piling loose soil over them. The gas is introduced under the film and allowed to stand, usually for about 24 h. Methyl bromide is a poisonous gas but the use of polyethylene film in this way allows safe handling as well as ensuring intimate contact between the soil and the fumigant for a suitably long period. Steam sterilisation is carried out in a similar manner, using 125 μm polyethylene or PVC film instead of the formerly used tarpaulin. The use of a thin flexible film has the advantage that it gives an indication when the soil has been completely steamed. This is because it 'balloons' above the soil surface only when the soil has reached the correct temperature.

Polyethylene film is also useful as a lining to small buildings in which seeds and similar items are fumigated against insect attack.

An unusual application for flexible PVC is in the form of a spiral strip in 3 m (9·8 ft) lengths. This is used in Australia for winding round the trunks of apple trees to protect them from rodents and from extremes of temperature.

21.3 FEEDSTUFFS AND CROP CONSERVATION

Plastics films find widespread use in this area as ground sheets, flexible bags, liners for storage facilities and silos. Polyethylene, PVC and nylon reinforced films are used for covering hay-ricks and as crop drying covers. When a non-reinforced film is used as a hay-rick cover it is sometimes covered with a net to prevent tearing of the film by the wind. It is still a cheaper method, however, than thatching or the use of cloth tarpaulins.

Plastics films make an appreciable contribution to the economic making of silage. Silage is cheaper to produce than hay on an equivalent food value basis. For a satisfactory product, the material to be ensiled should have an adequate level of sugars to ensure sterile fermentation, air must be removed as quickly as possible and fresh air must be prevented from entering subsequently. Carbon dioxide is evolved during fermentation and the film used in the silo should allow this to escape while still preventing the ingress of oxygen which leads to overheating and undesirable side reactions.

Tower silos are expensive and plastics films have brought about large savings in cost. One simple method is to build up a silage stack on 75 μm black polyethylene film until it is sufficiently high then place another sheet on top and seal it to the base one. Air is evacuated by means of a suitable vacuum pump.

Polyethylene film has also been used for the storage of potatoes, seeds, and winter plants and roots. There are many other ways in which plastics films help in crop conservation. One is the use of polyethylene film sleeves for the storage of apples while another is the use of polyethylene film to wrap banana stems before shipment. The film is wrapped round the stems before harvesting and tied at each end. Ventilation holes are made in the film to prevent the build up of carbon dioxide and ethylene which would affect ripening. The main reason for the use of polyethylene film is the protection it affords the fruit against mechanical damage. This latter is often due to friction between the stems causing blackening and scarring. The low frictional properties of the polyethylene film reduce the damage to a minimum.

An interesting method of seed conservation also depends on the use of a plastics film. A special fitment to the tractor during harvesting, lays the film on the ground alongside the tractor and the cut crop is turned over on to the film. Advantages are that the crop dries out quickly and moisture from the soil is prevented from rising to the crop. When drying is complete the crop is picked up with the film so that no seed is lost.

21.4 WATER CONSERVATION

Plastics films provide a cheap method of conserving water supplies. One widespread use is the making of temporary reservoirs, using black plastics film of about 250 μm thickness as a lining on the soil surface of suitable excavations. After bulldozing the soil to a suitable depth the film is laid, preferably on a sand layer in case the soil contains stones liable to puncture the film. Enough film is laid to extend over the sides of the reservoir so that weeds are prevented from growing at the waters edge. The cost of such a reservoir is very much less than that of a cement lined one, the cost ratio is roughly in a ratio of one cement to ten, based on a capacity of around 65 000 l (14 300 gallons). Large capacities are possible and in Canada two lagoons have been constructed with capacities of 27 million litres (6 million gallons) each. The question of seepage has been investigated and in the USA it has been shown that even film of only 150 μm thickness gives complete protection against seepage for over five years.

An even more striking use of polyethylene film in water conservation has been developed in the Sahara where the land is first excavated, then lined with film. Before the earth is replaced, perforated plastic pipes are placed near the base of the excavation. When the soil is replaced and plants have been placed in position, water is pumped through the perforated pipes at low pressure. The roots of the plants are attracted downward to the water and it has been shown that less water is required for plant growth using this method than if conventional irrigation methods were used.

21.5 REARING OF LIVESTOCK

A sheep house has been constructed from 250 μm polyethylene film sandwiched between two layers of nylon netting, spread over tubular steel arches. The sheep house is in Wales and is situated on an exposed hillside. The structure houses 150 sheep and cost only one eighth of the cost of a permanent sheep house.

A minor but interesting use of plastics film is in an automatic battery chicken system where the conveyor belts for transporting food and removing eggs and hen droppings are made from polyethylene terephthalate film. Another aid to animal husbandry is the use in Australia of polyethylene film coats which are fitted to newly sheared sheep to cut down deaths from cold.

21.6 BUILDINGS

Some uses for buildings have already been mentioned under the headings of 'Rearing of Livestock' and 'Growing' but there are also some other, more general, ones. Polyethylene film, for instance, with polypropylene or nylon netting support has been used in temporary roofing and walling for a number of farm buildings. Temporary shelters have also been constructed with the same material.

21.7 FUTURE GROWTH

In the UK, penetration by plastics films of the agricultural and horticultural market is still fairly low and a fair amount of future growth can be expected. The area of forced growing may not expand as quickly as some other areas because the climatic conditions of moderate temperatures, low light and high wind are more conductive to the use of permanent greenhouses. On the other hand, film for silage protection is likely to grow reasonably steadily because of the necessity for protection from our high rainfall. Water conservation uses, too, are more likely to grow in other parts of the world with a lower and more irregular rainfall.

22

Building and Construction

The building and construction industry is a very important one for plastics in general and accounts for about 25% of plastics consumption in the UK. The film portion of this is relatively small in tonnage (though not insignificant) but is, nevertheless, interesting.

One of the most important roles played by the plastics films used in building is that of a moisture or moisture vapour barrier. Thus, polyethylene film is extensively used as a lightweight tarpaulin for covering up materials and equipment on site. Apart from being cheaper than the normal tarpaulin, polyethylene is much more easily handled due to its lightness in weight, although one might have reservations about this in a high wind! This lightness in weight also means that the polyethylene film has to be firmly anchored when in use. There is one further advantage in the use of polyethylene and that is its transparency which enables stored goods to be identified without removal of the covering.

Even more important is the use of low density polyethylene film as a weatherproof covering for building in progress. This can range from a simple covering at unfinished windows for buildings which are otherwise almost complete, to 'tents' covering the whole building. Such protection enables building to be carried on under conditions which normally bring it to an expensive standstill. The particular advantage of a plastics film such as polyethylene in this application is again the transparency which means that shelter from the rain, etc., is supplied without blocking out the daylight. The film may be attached to scaffolding or may be made up on a light framework.

Whichever method is used, the effect of strong winds must be taken into account.

The above uses, important though they are, do not count as part of the building's construction. Polyethylene and PVC films are used in such integral applications, however, the most usual being the damp-proof course on which are laid traditional brick walls. The function of the damp-proof course is to prevent moisture from the ground making its way through the foundation of the building and up the walls. Bitumenised felt was formerly used but low density polyethylene or PVC films are both cheaper and easier to use. Wall top membranes may also be made from the same film materials, while a similar use, where the plastics film's moisture barrier properties are utilised, is in window flashings.

An extension of the idea of covering an uncompleted building with a film structure on scaffolding or some other frame structure is the use of inflated structures. These were developed initially for military shelters which were only required as temporary shelters but because of this they had to be light in weight and portable so that they could be taken down and re-erected elsewhere. From these simple beginnings, large and complex prefabricated inflatable buildings have been developed with air lock access. The materials are usually laminates, of films such as polyester and PVC, in order to give the required strength. These structures are kept rigid by continuously running low pressure compressors. They are useful as warehouses, construction shelters, exhibition shelters and as swimming pool or tennis court enclosures. The selection of the right material for a particular type of shelter will depend on the climatic conditions to which it will be exposed, the rigidity required, allowable cost of the structure and the size. Other problems include the aerodynamics of the structure, and a compromise will probably have to be made between internal space requirements and the shape best fitted for aerodynamic stability. The problem of snow loading is a factor in some climates while other factors to be considered are anchorage requirements, any special requirements for doors, heating and ventilation equipment and detailed design as it relates to the distribution of load.

A more permanent variation on the theme of inflatable shelters has also been developed, initially as a combat shelter. Essentially it is a shelter which is carried in a type of rucksack until a suitable foundation is found. The shelter is then inflated and later sets hard by a chemical reaction. It consists of polyurethane between two layers of low density polyethylene film. The polyurethane foams when the structure is inflated to shape, then hardens and sets so fixing the shape permanently. This type of building, or something very like it, will most likely be used in the initial exploration of the planets. Light weight is essential, of course, while the other advantage is the fact that the shelters take up little space until they are inflated ready for use. The polyurethane used to set the structure into its permanent shape also offers the very important bonus of heat insulation so that the structures are almost ideal for the purpose. These film and polyurethane inflatable structures might also have a future use, much closer to home, as shelters for disaster areas such as those hit by tornadoes or earthquakes. Hundreds of these could be flown to the spot and erected in a very short time because of their light weight and low space requirements.

A different application of plastics films in the construction industry (though one which again depends on their moisture vapour barrier properties) is in the shuttering of concrete. Concrete is improved both in strength and in its surface qualities by the use of curing membranes and mould linings to back whatever shuttering is used. The even moisture retention given by such membranes, normally made from either PVC or low density polyethylene films, prevents surface flaking of the concrete forms.

A similar philosophy underlines the use of low density polyethylene film as an aid to the curing of concrete. Low density polyethylene film is placed over the concrete, soon after pouring, to control loss of moisture and retain the heat of hydration. The curing of the concrete is thus more even and more easily controlled, giving a stronger and more uniform structure. The technique is not usually applied to the curing of concrete roads in the UK since a brush-marked surface is required by the Ministry of Transport to reduce skidding. If the film is put on early enough to control curing it will also smooth away the fine brush marks. Low density polyethylene film is used in concrete road construction in other ways including use as a slip layer and for sealing the formation of the road. The formation is the base down to which the overlying material is excavated and it is sealed in order to prevent seepage of water into, or loss of moisture from it, between the time it is uncovered and the time the sub-base is laid. The sub-base usually consists of crushed stones or cement-treated gravel. Sealing was previously carried out with bituminous products but 125 μm polyethylene film is just as efficient and is lower in total cost. The polyethylene slip layer is used to provide a break between the poured sub-base and the main concrete surface to prevent the two layers keying into each other. This allows slight relative movement between the concrete surface and the sub-base with temperature changes or earth movements, etc., and thereby reduces internal stresses in the construction and, consequently, the tendency to crack formation. It does not matter that the film slip layer may become punctured locally or lacerated by relative movements because it will already have served its purpose of preventing keying of the two layers. The film may be a simple layer of 125 μm thickness or may be two layers of 62 μm thick film, with or without a slip agent such as molybdenum sulphide, between them.

23
Plastics Papers

To some extent every plastics film can be looked on as a 'plastics paper' and, indeed, some plastics films have taken over markets previously held by paper. The polyethylene heavy duty sack, for instance, has almost completely replaced the multi-wall paper sack in the field of fertiliser packaging. However, what most people mean when they talk about plastics papers is the use of plastics to produce a film which has many of the visual and tactile properties of paper, particularly those papers used as writing or printing materials, or those used at retail store level as bags or wrappings.

These markets were comparatively untouched by competition from plastics until the late 1960s, mainly because of the low cost of paper. The situation has changed radically in recent years, for several reasons. One reason is that papers, generally, suffer from some technical limitation such as permeability to moisture vapour and gases, dimensional instability under conditions of varying humidity, fairly low tear strengths, lack of resistance to water, oils and greases. Specialised papers are available which possess superior properties in one or two respects while another way of overcoming these defects is to combine paper and plastics, as with polyethylene coated paper which can then be made into liquidproof cartons. Inevitably, these products are more expensive. However, there are many applications where a simple sheet having the properties of paper plus at least some of the properties of plastics, would be the ideal solution. This alone would not be reason enough to warrant an appreciable use of plastics as paper substitutes. There are also economic reasons, though. Paper prices have been rising steadily for some years and although the downward trend of plastics prices has now been reversed, the upward slope of plastics prices is generally not as steep as that for paper. It is often possible, therefore, for plastics to compete with paper on a straight cost per unit area basis. In some parts of the world there are additional factors which render plastics even more acceptable as paper substitutes. Japan, for instance, is now the third largest paper producer (behind the United States and Canada) and is increasingly dependent on imported pulp. In 1967, for example, the production of paper and board in Japan amounted to nearly 10 million tonnes of which 62% was made with new pulp. Of this amount, 7% (or almost half million tonnes) had to be imported. The position was made worse by the fact that 35% of the raw materials were obtained from waste paper. This is the highest percentage in the world and it is becoming increasingly difficult to maintain the manpower necessary to collect the waste paper. The net result is an increase in the dependence on pulp imports. One interesting statement made by the Japanese is that the value of imports of petroleum as a base raw material for the production of synthetic papers is only about a quarter to a half of the value of imports of the necessary pulp and timber for making the same amount of traditional paper, on an area basis, thus conserving valuable foreign exchange currency.

In many other countries these problems also exist, the difference being those of degree, rather than of kind. It seems logical, therefore, to expect similar developments in the field of synthetic papers, on a much more widespread scale in the future. Most of the common plastics materials have been investigated as paper replacements and the most interesting developments will be discussed in this chapter. Since we are concerned only with plastics films, however, the production of paper by spun-bonding techniques or from synthetic fibres (in the same way as paper is made from natural fibres) will not be included. This still leaves a wide range of plastics papers and classification is difficult. Classification by base polymer is not the answer because factors such as orientation, foaming or coating give completely different end-products. Division will be made, therefore, into coated and uncoated films and then, within these two classes, a division based on polymer in conjunction with any other operations such as orientation and foaming.

23.1 UNCOATED FILMS

23.1.1 HIGH DENSITY POLYETHYLENE FILM

A range of high density polyethylene films has been developed with thicknesses varying from 7 to 100 μm. The thinnest of these are similar to tissue papers in feel and appearance, the 20–30 μm film is similar to ordinary wrapping paper and the 40–100 μm film is used for supermarket carrier bags, tracing paper and water resistant envelopes. These are really three different fields and will be dealt with separately.

The thin (up to 10 μm) film is often referred to as microthin film or tissue-type film. Such film is made from high density polyethylene using blow-up ratios of from 4:1 to 7:1 and draw ratios of around 10:1. The film produced in this way has many of the properties of a fine greaseproof tissue paper. It has the look, feel and rustle of paper, for instance, and has good deadfold properties. Among the advantages that it has over paper are waterproofness, low gas and water vapour permeability, and good aroma retention. It is heat sealable, can be printed and is readily self-coloured. It can be made into bags either by heat sealing or by the use of adhesives.

Microthin high density polyethylene film is a competitor, then, to greaseproof paper (particularly thin greaseproof tissues). It is also a competitive material, both economically and technically, to bleached greaseproof papers, wet strength papers and vegetable parchments. From the price point of view this is normally true only on a cost per unit area basis rather than cost per unit weight. High density polyethylene tissues are very light in weight, being about a quarter of the weight of a paper of equivalent performance. There is nothing unfair in this comparison, of course, since cost per unit area is the only cost which has any meaning from the end-user's point of view.

As a wrapping material, high density polyethylene tissue has been used for a number of foodstuffs, including wet fish, meat, dried fruit, meat pies and the traditional Danish lunchpack, smorgasbröd. The use for meat pies is particularly interesting since it shows several technical advantages over the normally used greaseproof paper. Hot fat losses are reduced, release properties are better and the material does not embrittle and char. In the case of wet fish there are also technical advantages, including preservation of freshness and a reduction in odour penetration. Microthin high density polyethylene film has also been used as a boil-in-the-bag pack for uncooked boned hams because it provides a hygienic seal and no flavour is lost during cooking. There are also a number of non-food packaging uses and the wrapping of cut flowers is one good example of how to make use of the material's particular properties. The flowers are less prone to wilting through loss of moisture and the wrap itself does not wilt or drip water even after a long journey. It is also used for the wrapping of delicate instruments to keep them dust-free. The use of high density polyethylene overcomes problems such as corrosion and staining which are likely to occur with paper, unless special chloride and sulphur-free grades are used, which are expensive. There are a number of non-packaging uses too, including use as a lining material for trays in lipstick production, as a waterproof barrier between shoe soles and in-soles, and for stuffing the toe-caps of new shoes.

The slightly thicker film (20–30 μm) is also used as a wrapping paper. Its barrier properties are even better than those of the tissue type material because of its increased thickness and it is particularly suitable for the packaging of frozen meat. In Belgium, such film is being sold in packs of twelve sheets for household wrapping uses. It has the advantage over greaseproof paper that it can be washed and re-used. In Greece, this material has been attractively printed and then sold as a gift wrap. In the UK, bags made from high density polyethylene have replaced a major proportion of paper bags for retain use. Such bags are normally made from MG* sulphite or bleached Kraft but high density polyethylene has better tear strength and general all-round toughness. It is particularly good, of course, for wet or greasy products.

Film of 100 μm thickness has been used in Germany to make supermarket carrier bags which are free-standing by virtue of their stiffness. They thus possess the normal advantages of plastics without the limpness associated with the low density polyethylene carrier bag. Free standing carrier bags are important in supermarkets, of course, because of the time saving involved at check-out points.

The rate of growth in all these markets will be conditioned by the relative prices of paper and plastics but there is no doubt that the potential is great. The consumption of food wrapping paper in the UK alone is of the order of 160 000 tonnes/annum. Assuming that high density polyethylene films, on an equivalent area basis, would weigh about one quarter of the paper equivalent, this is a potential market for plastics of around 40 000 tonnes/annum.

23.1.2 CELLULAR POLYETHYLENE AND POLYPROPYLENE

Reducing the density of a plastics film by giving it a cellular structure

* MG = Machine Glazed. The paper is smooth on one side only due to having been dried by passing over a single, large diameter, heated metal cylinder.

allows the production of a film having a greater stiffness for a given weight of polymer. This follows from the fact that the stiffness of a film is proportional to the cube of its thickness. If, then, we reduce the density of the plastics film to half its normal value, then from the same weight of material we can produce film having twice the thickness of an unfoamed film. The stiffness of the foamed film will thus be two cubed (or eight) times that of the unfoamed equivalent. The reduction in density itself reduces the stiffness, of course, but this effect is only a linear one so that halving the density, halves the stiffness. The final effect of both these factors is thus an overall increase of four times the stiffness for a reduction in density of a half. For this, and other reasons, techniques for expanded plastics have been the subject of a great deal of development work. The latest plastics of commercial importance in this sphere are the polyethylenes and polypropylene. The manufacture of these expanded films has already been touched on but it should be noted here that in expanding the molten polymer the expanding agent, usually nitrogen, has to stretch it in order to create a cell. This stretching is dependent on the melt strength and hence on the temperature. A certain amount of control of cell size is thus possible and this controls strength and appearance. Density is usually reduced to between 0.6 and 0.7 g/cm^3.

The expanded films so produced have quite a few of the attributes of paper. They have the stiffness of paper, rather than of a plastics film, because of the effects discussed above. They also have the appearance and handle of paper. Their crease retention, while not as good as that of paper is better than that of the unfoamed film. These paper-like attributes are coupled with the advantages of plastics, including wet strength, dimensional stability, processability and heat sealability. In addition, their cellular nature gives them some heat insulation and cushioning properties which may be useful bonuses. The cellular films can be printed by the normal film printing methods and the results are enhanced by the silky surface sheen associated with these materials.

The increased stiffness of the cellular material enables even low density polyethylene to be used for making free-standing carrier bags. This, coupled with the unusual printing effects possible, make low density polyethylene a competitor in the 'pop-art' and other 'fashion' carrier bag fields. However, a reduction in strength makes it unsuitable for supermarket carriers except at uneconomic thicknesses. Other possible applications are record sleeves (where stiffness and cushioning are important) wrappers for frozen food (for good heat insulation) and refuse sacks (where the surface gives a better grip by the sack stand). Cellular polyethylene could also be used for instruction manuals where these have to be used in damp or greasy conditions, but this type of market is also open to competition by other coated and uncoated plastic papers. If blends of low density and high density polyethylene are converted into cellular film the resultant material is even stiffer and it can be thermoformed into trays or box inserts, with particular usefulness in the packaging of frozen foods.

Cellular polypropylene is rather more expensive than cellular polyethylene but it is sometimes used as the inner wrapper for tablets of toilet soap. One of the advantages of polypropylene for this application is its freedom from environmental stress cracking in contact with soaps and detergents. It also prevents interaction between the constituents of the soap and the printed outer wrapper.

23.1.3 POLYSTYRENE BASED 'PAPERS'

Polystyrene film is the basis for two papers produced in Japan known as 'Q'per' and 'Q'kote'. The latter is a coated material and will be dealt with later. The former is uncoated and can be transparent or rendered opaque by loading with pigment fillers. It also has a fine, porous surface layer produced by a solvent treatment. Other surface treatments include mechanical ones such as sand blasting. These surface treatments are known as 'paperising'. The porous surface layer gives an opaque appearance similar to that of paper because of light scattering. This porous layer is also important as an ink receptive layer. The appearance and other properties of the porous layer can be varied by altering the surface treatment conditions. The thickness of the porous layer for example, affects mainly the whiteness, density and opacity of the 'paper', while the pore size affects, in addition, smoothness, gloss and ink receptivity. Density varies from about 0.6 g/cm^3 for a translucent grade up to 0.8 g/cm^3 for a matt finish grade. The opacities of these different types range from 55% to 85% respectively. A type having a gloss surface is also available and has an intermediate density of 0.7 g/cm^3 and a high opacity (around 90%). The percentage gloss for the 3 different grades is 1% for the translucent, 6% for the matt and 40% for the glossy type. Another polystyrene based paper is 'Spiax' made in Japan by Sumatomi Chemical Co.

In addition to having the appearance, stiffness and deadfold properties of paper, these uncoated polystyrene materials have the characteristic rattle of paper. They also have excellent wet strength although they are not so good a barrier to moisture vapour as are the papers based on polyethylene and polypropylene. One interesting property is that hot stamping will render the paper partially transparent and this opens up possibilities in inkless printing.

Another method of producing synthetic paper-like materials is by heavily pigmenting the plastics film. This type of material has many of the properties of a coated art paper although special inks are needed for printing. Papers of this type can be based on a variety of polymers including high density polyethylene ('Polyart'—produced by BXL in England, 'Zexan'—produced by Mitsui, and 'UCAR'—produced by Union Carbide in USA) and polystyrene ('Printel'—produced by Sekisui Chemical Company in Japan). The general properties and applications of these papers are similar to art papers produced by coating methods and are dealt with in more detail under that heading.

23.1.4 POLYETHYLENE TEREPHTHALATE (PETP)

As mentioned in Chapter 7, this material is used in the manufacture of drafting materials for original drawings or for tracing. Where tracing is concerned PETP has the advantage of transparency over tracing paper. A more important advantage is its superior dimensional stability under conditions of changing humidity.

23.2 COATED FILMS

High class printing papers are produced by coating the surface of the paper with an emulsion consisting of a finely divided inorganic pigment (such as china clay) and a binder such as starch, casein or a synthetic latex. These surface coatings are formulated to give improved printability, including printing ink absorption and surface brightness. It was logical, then, to consider the coating of polymer films in order to give them similar surface properties to those of high class printing papers. Such synthetic printing papers would possess additional properties such as dimensional stability, strength and durability by virtue of their polymer substrates.

The main problem encountered was adhesion of the coating composition to the film, but this has now been overcome by use of solvent based priming coats where the substrate is particularly inert, as in the case of polypropylene. Such primers can be applied by air-knife coaters and then dried at room temperature. The film is then coated in the same way as paper and given similar finishes, e.g. supercalendering.

So far, four polymers have been used to any extent commercially as substrates for high class printing papers, namely polystyrene, high density polyethylene, polypropylene and PVC. These four offer the best combination of properties at an economic price. The general properties of these synthetic papers are dictated by the coating applied to the film and by the intrinsic film properties and it will be convenient to consider the various papers as a class with appropriate mention of the properties of a particular member when these differ significantly from the norm.

One paper, mentioned earlier in the chapter is 'Q'kote' which is manufactured in Japan. This is usually based on polystyrene but the trade name has also been used for synthetic coated papers based on a range of polymers, including stretched and oriented polypropylene, PVC, polyethylene and stretched/oriented polyethylene terephthalate. Another coated film is 'OJI-YUKA', made in Japan by Mitsubishi/Yuka.

One of the main advantages of using a plastics film base for a coated paper is that a much thinner base can be used. In Japan, for example, experimental papers have been produced consisting of a 40 μm thick base and a coating of 30 g/m^2 on both sides. Another, general, advantage of plastics film substrates is their dimensional stability under conditions of changing humidity. Printability of these papers is good and is generally comparable with that of the cellulose based printing papers. The use of similar printing inks and processes is also feasible because the surface structure of the synthetic papers is similar to those of their conventional counterparts.

The composition of the substrate affects the mechanical properties of the finished paper in matters such as tear strength, tensile strength, bursting strength, etc. In addition, the drying temperature after coating or after printing must be adjusted to take account of the composition of the substrate. PVC and polystyrene for example, cannot be dried at the same temperatures as can high density polyethylene and polypropylene.

Applications for synthetic art printing papers are broadly similar to those of traditional art papers with important additions due to their superior properties in the fields of dimensional stability, waterproofness, and chemical resistance. Thus, they can be used for manuals likely to be used in damp or greasy conditions. Examples here would include instruction manuals for equipment, and loose-leaf cookery books. They can also be used for making waterproof labels and for outdoor display materials. One interesting use for a synthetic art paper is the production of inlaid pictures on the face of plastics articles such as dishes and trays. The printed sheet is placed in the face of the injection mould and when the polymer is injected, the film label becomes an integral part of the moulding.

Synthetic printing papers as a whole are still dearer than the traditional ones by a factor of around 2–3, but their superior properties often make their use economic. Engineering data sheets, for instance, can often justify the extra cost by virtue of longer life and being able to dispense with a protective folder. Where data sheets are kept in a loose-leaf binder, there is very much less chance of their being lost because of tearing at the punched holes. Finally, costs are often less than would be expected because of the possibility of using thinner gauges of film.

Thin PETP (and, to a lesser extent, thin polypropylene) has been successfully used in the manufacture of carbon 'papers' and typewriter ribbons. PETP has the better resiliency (giving good recovery after indentation) and has a very high strength.

24

Film Tapes and Fibres

In Chapter 8 it was mentioned that one of the disadvantages of uniaxially stretched films was the fact that they tended to fibrillate and this militated against their use as film for packaging (or, for that matter, for building and horticultural uses). There is one vast field based on uniaxially stretched polyolefin films, however, and that is the film tape and fibre field.

Before studying this field of application in any detail, it will be helpful to define some of the terms used so that they may be referred to again without further explanation.

Film tape

Water bath or chill roll cast, or blown film that is split into narrow tapes (usually 2 or 3.5 mm wide and 50 μm or 28 μm thick, respectively) and stretched under heat. Stretch rates are of the order of 7:1. Stretched film tape or film yarn are equivalent terms.

Twisted film fibres

Water bath or chill roll cast or blown film that is split into wider tapes than the film tapes and again stretched under heat. Additionally, the tapes are then split by mechanical means into a coarse, fibre-like form. Stretch ratios are somewhat higher, being of the order of 10:1. Fibrillated film is an equivalent term.

Fine denier fibres

These result when the above process is taken still further, until a fine, soft, wool-like mass of fibres is obtained. Super fibrillated film is an equivalent term.

Denier

This is the weight, in grammes, of 9000 m of the fibre in the form of a continuous filament. The SI unit is the tex, which is the weight in grammes of 1000 m of the fibre.

Tenacity

This is a measure of tensile strength and is measured in mN/tex.

24.1 POLYMER CHOICE

Nylon and polyester fibres are normally too expensive for the types of market open to film fibres and tapes while another disadvantage in the case of film fibres is their reduced tendency to split, as compared with polypropylene and high density polyethylene. The choice between polypropylene and high density polyethylene is dependent on a number of factors. Polypropylene has a processing advantage because its tenacity is only slightly dependent on the melt flow index of the polymer so that relatively high melt flow grades may be used without appreciable loss of tenacity. High speed extrusion of the film is thus possible. Polypropylene stretched tapes also have a higher tenacity than those made from high density polyethylene even when low melt flow grades of the latter are used. The coefficient of friction of high density polyethylene is lower than that of polypropylene, too, and this can cause winding problems as well as stacking problems when the tapes are woven into sacks. In countries where high density polyethylene is used, it is usually its lower price that makes it attractive. Coextruded multilayer films are being investigated for some specific markets. One of these is for textured carpet yarns because of the latent self-crimping characteristics of the multilayer fibrillated yarns.

In the specific case of ropes, some interesting figures are given in *Table 24.1*. These show that polypropylene has only a marginally lower performance than nylon. However, nylon is distinctly more expensive than polypropylene, especially on a cost/volume basis. The cost differential between polypropylene and high density polyethylene is not very high and in any case the monofilament

Table 24.1 SOME PHYSICAL PROPERTIES OF ROPES

Material	*Rope diameter* 8 mm Weight (kg/100 m)	Breaking load (kg)	Breaking length* (m)	Ratio of Breaking length to manilla
Nylon (monofilament)	4·2	1350	32 000	3·2
Polypropylene	3·0	960	32 000	3·2
Polyethylene (monofilament)	3·15	700	22 000	2·2
Manilla	5·4	540	10 000	1·0
Sisal	5·4	490	9 000	0·9

*Breaking load divided by weight per metre.

extrusion method is more expensive than the stretched tape method. The real difference in cost, then, is even smaller and is insufficient to compensate for the poorer performance figures of high density polyethylene. As mentioned earlier polypropylene has advantages in processing speeds and strength when both polypropylene and high density polyethylene are converted by the stretched tape method. Polypropylene also fibrillates readily on twisting whereas high density polyethylene does not.

24.2 PROCESS AND EQUIPMENT

The process for producing stretched film tape consists of the following stages:

(1) Production of the film.

(2) Slitting of the film to tapes of predetermined widths.

(3) Heating and orientation of the tapes.

(4) Winding of the stretched tapes on to a suitable spool.

The preferred process is continuous but it may be separated into two stages whereby the film is wound on to a roll after extrusion. The roll is subsequently unwound, slit, stretched and wound onto spools. The biggest advantage of the two stage process is that the ultimate line speed is not governed by the extruder speed so that both extruder and stretcher can be run at optimum speeds. A wide range of equipment is commercially available for each of the above steps but the make up of an optimum complete line will depend on a number of factors including size of plant, equipment available, cost of equipment, previous technology and the precise products to be produced.

24.2.1 FILM EXTRUSION

All three techniques mentioned in Chapter 8, namely, slit die extrusion with chill roll casting, slit die extrusion with water bath quenching and blow extrusion are suitable for the production of the basic film. There are, however, advantages and disadvantages attached to each one of these techniques.

Slit die film systems are generally reckoned to give higher line speeds since the film can be more efficiently cooled. Closer gauge control is also possible. Disadvantages include a greater amount of scrap due to the edge thickening which occurs in flat film extrusion. This is not, however, a factor when the film tapes are to be used for the manufacture of ropes since the thicker tapes can be used as part of the rope's central core. Of the two slit die methods, chill roll casting has the tighter control of film thickness but differential cooling of the two sides of the film makes it a less efficient process for thick film extrusion. Disadvantages of the water bath quenching method include a higher variation in thickness than in the chill roll casting method and the carry-over of water. Devices for the removal of excess water have, therefore, to be included. Possible methods include the incorporation of vacuum/heat dryers in the line.

Blown film equipment is cheaper than that for slit die extrusion. An important characteristic of blown polypropylene film is that some lateral orientation occurs during the blowing process. As this is never completely eliminated during the subsequent longitudinal orientation the final film tapes have a reduced tendency towards self-splitting. The strength is also reduced somewhat and the film has a softer feel than cast film. The softer feel is desirable in tapes for weaving and fibrillation, and especially for fabrics from fibrillated film because of the enhanced consumer appeal. The reduced self-splitting characteristics are also desirable, both for weaving (which is natural) and for fibrillation because reduced splitability leads to a more controlled splitting, at least when a pin roller unit is used (see later section on fibrillation) and less post-fibrillation.

24.2.2 FILM SLITTING

The film is slit into tape of sufficient widths so that the stretched and oriented tapes will be of the required denier. Slitting is usually carried out using a slitter bar containing razor-like blades. During slitting the film should be under controlled tension for accurate cutting and this is best achieved by siting the slitter bar close to a roll or drag bar.

24.2.3 HEATING AND ORIENTATION

After slitting, the tapes are led into a pull-roll stand (or godet unit). This may consist of 3, 5 or 7 power driven rolls together with a nip roll, covered with neoprene or urethane rubber to prevent slippage. From the godet unit the tapes enter an oven, usually about 3–4 m (9·8–13·1 ft) long, fitted with accurate control of temperature and with special regard to uniformity of heating. Circulating hot air ovens are the most usual but infra-red heaters are also used and hot-plates (heating by contact) have been developed. At the other end of the heating unit is a second godet unit operating at a higher speed than the first. The ratio of the speeds of the first and second godet units is the draw ratio (7:1 for weaving tapes and 10:1 for tapes for fibrillating). Where tapes are to be used for woven carpet backing there may be sufficient heat generated, during the weaving process, to cause tape shrinkage unless the tapes are previously heat set. This relaxation and heat setting is carried out by running them through a second oven followed by a third godet unit.

Later types of stretching equipment are based on hot roll techniques. Blown or cast film is stretched in its full width on a unit consisting of pre-heated rolls, stretching rolls and annealing rolls. It is then wound up as wide stretched film which can be slit into tapes. Because the stretching is controlled over a very short distance only a slight reduction in film width occurs and thus a larger number of tapes can be produced from a given film width than with conventional methods based on slitting before stretching. Advantages of hot roll stretched polypropylene film include:

(1) High output.

(2) Low splitting tendency.
(3) High tenacity.

One disadvantage is that output is limited by the high initial film thickness required. This is because the film thickness after stretching is directly proportional to the stretch ratio used. With a typical stretch ratio of 10:1 the initial film thickness would be approximately 400–500 μm and adequate cooling and pre-heating of such films is difficult.

24.2.4 WINDING

The winding of tapes is more difficult than the winding of natural or synthetic monofilaments because of the added complication of possible twisting. This can reduce the tensile strength by 15–20% and must, therefore, be avoided. Another critical factor is tension which must be kept constant. Constant tension winders are available but are expensive. Constant torque winders are cheaper but require manual adjustment by the machine operator as the diameter of the wound bobbin increases.

24.3 FIBRILLATION

The film tape, manufactured as outlined earlier, can be fibrillated by mechanical splitting in a controlled manner to give a range of fine denier fibres. The three basic systems for fibrillating are:

(1) Pin-roller system.
(2) Castellated die system.
(3) Embossed roller system.

24.3.1 PIN-ROLLER SYSTEM

The equipment consists of a roller which revolves at a peripheral speed higher than the linear speed of the film and which can incorporate up to approximately 30 000 needles in a series of bars; an assembly of rollers for adjusting the contact area of the tapes; and draw off rollers for adjusting the tension. As they revolve the needles split the tapes, forming a regular or irregular network pattern. The size of this network can be adjusted by altering the roller speed and the contact arc.

24.3.2 CASTELLATED DIE SYSTEM

In this system the usual flat die is replaced with three flat dies each having 35 profiled, castellated slots through which a total of 105 tapes are extruded. The profiles of each slot produce 25 ribs in each tape. Stretching of these ribbed tapes shears them into 25 single strand fibres.

24.3.3 EMBOSSED ROLLER SYSTEM

Film, either above or below its crystalline melting point, is brought into contact with profiled rolls. The pattern is then embossed under slight pressure on to the film. Subsequent stretching of the film produces separated, continuous filaments corresponding to the particular profile used. The method is said to offer to the end user close control over properties.

24.4 SACK MANUFACTURE

Polypropylene tapes can be woven more readily on most looms than natural fibres, such as jute, because the polypropylene tapes are smooth and so do not snag. Circular weaving, flat weaving, or knitting techniques can all be used with polypropylene tapes. On flat looms speeds of around 20 sacks/h are possible, which is equivalent to the rate achieved for jute sacks. If circular looms are used, however, the rate for polypropylene film tape sacks is claimed to be well over 60 sacks/h. One criticism of circularly woven polypropylene sacks, however, is that they are difficult to stack. This arises from the 'pillow' shape which such a sack tends to assume when it is well filled and it is thus advisable not to over fill.

Closing of woven sacks can be achieved by heat sealing although there are difficulties. It is more usual to close the sacks by conventional techniques such as stitching. This has the advantage that the user does not have to invest in new equipment.

24.4.1 ADVANTAGES OF POLYPROPYLENE SACKS

Woven polypropylene sacks can compete in price with jute sacks but, more important still, is the question of price stability. Because of annual fluctuations in the jute crop there are corresponding fluctuations in price and these can be appreciable. The price of polypropylene is, by comparison, extremely stable and it is possible to plan ahead with more certainty.

Woven polypropylene sacks are appreciably lighter than jute sacks. A jute sack weighing 1 kg (2·2 lb), for instance could be replaced by a polypropylene one weighing only about 0·2 kg (0·44 lb). The greatest benefit here lies in reduced freight charges for empty sacks. The light weight is not an unmixed blessing, however, for some products, such as animal feeding stuffs, are sold on gross weight (i.e. including sack weight). Because the polypropylene sack is lighter the gross weight has to be made up by an increase in weight of the product.

An important advantage of the polypropylene sack is its resistance to rotting in damp conditions. This is particularly important in tropical areas where rotting of jute sacks can occur in a matter of weeks. Polypropylene sacks are also resistant to insects and although jute sacks can be treated to improve their water repellency and insect resistance, this adds to their cost. Polypropylene sacks can be made to withstand prolonged exposure to sunlight by the incorporation of ultra-violet stabilisers in polypropylene during the extrusion process. Some idea of the outdoor performance of the two types of sacks can be gained from the fact that a natural fibre yarn exposed to high ultra-violet light conditions loses 80% of its tenacity in two to three months, whereas an ultra-violet stabilised polypropylene stretched tape (either pigmented or unpigmented) loses only 50% of its tenacity in a period of three to four months. Polypropylene sacks are also superior to jute sacks in their resistance to chemical attack.

Polypropylene sacks are better able to retain dusty products, possibly because jute yarns are round or oval in section and this allows fine particles to escape between the yarn interstices. In the case of flat tapes, such as polypropylene film tapes, the load of the sack contents locks the intersecting tapes and appreciably reduces the escape of fine particles.

Polypropylene sacks are tougher and more resistant to breakage by impact. Jute sacks containing 91 kg (200 lb) of product have been tested to withstand about 7 drops before bursting occurred. Polypropylene sacks, containing the same weight of product, withstood 15–16 exactly similar drops before bursting.

The coefficient of friction of jute is higher than that of polypropylene but the incorporation of anti-slip compounds and variations in the sack weave can largely overcome this problem and the stacking characteristics for polypropylene sacks can be entirely satisfactory.

Polypropylene does not impart any odour or flavour to food products while another advantage in the field of food packaging is the absence of contamination via loose fibres. Another product where contamination by jute fibres is a serious problem is wool. The packaging of bales of wool in Australia is already largely in polypropylene and high density polyethylene and results have been very satisfactory.

The general appearance of polypropylene sacks is very attractive. Four colour printing is already in commercial use and, in addition, the tapes themselves can be manufactured in almost any colour.

24.5 APPLICATIONS

The use of polypropylene fabrics for the packaging of bales of wool has already been mentioned. Uses for sacks include the packaging of potatoes and other horticultural produce, grain and animal feeding stuffs. In the case of grain, the ventilation is important since condensation inside the sack would lead to fermentation of the grain. In addition, a woven sack is necessary because of the practice of sampling through the sack without opening it. This obviously has to be done without damage to the sack. Additional applications in tropical countries include the packaging of coffee and sugar, while possibilities exist in the packaging of fish, cotton and groundnuts. In general, woven polypropylene sacks will find applications where the advantages of plastics are required but where a normal film sack is unacceptable for reasons of ventilation, sampling, etc. Much of the impetus for the growth of polypropylene sack manufacture in the USA was supplied by the use of polypropylene for military sandbags in the jungles of Vietnam. The main factor here, of course, was the resistance of such sacks to the hot and humid conditions so damaging to jute sacks. Insufficient ultra-violet stability in the original polypropylene sacks was later overcome with more efficient ultra-violet stabilisers.

The potential markets for sacks are already vast and are growing rapidly. Both the USA and the USSR, for instance, use about 400 million sacks/year. South and East Africa require about 145 million/year; the UK 65 million; Germany 45 million; and Italy 20 million.

Because of the trends towards bulk delivery, the demand for sacks may be static or even decreasing slightly in countries such as the UK but in the developing countries the demand is increasing due to new and expensive agricultural products needing efficient protection and the long-term prospects for plastic woven sacks looks very bright indeed.

If an impermeable sack is required it is possible to laminate woven polypropylene fabric to low density polyethylene film. By varying the thickness of the film and the type of weave used in the polypropylene fabric, a wide range of sacks can be produced. At one end of the scale one can produce a closely woven sack with just enough low density polyethylene for waterproofness, while at the other end we can get what is essentially a film sack supported by an open weave polypropylene fabric.

Woven stretched products also include decorative fabrics, carpets backing and industrial fabrics such as filtercloths, belts and tarpaulins.

For example, woven polypropylene tape fabrics are successfully established for the primary backings of tufted carpets since they overcome the technical and economic disadvantages of natural fibre backings. The latter are not rotproof, have a low dimensional stability and tend to break or deflect the tufting needles, particularly during high speed, fine-gauge tufting. Woven polypropylene fabrics are much lighter, are rotproof and moisture proof and give a higher pile than can be obtained with jute backings.

Tufted carpet backing is a large market for woven film tapes. A high degree of penetration has already been achieved in the American and European tufted carpet industries and a similar backing is being used for needle punched 'indoor–outdoor' carpeting where the carpet pile is itself made from polypropylene. The polypropylene tapes are fibrillated and chopped into staple lengths for this application. Fibrillated stretched tape is also being used as face yarn for tufted carpets. The potential market for polypropylene carpet backing in the UK can be judged from the fact that during the three years 1964–1966, an average of over 100 000 tonnes/annum of jute yarn and jute fabric went into floor covering of one sort or another. Although polypropylene woven fabrics are lighter in weight than jute, the total is still appreciable.

Film fibres (fibrillated film) can be considered as a replacement for sisal. The main areas for takover are agricultural baler twine, packaging twine and ropes. Baler twine is a large potential outlet for film fibres and has already been penetrated to some extent. The use of polypropylene baler twine has the advantage of giving a rotproof twine as well as being waterproof and not harmful to the digestive systems of farm animals. Polypropylene has a breaking load that is slightly lower than that of sisal but has a much greater strength to weight ratio. The addition of ultra-violet stabilisers to the polypropylene gives a twine with a longer life than sisal twine. Polypropylene twine has a knot strength comparable to that of sisal and can be used on normal twine dispensing equipment with a minimum of modification. The polypropylene twine can, of course, be self-coloured and this opens up several possibilities for the use of twine for identification purposes. The arguments concerning price stability and continuity of supply also apply in the case of twine.

Agricultural baler twine is a price sensitive commodity and is not so much sold on technical reasons as is packaging twine. In the latter case, superior performance can usually command a price premium and it is for this reason that film fibres have penetrated the packaging twine market to a greater extent that the baler twine one.

The other large market for film fibres is rope. The natural fibres normally used for ropes are sisal and manilla hemp. Manilla hemp is the strongest of all vegetable fibres and thus is used for the manufacture of heavy rope and cordage for marine, industrial and farm use. It is still not as strong as polypropylene, however, as can be seen from *Table 24.1*. Polypropylene film fibre ropes are lighter in weight than manilla hemp or sisal ones and will float on water because the specific gravity of polypropylene is less than one. In addition, they are not subject to moisture pick-up with consequent increase in weight. These points are particularly important in the case of marine ropes.

The newest developments lie mainly in the field of fine fibres or super fibrillated film. Polypropylene fine fibres are very soft, very light in weight and very strong, with excellent abrasion resistance. Outlets include knitting or sewing threads, while in staple form they can be used as reinforcement in non-woven disposables.

25 Degradability, Recycle and Re-use

The concept of plastics films that will degrade easily when finally discarded promises advantages, whether they end up on a disposal tip or are thrown away to become litter. Plastics films will normally take two to five years to degrade in the open air. The inertness and durability of plastics films are, of course, two of their strong points when they are considered as packaging materials and no easy way has yet been found to make them fully satisfactory during use and easily destructible after use.

One suggestion that has been made is to render the plastics film biodegradable, i.e. degradable by biological action, such as bacterial attack. This could have extremely undesirable consequences if a successful bacterial strain were developed. Since, to be successful, they would have to be capable of living in the soil, they could well attack plastics water pipes or underground cables, with devastating results.

The action of sunlight has also been suggested as a trigger for causing the degradation of plastics. Additives have been developed which cause plastics to break down in the presence of UV light. The danger in this approach lies in controlling the exact time at which the breakdown will begin. Premature degradation while the film was in use would be disastrous. The advantages, too, become more nebulous on closer scrutiny. Film on the disposal tip is soon covered by earth or by a fresh load of rubbish and there may, therefore, be only a short exposure to sunlight. The problem of litter may be partly solved by this approach but even so there would still be an appreciable time before the film disappears completely. One application for UV degradable film is in the mulching of banana shoots in Israel as was mentioned in Chapter 21.

A safer way of achieving degradability, namely, by the use of starch-filled polyethylene, was mentioned in Chapter 2. Here, the film does not degrade until it is in contact with soil-based enzymes which destroy the starch filling.

There is, however, one argument against all degradable plastics that is particularly relevant today—the plastics material is completely lost. Better solutions, and ones that are in line with the need to conserve the earth's raw material resources, are those of recycle and re-use. Plastics films present a variable picture from the resource use point of view. On the one hand we have polypropylene, polystyrene, PVC and the polyethylenes (obtained basically from crude oil which is a non-renewable resource) while on the other hand we have cellulose acetate and regenerated cellulose which utilise renewable resources, such as trees. Even in the case of the latter materials there are arguments in favour of re-use or recycling. Their manufacture relies heavily on the use of wood pulp from overseas and so their recycling would help to improve the UK's balance of payments position.

At this point it may be useful to define the terms re-use and recycle as they are sometimes used interchangeably, leading to confusion. Re-use means using the article in an unchanged condition, as in the refilling of a plastics bottle. Recycling may be direct or indirect. Direct recycling means that the basic materials are recovered and can then be converted into other products. The repulping of paper and board with subsequent moulding to produce pulp containers is one example of this.

Indirect recycling is any process wherein the basic material is decomposed and the decomposition products re-enter one or other of the natural cycles that occur continuously in the environment. The incineration of plastics as, for example, polystyrene and the polyolefins leads mainly to the production of carbon dioxide which then becomes part of the carbon/oxygen natural cycle. Most natural cycles are extremely long as, for example, the one that originally produced the crude oil from which we now obtain our plastics raw materials. This was of the order of millions of years and is an impractical one from the point of view of plastics recovery! It should be noted that what is usually referred to as waste disposal is normally a form of indirect recycle. The rusting of tinplate containers, for instance, in land in-fill eventually produces deposits of iron compounds, similar to those from which the original metal was smelted, although it would take hundreds of years to produce economically viable concentrations. Similarly, as mentioned earlier, the incineration of plastics is a form of indirect recycle. Waste materials never disappear entirely, they merely recycle out of sight. As we have seen, however, most indirect recycling involves long periods of time for the build-up part of the cycle while the time taken for the human race to use the resultant products is becoming increasingly short. The cycles soon get horribly out of balance, with the result that the earth's mineral resources are approaching exhaustion in some cases.

The question then arises—if re-use or direct recycle is seen to be so necessary, why isn't it practised more widely and what can be done to improve the position? There are technical problems but the biggest problem is economic viability. Even with paper and board where a widespread system for collection and re-sale already exists, there is scope for a larger percentage of recovery but the financial incentives are just not high enough to justify the extra costs involved in expansion. When we come to consider plastics the economics are complicated by the fact that extensive sorting usually has to be carried out. Not only are there different plastics but there are various grades within each type. So far there are no really big operators able to collect and sort on an economic scale. Various minor exercises have been carried out, such as the collection of polystyrene drinking cups from industrial canteens and low density polyethylene fertiliser sacks from farms. Both of these end-uses operate on a fairly standard polymer grade and so they do not involve problems of sorting. Even so, they remain relatively small scale operations.

There are ways of recycling plastics other than by shredding, melting, extruding and granulating to produce pellets of pure polymer. Some of these processes do not depend on separation of any mixed plastics. The mixed plastics stock is chipped (or shredded in the case of film), heated and then rolled under pressure. Items made from this heated and compressed material include pallets and panels for agricultural out-buildings where stresses are fairly low. A similar process was also developed utilising plastics-coated paper, a material that often causes problems during normal repulping operations. Once again, however, it is the question of economics which is so difficult to solve. It may be that we have got to wait until the earth's resources have been depleted still further before some of these recycling processes are considered to be economically viable.

One of the reasons for the existence of so much mixed waste has already been considered, namely, the difficulties involved in segregating plastics films, paper, cellulose film and aluminium foil, but another factor is the trend towards the greater use of composite materials, particularly in packaging. With the thousands of laminates and plastic-coated materials in use today the problem of separation is correspondingly large.

One other method of recycling is known as pyrolysis and consists in heating the plastic waste in the absence of air (or with very limited quantities of oxygen). Under such conditions organic materials break down to give simple organic liquids and gases. These are then collected and separated to give valuable feedstocks for the chemical industry. If these feedstocks are used for the manufacture of further plastics we will have achieved a true direct recycling process but equally valuable end products may be produced without affecting the basic argument. It is not even necessary that plastics should be segregated from other waste as the many other organic materials in collected wastes can also be pyrolised. Research continues on this process and some small-scale plants are in operation in the U.S.A. but costs are, at present, high.

FURTHER READING

ANON., Modern Plastics International, 36 (February 1974)

BRIDGEWATER, A. V. and MUMFORD, C. J., *Waste Recycling and Pollution Control Handbook*, George Godwin (1980)

KAMINSKY, W., MENZEI, J. and SINN, H., Conservation and Recycling, 1(1), 91 (1977)

STAUDINGER, J. J. P., *Plastics and the Environment*, Hutchinson, London (1974)

ZERLAUT, G. A. and STAKE, A. M., *Recycling and Disposal of Solid Waste*, Noyes Data Corporation (1975)

26
Plastics and Energy

Few people will dispute that an energy crisis exists although opinions differ as to its extent and on the measures to be taken to combat it. One school of thought seems to hold the view that packaging is the villain of the piece and, in particular, plastics packaging, which is seen as using badly needed energy in its various manufacturing processes as well as the oil used directly as a feedstock, with a consequent further loss of an energy source.

The argument that plastics packaging is a major factor in the depletion of oil reserves does not stand up under examination. The greatest use for crude oil is as a source of fuel (petrol, kerosene or fuel oil) or for related uses such as lubricating oils. Only about 10% is used for the manufacture of all types of chemicals and of this an equally small percentage is used for all plastics. Taking the matter still further, about 25% to 30% of plastics are used in the packaging industry.

The real problem, then, is to find alternative energy sources. The existing reserves of oil, coal and natural gas would last for hundreds of years as chemical feedstocks, so giving the world a more than adequate breathing space to develop new sources.

One of the latest estimates of world energy sources is that given in the 1979 Ford Foundation Report entitled *Energy—The Next 20 Years* (see *Table 26.1*).

At present rates of energy usage this would last for about 90 years (or 900 years if used only as chemical feedstocks).

Apart from this, attention is being increasingly directed towards a

Table 26.1. WORLD ENERGY RESERVES IN 1979

	Million tonnes of coal equivalent
Solid fuels	648 000
Crude petroleum	136 000
Natural gas	94 000
	878 000

wide range of renewable resources. For convenience the name 'Biomass' has been given to renewable resources with a biological basis (e.g. trees, seaweed, field crops, sewer waste, etc.).

One important monomer is ethylene and this can be derived from ethanol, via the fermentation of sugars from plant juices, cellulose and starch. Methanol can be made from agricultural residues or from municipal solid waste and is important because it is the source of formaldehyde. The major barriers at present are economic rather than technical but this will change as the non-renewable sources become even scarcer. The conversion of Biomass into feedstocks and polymers of interest is outlined in *Figure 26.1*.

The arguments concerned with the energy used in the manufacture of plastics packaging are much more difficult to unravel. The total energy used is a complex quantity, and if comparisons are to be made the following factors must be taken into account.

(a) energy required to obtain the raw material;
(b) energy value of the raw material;
(c) transportation energy to deliver the raw material to the processing plant;
(d) processing energy;
(e) transportation energy to deliver containers to the filling plant;
(f) transportation energy to deliver the filled containers to the retail outlet;
(g) energy used for post-consumer disposal of containers.

Studies based on the above energy parameters have been carried out in the USA during an evaluation of the environmental impact of various competing packages. The conclusion reached was that plastics containers are not notably more or less energy intensive than other containers.

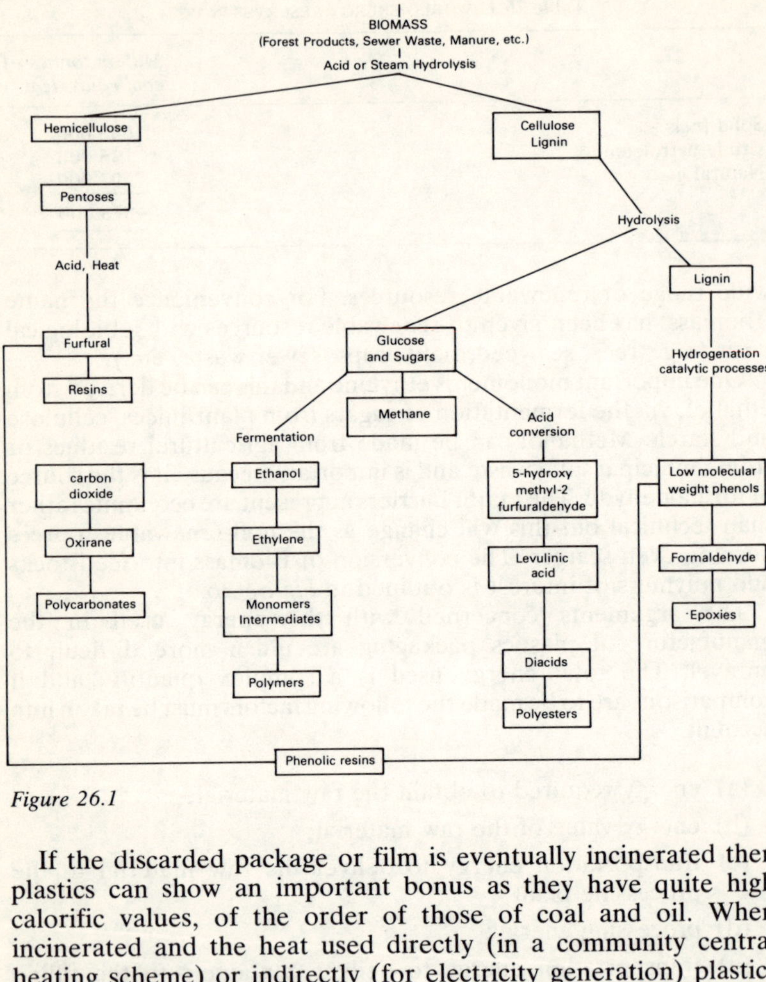

Figure 26.1

If the discarded package or film is eventually incinerated then plastics can show an important bonus as they have quite high calorific values, of the order of those of coal and oil. When incinerated and the heat used directly (in a community central heating scheme) or indirectly (for electricity generation) plastics are, in effect, 'paying back' part of the energy used in their manufacture.

So much, then, for the comparison of plastics with other packaging materials. There is, however, another factor that applies to all packaging, namely, the energy used in the manufacture of the product to be packaged. If most products were to be sold unpackaged there could be a considerable loss during distribution from factory to consumer. This is particularly true in the case of foodstuffs. Agriculture itself is an energy-intensive operation from tilling through harvesting to drying. Fertilisers, weed-killers and insecticides also consume energy in their manufacture. This energy is wasted unless the food reaches the consumer in an edible condition.

It is obvious that the energy equation is not a simple one and it would seem that plastics films can save as much energy as is consumed in their manufacture.

There will, presumably, always be criticism but once it moves from the general to the particular it can be rebutted in most cases.

FURTHER READING

CADMAN, M. H., Energy—A World Perspectivc, 'Full Circle', Institute of Packaging Seminar hcld on 7–9 October 1980 in Norwich, England, Paper no. 1

SMITH, D. J., Energy—Policy, Politics and Packaging Development, Idem, Paper no. 2

Appendix A

Properties of Plastics Films

Material / *Property*	*Low density polyethylene*	*High density polyethylene*	*Cast polypropylene*	*Oriented polypropylene*	*Rigid PVC*	*Plasticised PVC*	*PVDC*	*Oriented polystyrene*	*Regenerated cellulose (MS Grade)*	*Cellulose acetate*	*Nylon 11*	*EVA*	*Ionomer*	*Polycarbonate*	*Polyester*
Yield (m^2/kg) (for 25 μm film)	42.6	41.2	44.0	44.0	28.4	27.0 –30.5	23.4	36.9	27.4	29.8	38.3	41.9	41.7 –42.1	32.3	28.4
Tensile strength (MN/m^2)	8.6–17.3	17.3 –34.6	41.5	165 –170	45.0 –55.3	27.6 –34.5	48.4 –138	62.2 –82.7	48.4 –110	48.5 –82.7	69.0 –96.8	13.8	17.3 –24.2	58.8	175
Elongation at break (%)	500	300	300	50–75	120	14–200	20–40	20	15–25	15–45	250 –400	650 –800	300 –400	75	70–100
Tear strength (Elmendorf) (g/25 μm)	200–300	20–60	50	5–10	—	—	10–30	4–20	2–10	2–15	400 –500	50 –100	20–80	10–16	30
Burst strength (Mullen) (kN/m^2) (for 25 μm film)	330	—	—	—	205 –275	—	205 –485	—	138 –155	205 –615	415	—	—	170 –240	310 –415
Water vapour transmission (g/m^2/day) (for 25 μm film at 90% R.H. and 38°C)	15–20	5	10–12	7	30–40	15–40	1.5 –5.0	70 –150	5–15	100 –320	40 –80	50–60	25–35	77–93	25–30
Oxygen permeability (cm^3/m^2/day/atm) (for 25 μm film)	6 500 –8 500	1 600 –2 000	3 700	2 000 –2 500	150 –350	—	8–25	4 500 6 000 –	670 (when dry)	2 000 –3 000	500	11 000 –14 000	6 000 –7 000	4 500	40–50
Carbon dioxide permeability (cm^3/m^2/day/atm) (for 25 μm film)	30 000 –40 000	8 000 –10 000	10 000	7 500 –8 500	450 –1 000	—	50	13 000	985 (when dry)	15 500	1 900	40 000 –50 000	6 000 – 7000	27 000	300–350
Resistance to oils and greases	Some oils cause swelling	Good	Ex	Ex	Ex	Depends on exact formulation	Ex	V Good	Ex	Ex	Ex	Ex	VGood	Good	Ex

Appendix B
Identification of Film Materials

The most reliable method of identifying a plastics film is by infra-red spectroscopy but the equipment is expensive and requires skilled operation. There are many simple tests, however, and if they are applied in combination surprisingly good results can be obtained, with experience. If the material to be identified is a laminate then it is necessary to separate it into its component parts before applying any tests.

B.1 SEPARATION OF LAMINATE PLIES

This is best carried out by refluxing the sample with a suitable solvent. The choice of solvent is dictated by the adhesive used for laminating but if this information is not available the best solvent must be found by experiment.

Table B.1 shows some of the commonest laminating adhesives together with possible solvents.

B.2 PHYSICAL PROPERTY TESTS

Films can be separated into two classes on the basis of their extensibility. A sample strip should be taken in the hands and

Table B.1

Adhesive	*Solvent(s)*
Wax	Acetone, carbon tetrachloride, toluene
Starch	Water
Casein latex	An emulsion of Duponol (obtainable from DuPont Chem. Company), toluene, ammonia and water
Vinyl (PVA, PVAc, etc.)	Acetone, carbon tetrachloride, toluene
Low density polyethylene	Toluene (at $>60°C$)

stretched gently, but firmly, noting the approximate amount of stretch. Extensible films (classified as those stretching more than 100%) include low density polyethylene, high density polyethylene, polypropylene, stretch PVC, polyvinyl alcohol and rubber hydrochloride. Non-extensible films include cellulose acetate, polyester, regenerated cellulose, nylon, oriented polystyrene and polyvinylidene chloride.

Stiffness is another property which can often give some information as to the material used. Low density polyethylene, polyvinylidene chloride some vinyls and some grades of rubber hydrochloride are limp and soft to the touch and make very little noise when crumpled. Regenerated cellulose, cellulose acetate, nylon, polypropylene, polyester, rigid PVC and some grades of rubber hydrochloride are stiffer, harder to the touch and make more noise when crumpled. Oriented polystyrene is even stiffer and when rattled it has a metallic ring.

Tear strength, or rather, tear propagation resistance, can also yield useful information. The film should be nicked with a knife or razor blade and then torn. Materials with a low tear strength include regenerated cellulose and cellulose acetate while nylon and polyester are slightly higher. Materials with a high resistance to tear propagation include PVC, polyvinylidene chloride, low density polyethylene, and rubber hydrochloride.

B.3 BURNING TESTS

Two observations may be made with the aid of a flame, namely, whether the film burns after removal of the flame and what kind of odour is given off. *Table B.2* gives the behaviour of some common films when held in a flame for about ten seconds and then withdrawn.

Table B.2

Film	*Continue burning*	*Odour*
Cellulose acetate	Yes	Acetic acid
Cellulose acetate/butyrate	Yes	Acetic acid and rancid butter
Nylon	No	Burning hair or feathers
Polyethylene	Yes	Burning wax
Polypropylene	Yes	Burning wax
Polyester	Yes	Sweet, esterlike
Polystyrene	Yes (gives sooty specks)	Styrene
Polyvinyl chloride	No	Acrid
Polyvinylidene chloride	No	Acrid
Regenerated cellulose	Yes (does not melt)	Burning paper
Rubber hydrochloride	No	Acrid
Polycarbonate	No	

B.4 DENSITY

Although films do vary in density, there is a good deal of overlap so that this property is of limited use for identification purposes. The only real distinction which can readily be made is between the polyolefin films (polyethylene, polypropylene and poly 4-methyl pentene-1) and the rest. This is because the polyolefines all have densities less than unity and so float on water. Even this test can be misleading if the film sample is creased and so retains small air bubbles. The addition of a small amount of a wetting agent to the water is also essential.

B.5 CHEMICAL TESTS

B.5.1 SOLUBILITY

A systematic scheme which makes use of the different solubilities of the common film materials is given in a PIRA Technical Reference Note and is reproduced in *Table B.3* by kind permission of PIRA (The Research Association for the Paper and Board, Printing and Packaging Industries). Before using the table, tests must be made for regenerated cellulose films (coated or plain) and these are then identified by means of *Table B.4*. The tests for regenerated cellulose film are as follows:

Table B.3

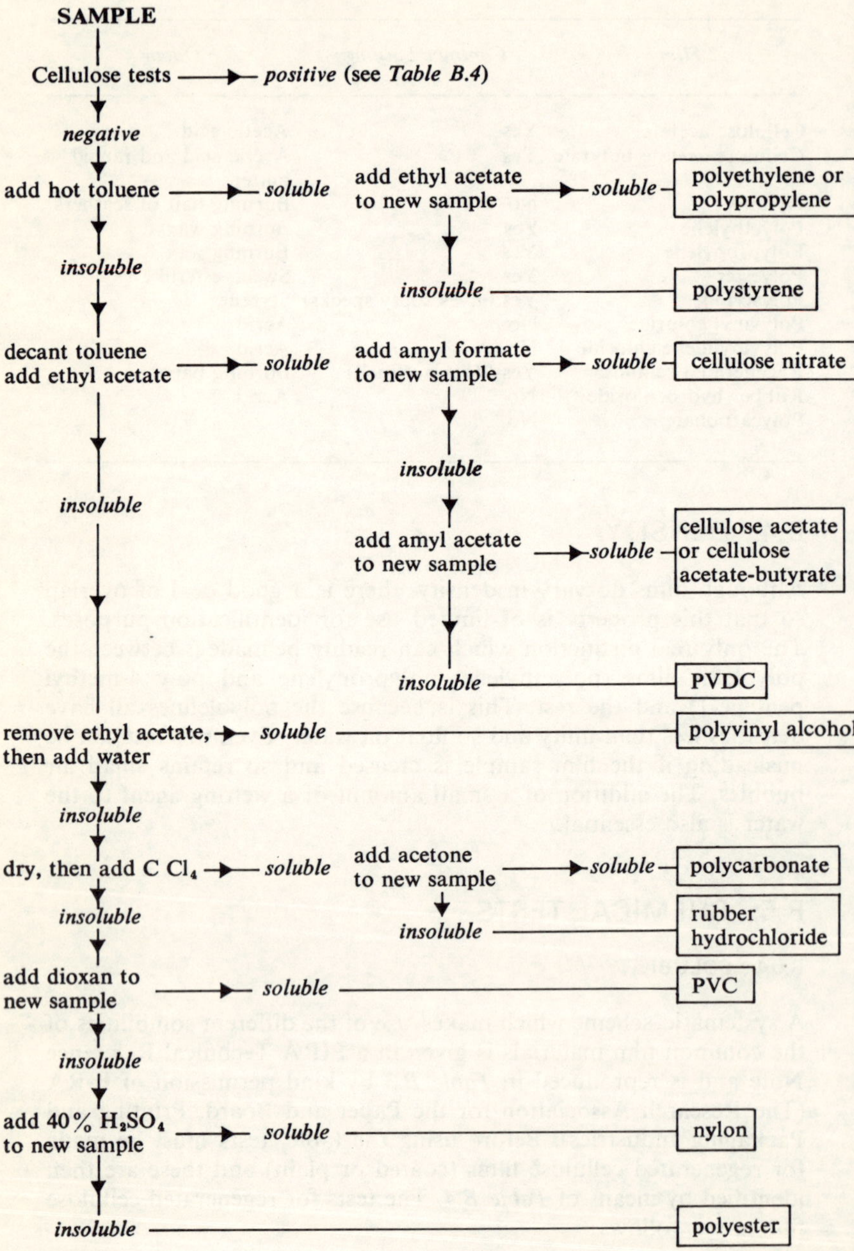
SAMPLE
Cellulose tests
positive (see Table B.4)
negative
add hot toluene
soluble
add ethyl acetate to new sample
soluble
polyethylene or polypropylene
insoluble
insoluble
polystyrene
decant toluene add ethyl acetate
soluble
add amyl formate to new sample
soluble
cellulose nitrate
insoluble
insoluble
add amyl acetate to new sample
soluble
cellulose acetate or cellulose acetate-butyrate
insoluble
PVDC
remove ethyl acetate, then add water
soluble
polyvinyl alcohol
insoluble
dry, then add C Cl4
soluble
add acetone to new sample
soluble
polycarbonate
insoluble
insoluble
rubber hydrochloride
add dioxan to new sample
soluble
PVC
insoluble
add 40% H2SO4 to new sample
soluble
nylon
insoluble
polyester

(1) Stretch test—not extensible.

(2) Burning test—odour of burning paper.

(3) Tear test—tear strength very low.

(4) Apply hot match head to the film—no hole melted in the material; denotes cellulose.

Other simple chemical tests which may be used are as follows.

B.5.2 BEILSTEIN TEST FOR HALOGENS

Heat a length of clean copper wire in a non-luminous bunsen flame until the green colour vanishes. Touch the test film with the hot wire and reheat in the bunsen flame. A green colour denotes the presence of halogens. Thus, PVC, polyvinylidene chloride and rubber hydrochloride (containing chlorine) and PTFE or PTCFE (containing fluorine) will all give a positive reaction to the Beilstein test. Some coated cellulose films and coated, oriented polypropylene films may also give a positive reaction because of the presence of polyvinylidene chloride in the coating.

B.5.3 GRIESS TEST FOR NITROGEN

Heat a sample of the film with manganese dioxide in a combustion tube for three minutes then hold over the tube mouth a piece of filter paper moistened with Griess reagent. A pink or red coloration denotes the presence of nitrogen. Griess reagent is made up as follows: Solution A: 1 g sulphanilic acid is dissolved in 100 cm^3 of 30% acetic acid.

Solution B: 0.03 g naphthylamine is boiled in 70 cm^3 of water. The colourness supernatant liquor is decanted and mixed with 30 cm^3 of glacial acetic acid.

Equal quantities of solutions A and B are mixed as required.

A positive test for nitrogen is given by the polyamides (nylons) and by the nitrocellulose coatings on moisture-proof grades of regenerated cellulose film.

B.5.4 WECHSLER TEST

This test differentiates between PVC and polyvinylidene chloride which both give positive reactions to the Beilstein test. A mixture of pyridine and a methanol solution of potassium hydroxide is applied to the film. PVC and vinyl chloride/vinyl acetate copolymers give a pale brown colour while polyvinylidene chloride gives a dark brown to black colour. Rubber hydrochloride does not react at all.

B.5.5 LIEBERMANN–STORCH–MORAWSKI TEST

This test is a good one for distinguishing between the four films containing chlorine, namely, PVC, vinyl chloride/vinyl acetate, rubber hydrochloride and polyvinylidene chloride (including PVDC coated cellulose films).

A small piece of film is placed on a spot plate and covered with a few drops of acetic anhydride. One drop of concentrated sulphuric acid is added so that it enters the liquid. The colours on the film and in the liquid are observed over a period of half an hour.

PVC gives a blue coloration (slow).
VC/VA copolymers give a slow change from green to blue to brown.
PVDV gives a yellow coloration (slow).
Rubber hydrochloride gives no coloration.

B.6 INSTRUMENTAL TESTS

Ultra-violet and infra-red spectrophotometry can be used to identify films by comparing the ultra-violet and infra-red absorption spectra of the unknown film with spectra previously obtained on known materials.

Table B.4

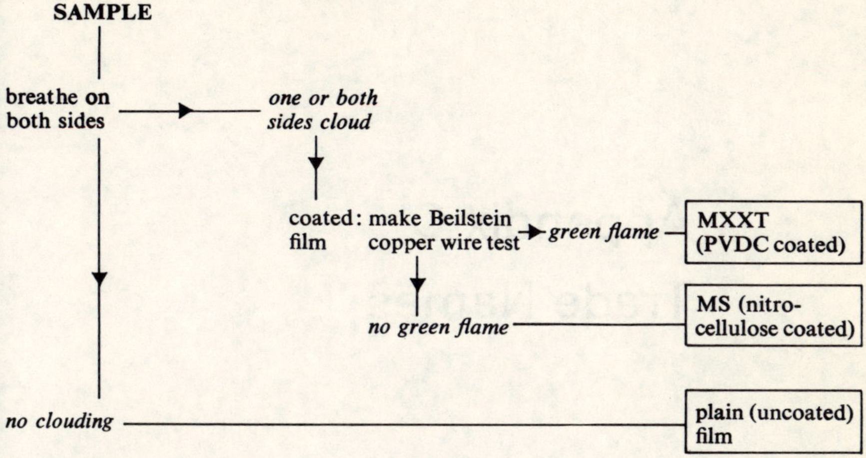
SAMPLE
breathe on both sides
one or both sides cloud
coated film: make Beilstein copper wire test
green flame
MXXT (PVDC coated)
no green flame
MS (nitro-cellulose coated)
no clouding
plain (uncoated) film

Appendix C
Trade Names

Trade name	*Type of material*
Abselex P	Acrylonitrile butadiene styrene
Acelon	Cellulose acetate
Acetophane	Cellulose acetate
Aerozote	Expanded polyethylene
Afubel	Cellulose acetate
Alathon	Polyethylene
Alathon/EVA	Ethylene-vinyl acetate co-polymer
Alkathene	Polyethylene
Alloprene	Chlorinated rubber
Amerplas	Varied thickness polyethylene
Austrophan	Cellulose
Azopal	Unplasticised PVC
Beneron	Acrylonitrile butadiene styrene
Bepsatene	Polyethylene
Bexfilm	Cellulose acetate and triacetate
Bexfoam	Expanded polystyrene
Bexoid	Cellulose acetate
Bexphane P	Biaxially oriented polypropylene
Bexphane S	Biaxially oriented polystyrene
Bexthene	Polyethylene
Britainphane	Cellulose
Cabulite	Cellulose acetate/ butyrate
Capran	Nylon 6
Cautex	PVC
Celawrap	Cellulose acetate
Cellidor	Cellulose acetate and acetate/ butyrate
Cellit	Cellulose acetate and acetate/ butyrate
Cellolam	Wax laminated cellulose
Cellon	Cellulose acetate
Cellophane	Cellulose
Cellothene	Polyethylene coated cellulose
Celluloid	Cellulose nitrate
Chromalon	Metallised PVC
Cinabex	Cellulose acetate
Ciponyl	Polyamide (nylon)
Cipoviol	Polyvinyl alcohol
Clarifoil	Cellulose acetate
Clarothene	Ethylene co-polymer
Cournova	Polypropylene
Craylene	Plasticised PVC
Craylon FC	Unplasticised PVC

Trade name	*Type of material*
Craytherm	Unplasticised PVC
Cryovac L	Modified polyethylene
Cryovac Y	Biaxially oriented polypropylene
Cryovac S	Vinylidene chloride vinyl chloride co-polymer
Crystophane	PVC
Danathene	Polyethylene
Densothene	HD Polyethylene
Diolon	Polyamide-polyolefine laminate
Diomex	Polyester-polyolefine laminate
Diophane	Cellulose
Dioplex P	Polyethylene-polypropylene laminate
Dioplex PS	PVDC coated polypropylene-polyethylene laminate
Diothene	Low density polyethylene
Drakatileen	Polyethylene
Drakavinyl	PVC
Durethan	Nylon 6
Dylan	Polyethylene
Embafilm	Cellulose acetate
Embafilm B	Cellulose acetate/butyrate
Enkothene	Polyethylene
Ethocel	Ethyl cellulose
Ethulon	Ethyl cellulose
Europhane	Cellulose
Fablonex	Polyester-PVC laminate
Fablonyl	Polyamide
Fablothene	Polyethylene
Fertene	LD Polyethylene
Filmco	PVC
Finoplas	Modified HD Polyethylene
Flovic	Vinyl chloride-acetate co-polymer
Forticel	Cellulose propionate
Fortiflex	HD Polyethylene
Franklite	Polyethylene
Garmil	Polyethylene
Garpac	Polyethylene
Genopak	PVC
Genotherm	PVC
Ger-pak	Polyethylene
Guttagena	PVC
Hagaten	Polyethylene
Hilex	HD Polyethylene
Hostaform C	Acetal co-polymer
Hostalen G	HD Polyethylene
Hostalen PP	Polypropylene
Hostaphan	Polyester
Hypak R	Polyamide
Iolon	Ionomer
Kel F	Polytrifluorochloroethylene
Krehalon	Vinyl chloride-vinylidene chloride co-polymer
Lacqtene	Polyethylene
Lactophane	Cellulose
Laythene	Polyethylene
Lexan	Polycarbonate
Lumaline	Metallised cellulose acetate
Lumarith	Cellulose acetate
Luparen	Polypropylene
Lupolen	Polyethylene
Makrolon	Polycarbonate
Meculon	Metallised polyester
Melinex	Polyester
Metalon	Metallised polyester-PVC laminate
Metathene D	Nitrocellulose coated cellulose
Metathene X	PVDC coated cellulose
Metathene M	Polyester
Metathene MX	PVDC coated polyester
Metathene P	Oriented polypropylene
Metathene PPX	PVDC coated oriented polypropylene
Milwrap	Polyethylene
Mipolam	Plasticised PVC
Mirralon	Metallised cellulose acetate
Mirro-brite mylar	Metallised polyester

Trade Name	*Type of Material*
Montothene	Ethylene-vinyl acetate co-polymer
Moplefan	Polypropylene
Mylar	Polyester
Mylothene	Polyester-poly ethylene laminate
Nataraj	PVC
Natene	Polyethylene
Nichiray	Biaxially oriented polyamide
Pentathene	Polyethylene
Pevalon	Polyvinyl alcohol
Phriphan	Cellulose
Phrithen	Polyethylene
Plastokarbon	Polycarbonate
Plastotene	Polyethylene
Plastothen	Polyethylene
Plastylene	Polyethylene
Platilon C	Nylon 6
Pliofilm	Rubber hydrochloride
Polyart	Modified polyethylene
Polycell	Polyethylene coated cellulose
Polyflex	Biaxially oriented polystyrene
Polytherm	Unplasticised PVC
Pony	PVC
Porvic	Microporous PVC
Propafilm C	PVDC coated biaxially oriented polypropylene
Propafilm O	Biaxially oriented polypropylene
Propathene	Polypropylene
Propophane	Polypropylene
Propophane EPI	Ethylene-propylene co-polymer
Propylex	Polypropylene
Prylene	Polypropylene
Ralsin	Nylon 11
Rayofilm	Cellulose
Rayophane	Cellulose
Rayoscine	Cellulose
Resinite	PVC
Reynolon	PVC
Rhodialine	Cellulose acetate
Rhodopas	PVC/acetate
Rhodophane	Cellulose acetate

Trade Namc	*Type of Material*
Rholene	HD polyethylene
Rilsan	Nylon 11
Santoforme F	Expanded flexible cellular polystyrene
Saran	Vinylidene chloride-vinyl chloride co-polymer
Saranex	Polyethylene-PVDC-polyethylene laminate
Sarolene H	HD polyethylene
Sarolene 63P	Polypropylene
Sarovyl	PVC
Saroy SP61	Polystyrene
Sicofoil A	Cellulose acetate
Sicofoil N	Cellulose nitrate
Sidaplax	Biaxially oriented polystyrene
Sidathene	Polyethylene
Silpak	Polyethylene
Styrafoil	Biaxially oriented polystyrene
Suprathen	LD polyethylene
Supronyl	Polyamide
Suprotherm C	PVC
Surlyn	Ionomer
Synthene	Nylon-polyethylene laminate
Tectophane	Cellulose
Tenasco	Rayon reinforced polyethylene sheeting
Tenofol	Polyethylene
Termex	Polyester
Terphane	Polyester
Tissuthene	Ultra thin HD polyethylene
Trespaphan	Polypropylene
Triacel	Cellulose acetate and acetate/butyrate
Trithene	Polytrifluoro-chloroethylene
Trosifol	Polyvinyl butyral
Trovilon	PVC
Trovitherm	Unplasticised PVC
Trycite	Oriented polystyrene
Tuftane	Polyurethane

Trade Name	*Type of Material*
Tullon	Polyamide
Ukayphane	Cellulose
Ultrathene	EVA co-polymer
Utilex	Cellulose acetate
Valeron	Cross laminated biaxially oriented polyethylene
Velbex	PVC
Ventoplas	Perforated polyethylene
Viniplas	PVC

Tradc Name	*Type of Material*
Vinolyte	Vinyl chloride-vinyl acetate co-polymer
Viscacelle	Cellulose
Viscophane	Cellulose
Vistal	Polyethylene
Vitafilm	PVC
Vitrone	PVC
Vondafol PC	Polycarbonate
Vondafol PPO	Polyphenylene oxide
Vybak	PVC
Zendel	Polyethylene

Index